Lecture Notes in Control and Information Sciences

Volume 478

This series aims to report new developments in the fields of control and information sciences—quickly, informally and at a high level. The type of material considered for publication includes:

1. Preliminary drafts of monographs and advanced textbooks
2. Lectures on a new field, or presenting a new angle on a classical field
3. Research reports
4. Reports of meetings, provided they are
   (a) of exceptional interest and
   (b) devoted to a specific topic. The timeliness of subject material is very important.

More information about this series at http://www.springer.com/series/642

Fabien Lauer · Gérard Bloch

Hybrid System Identification

Theory and Algorithms for Learning Switching Models

Springer

Fabien Lauer
LORIA
Université de Lorraine, CNRS
Nancy, France

Gérard Bloch
CRAN
Université de Lorraine, CNRS
Nancy, France

ISSN 0170-8643 ISSN 1610-7411 (electronic)
Lecture Notes in Control and Information Sciences
ISBN 978-3-030-00192-6 ISBN 978-3-030-00193-3 (eBook)
https://doi.org/10.1007/978-3-030-00193-3

Library of Congress Control Number: 2018954018

This Springer imprint is published by the registered company Springer Nature Switzerland AG
The registered company address is: Gewerbestrasse 11, 6330 Cham, Switzerland

Preface

This book is about *hybrid system identification*, a field concerned with the theory and practice of building mathematical models of dynamical systems switching between different operating modes, from experimental observations. Mathematical models have always been of primary importance in many branches of science and especially in automatic control, where they constitute the first building block of most analysis and control schemes. However, in many cases, such models cannot be easily built from first principles or contain unknown parameters. System identification precisely aims at estimating these parameters or, more generally, the models, from experimental observations of the system behavior, usually taking the form of input–output data. When the system under consideration involves multiple operating modes or discontinuities, this usually calls for the estimation of hybrid models that can precisely take these particularities into account.

While system identification is now a well-established field, starting in the 1960s with the availability of digital measurements, the subject of hybrid system identification has only been more recently developed, with most of the activity spanning only the last 15 years or so. Nonetheless, it already served in many applications in automotive control, computer vision, or even system biology, and numerous advances were made in that period. Therefore, we felt it was time for a book dedicated to this topic which could introduce new researchers to this passionating field with many ramifications. Indeed, hybrid system identification lies at the intersection of a number of disciplines, including control theory (for the analysis of switched dynamical systems), system identification (for the estimation of models of dynamical systems), and machine learning (for the classification of the data into the different operating modes and the estimation of nonparametric models). Optimization and computational complexity also play important roles here, perhaps much more pronounced than in traditional system identification. This book covers all the necessary background material on these fields in order to grasp the main issues and understand the rationale behind the main methods of hybrid system identification. After these overviews, we emphasize what makes the identification of hybrid systems so different from classical system identification and how much more complex the task can be. Then, practical methods leveraging the complexity bottleneck are presented and the book ends with some more advanced material, including a chapter on nonlinear hybrid systems with strong connections to machine learning. Throughout the book, examples and figures are given to illustrate the key points and we tried to keep the presentation accessible to most researchers or graduate students in a control-related field, even when discussing, e.g., computational complexity or machine learning topics.

It is worth noting that the keywords "hybrid system" may have multiple meanings in science (and elsewhere), which can introduce some confusion regarding the subject of this book. Here, we define hybrid systems as systems switching between different operating modes, which amounts in technical terms to systems combining continuous and discrete *states*, the discrete state indicating the current operating mode. But, as an example, a growing community uses the keyword "hybrid" to describe systems that combine continuous-*time* and discrete-*time* behaviors, which is a completely different topic. We believe that the terminology that we chose for this book is the dominant one in the control community. In particular, it is in line with the terminology used for the names of Technical Committees (TC) of the two major control societies (the IFAC TC 1.3 "Discrete Event and Hybrid Systems" and the IEEE Control Systems Society TC "Hybrid Systems"), relevant journals (e.g., "Nonlinear Analysis: Hybrid Systems"), and conferences (e.g., "Hybrid Systems: Computation and Control" or "Analysis and Design of Hybrid Systems").

Besides, using "switched system" instead of "hybrid system" would have introduced another confusion. Indeed, we here precisely divide "hybrid systems" into the "switched" and "piecewise" subclasses, the first referring to arbitrarily switched systems and the second one to systems switching according to a partition of the regression/state space.

Our interest in hybrid system identification can be traced back to 2007, while Fabien was in the middle of his Ph.D. supervised by Gérard, and we were developing machine learning techniques for system identification. At that time, Fabien attended the 2nd HYCON Ph.D. School on Hybrid Systems in Sienna, Italy, and came back with early thoughts on a new algorithm mixing ideas from support vector regression and the algebraic approach of Sect. 6.1.1. Then, Fabien met with René Vidal in 2008 while presenting this work at the HSCC conference in St. Louis, MO, USA. This initiated our collaboration with the main contributor to the algebraic approach, with whom we generalized the algorithm to what is now known as the continuous optimization approach of Sect. 6.1.2 and drafted its nonlinear extension presented in Sect. 9.1.

We further worked on this nonlinear extension with our former Ph.D. student Van Luong Le, with whom we also developed a number of other techniques, including the sparsity enhancing selective ℓ_1-minimization sketched in Sect. 6.2.3.2, the method to learn piecewise smooth models via convex optimization exposed in Sect. 7.2 and the sparsity-based method for nonlinear hybrid systems described in Sect. 9.2. The later is largely inspired by the work of Laurent Bako, who introduced the error sparsification method of Sect. 6.2.3.

Beyond many interesting conversations on sparsity and hybrid system identification, Laurent also deserves special thanks as this is with him that we first discussed the idea of writing a book on this topic in 2011 and drafted a preliminary table of contents in 2013. However, we all had other duties and the project did not go much forward during some time. In 2016, we picked it up again only the two of us; this was easier since we were both working in Nancy. So we could work on a more regular basis and ended up actually writing this book, after some rethinking of the table of contents.

During these years, we benefited from interactions with other colleagues that worked with us on hybrid system identification or related topics. In particular, we had the chance to work with Hoai An Le Thi and Pham Dinh Tao on DC programming techniques and with Henrik Ohlsson on sparse recovery issues.

Fabien also warmly thanks Christoph Schnörr for hosting him during his postdoc at the University of Heidelberg and Yann Guermeur at the LORIA for providing him the necessary means and freedom to work on this project.

Finally, writing such a book is rather time consuming, and we thank our families. Fabien thinks of Caroline, Adam and Noémie, and Gérard of Patricia, Antonin and Etienne, the latter even if the book subject is far from laser–matter interactions and Ultrafast Phenomena.

Nancy, France
July 2018

Fabien Lauer
Gérard Bloch

Contents

1 **Introduction** . . . 1
  1.1 What are Hybrid Systems? . . . 1
    1.1.1 Dynamical Systems . . . 1
    1.1.2 Hybrid Systems . . . 3
  1.2 What is System Identification? . . . 5
  1.3 Applications . . . 6
  1.4 Outline of the Book . . . 8
  References . . . 11

2 **System Identification** . . . 15
  2.1 Input–Output (I/O) Models . . . 16
    2.1.1 Models, Predictor, and Prediction Error . . . 16
    2.1.2 Parameter Estimation . . . 22
    2.1.3 Optimization . . . 24
  2.2 State-Space (SS) Models . . . 28
    2.2.1 Models and Properties . . . 28
    2.2.2 Parameter Estimation . . . 34
  2.3 Recursive Identification . . . 39
  2.4 Nonlinear System Identification (★) . . . 41
    2.4.1 Parametric Models . . . 41
    2.4.2 Nonparametric Models . . . 42
  2.5 Model Selection and Assessment . . . 52
    2.5.1 Model Assessment . . . 52
    2.5.2 Model Selection . . . 53
  Notes . . . 55
  References . . . 56

3 Classification 59
3.1 Discrimination 60
3.1.1 Binary Linear Classification 61
3.1.2 Multi-class Problems 68
3.1.3 Nonlinear Classification (★) 70
3.2 Clustering 71
Notes 72
References 74

4 Hybrid System Identification 77
4.1 Hybrid System Models 78
4.1.1 State-Space Versus Input–Output Models 78
4.1.2 Linear Versus Nonlinear Submodels 79
4.1.3 Piecewise Smooth Versus Arbitrarily Switched Systems 82
4.2 Identification Problems 86
4.2.1 Hybrid System Identification with Unknown Mode 86
4.2.2 The Trade-Off Between the Number of Modes and the Error 87
4.2.3 Fixing the Number of Submodels 88
4.2.4 Fixing a Bound on the Error 90
4.2.5 Hybrid Model Assessment 93
4.3 Other Related Problems (★) 95
4.3.1 Nonlinear System Identification 95
4.3.2 Subspace Clustering 96
Notes 99
References 100

5 Exact Methods for Hybrid System Identification 103
5.1 Straightforward Solutions 104
5.1.1 Switching Regression with Fixed s 105
5.1.2 Bounded-Error Estimation 106
5.1.3 Piecewise Affine Regression with Fixed s 108
5.2 Hardness Results (★) 109
5.2.1 Basics in Computational Complexity 109
5.2.2 Hardness of Switching Regression 111
5.2.3 Hardness of PWA Regression 116
5.2.4 Hardness of Bounded-Error Estimation 118
5.3 Polynomial-Time Algorithms for Fixed Dimensions 118
5.3.1 PWA Regression with Fixed s and d 119
5.3.2 Switching Regression with Fixed s and d 122
5.3.3 Bounded-Error Estimation with Fixed d 124

5.4 Global Optimization with Branch-and-Bound 126
5.4.1 Switching Regression 127
5.4.2 Bounded-Error Estimation 132
5.4.3 PWA Regression 134
5.5 The Need for Approximation Schemes/Heuristics 137
Notes 138
References 139

6 **Estimation of Switched Linear Models** 141
6.1 Fixed Number of Modes 142
6.1.1 Algebraic Method 142
6.1.2 Continuous Optimization Approach 145
6.1.3 Block-Coordinate Descent Approach 148
6.2 Free Number of Modes 151
6.2.1 Bounded-Error Approach 151
6.2.2 Block-Coordinate Descent Approach 154
6.2.3 Error Sparsification Method 155
6.2.4 Parameter Sparsification Method 159
Notes 164
References 165

7 **Estimation of Piecewise Affine Models** 169
7.1 From Switched Affine Models to PWA Models 170
7.2 From Nonlinear Models to PWA Models 171
7.2.1 Limitation of the Classical Regularization Schemes 172
7.2.2 Local Regularization 172
7.2.3 Learning Smooth Models of PWA Functions by Convex Optimization 173
7.2.4 Recovering PWA Models 175
7.3 From Local Models to PWA Models 175
7.3.1 Fitting Local Models over Neighborhoods 177
7.3.2 Controlling the Variations of the Local Models 178
7.3.3 Clustering the Local Models 179
Notes 181
References 181

8 **Recursive and State-Space Identification of Hybrid Systems** 183
8.1 Input–Output Models 184
8.1.1 Parallel Identifiers 184
8.1.2 Model Tracking 189
8.2 State-Space Models 192
8.2.1 Problem Statement 193
8.2.2 Off-Line Approaches 194

Notes 201
References 202

9 Nonlinear Hybrid System Identification (★) 205
9.1 Continuous Optimization Approach for Switched Nonlinear Models 206
9.1.1 Overall Procedure 206
9.1.2 A Few Methods for Structure Selection 208
9.2 Error Sparsification Approach for Switched Nonlinear Models 213
9.2.1 Choice of the Regularizer 214
9.2.2 Sparsity Versus Compressibility 215
9.3 Sum-of-Norms Approach for Piecewise Smooth Models 218
9.3.1 Controlling the Variations of Nonlinear Local Models 219
9.3.2 Clustering Local Models in the RKHS 221
Notes 224
References 224

10 Outlook 227
10.1 Other Forms of Hybrid Models 228
10.1.1 Input–Output Models 228
10.1.2 Nonlinear Submodels 228
10.1.3 Continuous-Time Models 229
10.2 Computational Complexity and Exact Methods 229
10.3 Statistical Guarantees 229
10.4 Model Selection 230
10.5 Applications 230
References 231

Appendix A: Basics of Probability 233

Appendix B: Basics of Linear Algebra 239

Index 251

Notations

General Notations

Scalars are written in default and lowercase letters, e.g., a, unless otherwise specified.

Vectors are column vectors written in boldface and lowercase letters, e.g., $\boldsymbol{a}$, unless otherwise specified. For a vector $\boldsymbol{a}$, the nth entry is denoted a_n. All inequalities between vectors are meant entrywise.

Matrices are written in boldface and uppercase letters, e.g., $\boldsymbol{A}$, unless otherwise specified. For a matrix $\boldsymbol{A}$, the entry in the nth row and the pth column is denoted A_{np}.

Sets

$\mathbb{N}$	The set of natural numbers
$\mathbb{Q}$	The set of rational numbers
$\mathbb{R}$	The set of real numbers
$\mathbb{A}^n$	The set of n-dimensional vectors, with entries in the set $\mathbb{A}$
$\mathbb{A}^{n\times p}$	The set of $n \times p$-dimensional matrices, with entries in the set $\mathbb{A}$
$[n]$	The set of integers from 1 to n
$\mathcal{Y}^{\mathcal{X}}$	The set of functions from the set $\mathcal{X}$ into the set $\mathcal{Y}$
$\in$	Set membership; e.g., $x \in \mathbb{A}$ means that x is an element of $\mathcal{A}$
$\cup$	Set union, e.g., $\mathcal{A} \cup \mathcal{B} = \{x : x \in \mathcal{A} \,\text{or}\, x \in \mathcal{B}\}$
$\cap$	Set intersection, e.g., $\mathcal{A} \cap \mathcal{B} = \{x : x \in \mathcal{A} \,\text{and}\, x \in \mathcal{B}\}$
$\subseteq$, $\subset$	Set inclusions, e.g., $\mathcal{A} \subseteq \mathcal{B} \Leftrightarrow \forall x \in \mathcal{A}, x \in \mathcal{B}$, while $\mathcal{A} \subset \mathcal{B} \Leftrightarrow \mathcal{A} \subseteq \mathcal{B}$ and $\mathcal{A} \neq \mathcal{B}$
$\setminus$	Set difference, e.g., $\mathcal{A} \setminus \mathcal{B} = \{x \in \mathcal{A} : x \notin \mathcal{B}\}$
$\lvert\mathcal{A}\rvert$	Cardinality of the set $\mathcal{A}$

Vectors and Matrices

1 Vector or matrix of dimension clear from context or written in subscript with all entries equal to 1
0 Vector or matrix of dimension clear from context or written in subscript with all entries equal to 0
I Identity matrix of dimension clear from context or written in subscript with main diagonal entries equal to 1 and all other entries equal to 0

Operations

$\dot{\boldsymbol{a}}$	Time derivative of the vector $\boldsymbol{a}$
$\boldsymbol{a}^\top, \boldsymbol{A}^\top$	Transpose of the vector $\boldsymbol{a}$ or matrix $\boldsymbol{A}$; see (B.3)
$\lvert a \rvert$	Absolute value of the scalar a
$(a)_+, (a)_-$	Positive and negative parts of a, $(a)_+ = \max\{0, a\}$, $(a)_- = \min\{0, a\}$
$(\boldsymbol{a})_+, (\boldsymbol{a})_-$	Positive and negative parts of the vector $\boldsymbol{a}$, defined entrywise
$\lceil a \rceil$	Ceiling function returning the smallest integer greater than or equal to a
$\delta_{n,p}$	Kronecker delta which is 1 if $n = p$ and is 0 if $n \neq p$
$\det(\boldsymbol{A})$	Determinant of the (square) matrix $\boldsymbol{A}$; see (B.23)
$\mathrm{diag}(\boldsymbol{a})$	Diagonal matrix with the entries of $\boldsymbol{a}$ on the diagonal; see (B.6)
$e^{\boldsymbol{A}}$	Exponential of the (square) matrix $\boldsymbol{A}$; see (B.10)
$\mathbb{E}$	Mathematical expectation; see (A.2), (A.3)
∇f	Gradient of the scalar-valued function f; see (B.17)
$\nabla^2 f$	Hessian of the scalar-valued function f; see (B.18)
inf	Infimum
$\mathrm{D}\boldsymbol{f}$	Jacobian of the vector-valued function $\boldsymbol{f}$; see (B.18)
log	Natural logarithm
max	Maximum
min	Minimum
$\mathcal{O}(g)$	$f = \mathcal{O}(g)$, asymptotic notation meaning that, for $f : \mathbb{N} \to \mathbb{N}$ and $g : \mathbb{N} \to \mathbb{N}$, $\exists n_0 \in \mathbb{N}$, $c \in \mathbb{N}$, such that $\forall n \geq n_0, f(n) \leq cg(n)$
$\mathbb{P}$	Probability measure; see Appendix A.1
$\mathrm{rank}(\boldsymbol{A})$	Rank of the matrix $\boldsymbol{A}$; see (B.22)
$\mathrm{sign}(a)$	Sign of the number a, $\mathrm{sign}(a) = 1_{a \geq 0} - 1_{a<0}$
sup	Supremum
$\mathrm{tr}(\boldsymbol{A})$	Trace of the (square) matrix $\boldsymbol{A}$; see (B.4), (B.35)
$\mathrm{vec}(\boldsymbol{A})$	Vectorization of the matrix $\boldsymbol{A}$; see (B.12)
$\mathbb{1}$	Indicator function, $\mathbb{1}$ is $\mathbb{1}$ if the boolean expression A is true and 0 otherwise
$\binom{n}{p}$	Binomial coefficient, $\binom{n}{p} = \frac{n!}{p!(n-p)!}$

$\otimes$	Kronecker product; see (B.11)
$\langle \cdot, \cdot \rangle_{\mathcal{V}}$	Inner product in the space $\mathcal{V}$ ($\mathcal{V}$ can be omitted when clear from context); see Appendix B.2.3

Norms

$\|\boldsymbol{a}\|_p$	ℓ_p-norm of the vector $\boldsymbol{a}$; see (B.1)
$\|\boldsymbol{a}\|_0$	ℓ_0-pseudo-norm of the vector $\boldsymbol{a}$; see (B.2)
$\|\boldsymbol{A}\|_F$	Frobenius norm of the matrix $\boldsymbol{A}$; see (B.5)

Specific Notations

As a subscript, the letter j always refers to a mode or submodel index for hybrid models or a category index in classification.

As a subscript, the letter k always refers to the discrete time or a data point index.

$\hat{\boldsymbol{a}}$	Estimate of $\boldsymbol{a}$
CE	Classification error rate; see (4.27)
d	Dimension of $\boldsymbol{x}$, the regression vector or input vector of classifiers
$\mathcal{D}$	Data set of regressor–output pairs; see (2.24)
e, $\boldsymbol{e}$	Prediction error; see (2.6, 2.25, 2.50)
ε, $\boldsymbol{\varepsilon}$	Noise term in I/O models
f	Real-valued function, model to be estimated
$\mathcal{F}$	Model space (a real-valued function class)
FIT	Normalized criterion of fit; see (2.83)
$\boldsymbol{\phi}$	Nonlinear mapping to feature space; see (2.75)
g	Classifier, function with output in a finite set
$\boldsymbol{g}$	Parameter vector of a linear classifier
$\mathcal{G}$	Set of classifiers (an integer-valued function class with finite co-domain)
$\tilde{g}_i$	Coefficients of the impulse response of a SISO system; see (2.1)
$\tilde{h}_i$	Coefficients of the disturbance impulse response; see (2.2)
$\mathcal{H}$	Reproducing kernel Hilbert space; see Definition 2.2
j	Index of mode, submodel, or category, with $j \in \mathcal{Q}$
J	A cost function to be minimized
k	Discrete time or data point index
K	Kernel function; see Definition 2.1
$\boldsymbol{K}$	Kernel matrix with $K_{ik} = K(\boldsymbol{x}_i, \boldsymbol{x}_k)$; see (2.65)
ℓ	Loss function; see (2.27)
λ	Regularization hyperparameter
$MNPE$	Mean normalized parametric error; see (4.26)
MSE	Mean squared error; see (2.82)

n_a	Number of lagged outputs in the regression vector (SISO systems)
n_b	Number of lagged inputs in the regression vector (SISO systems)
n_d	Pure delay (SISO systems)
$n_\mathbf{x}$	Number of states, dimension of the state vector
n_u	Number of inputs, dimension of the input vector
n_y	Number of outputs, dimension of the output vector
N	Number of (training) data points
NPE	Normalized parametric error; see (2.80)
Ω	Continuous state-input domain
q	Discrete state, index of mode or category, with $q \in \mathcal{Q}$
$\mathcal{Q}$	Set of mode or category indexes, of cardinality s (usually $\mathcal{Q} = [s]$)
q	Forward shift operator
R	Regularizer; see (2.63)
s	Number of modes and submodels or number of categories in classification
t	Continuous time
$\boldsymbol{\theta}$	Parameter vector
$u, \boldsymbol{u}$	System input
v	Disturbance term in SISO I/O models
$\boldsymbol{v}$	Measurement noise term in SS models
$\boldsymbol{w}$	State noise term in SS models
$\mathbf{x}$	(continuous) State vector
$\boldsymbol{x}$	Regression vector or input vector of classifiers
$\boldsymbol{X}$	Observation matrix composed of regression vectors as $\boldsymbol{X} = [\boldsymbol{x}_1 \ \ldots \ \boldsymbol{x}_k \ \ldots \ \boldsymbol{x}_N]^\mathrm{T}$
$\mathcal{X}$	Regression domain or input space for classification
$\mathcal{X}_\phi$	Feature space
$y, \mathbf{y}$	System output
$\mathcal{Y}$	Output domain
$\boldsymbol{y}$	Target vector defined as $\boldsymbol{y} = [y_1 \ \ldots \ y_k \ \ldots \ y_N]^\mathrm{T}$

Abbreviations

AFMM	Adaptive Forgetting through Multiple Models (algorithm); see Sect. 8.1.2.2
ARMAX	AutoRegressive Moving Average with eXogenous input (model); see (2.16)
ARX	AutoRegressive with eXogenous (eXternal) input (model); see (2.9)
BJ	Box–Jenkins (model); see (2.22)
FIR	Finite Impulse Response (model); see (2.7)
FVS	Feature Vector Selection; see Sect. 9.1.2.2
HDC	Hybrid Decoupling Constraint; see (6.2)

I/O	Input–Output (model); see (1.2, 1.4)
KPCR	Kernel Principal Component Regression; see Sect. 9.1.2.3
MCS	Multi-level Coordinate Search (minimizing algorithm); see Sect. 6.1.2.1
ME	Minimum-of-Errors (problem); see (4.21, 6.8)
MILP	Mixed-Integer Linear Program; see (5.51)
MIN PFS	Minimum Number of Feasible Subsystems Problem; see Sect. 6.2.1
MINLP	Mixed-Integer NonLinear Program; see (4.19, 8.13)
MIQP	Mixed-Integer Quadratic Program; see (5.50)
NLMS	Normalized Least Mean Squares (algorithm); see (2.59)
OE	Output Error (model); see (2.18)
pdf	probability density function
PE	Product-of-Errors (problem); see (6.9, 9.1)
PEM	Prediction Error Method; see (2.26)
PWA	PieceWise Affine (model); see Sect. 4.1.3.2
PWARX	PieceWise ARX (model); see Sect. 4.1.3.2
PWS	PieceWise Smooth (model); see Sect. 4.1.3.2
RKHS	Reproducing Kernel Hilbert Space; see Definition 2.2
RLS	Recursive Least Squares (algorithm); see (2.60)
s.t.	subject to
SARX	Switched ARX (model); see Sect. 4.1.3.1
SHS	Stochastic Hybrid Systems; see Sect. 1.1
SISO	Single-Input, Single-Output (model)
SS	State-Space (model); see (1.1, 1.3, 2.34, 2.47)
SVD	Singular Value Decomposition; see (B.2.8)
SVM	Support Vector Machine; see Sect. 3.1.1.2
SVR	Support Vector Regression; see (2.78)
vs.	versus
w.r.t.	with respect to

List of Figures

Fig. 1.1 Two examples of PWA function: a saturated affine function and one with a dead zone . . . 4
Fig. 1.2 Roadmap of the book . . . 10
Fig. 2.1 A one-dimensional convex function . . . 26
Fig. 2.2 Empirical and expected risks vs. the model complexity . . . 44
Fig. 2.3 Effect of the regularization parameter λ on kernel ridge regression. . . . 49
Fig. 2.4 The ϵ-insensitive loss function . . . 52
Fig. 3.1 Optimal separating hyperplane . . . 64
Fig. 3.2 Influence of C on the margin of the support vector machine classifier . . . 65
Fig. 3.3 Multi-class problem with 3 categories. . . . 69
Fig. 3.4 K-means Algorithm 2 in action. . . . 73
Fig. 4.1 Partition of the regression space into four convex polyhedra in dimension 2 . . . 83
Fig. 4.2 A one-dimensional switching linear regression problem with two modes . . . 89
Fig. 4.3 A one-dimensional PWA regression problem with two modes . . . 91
Fig. 4.4 A one-dimensional switching regression problem with two modes tackled by a greedy bounded-error approach . . . 92
Fig. 4.5 A one-dimensional PWA regression problem with two modes tackled by a generic nonlinear regression method. . . . 96
Fig. 4.6 Illustration of the difference between regression and subspace estimation (a) . . . 98
Fig. 4.7 Illustration of the difference between regression and subspace estimation (b) . . . 98
Fig. 5.1 Reduction of the partition problem to a switching regression problem. . . . 113
Fig. 5.2 Reduction of the partition problem to a PWA regression problem . . . 117

Fig. 5.3 One-dimensional example of binary PWA regression solved exactly by Algorithm 3 121
Fig. 5.4 Illustration of Proposition 5.1 for $s = 2$ and $d = 1$ 123
Fig. 5.5 Illustration of Lemma 5.2 for $d = 1$ 125
Fig. 5.6 Illustration of the splitting procedure when $s = 2$ and $d = 1$ 129
Fig. 5.7 Hinge functions. 134
Fig. 6.1 Illustration of the continuity of the pointwise minimum of several continuous functions. 146
Fig. 6.2 Probability of success of the k-LinReg Algorithm 151
Fig. 6.3 Exponentially decreasing ϵ-insensitive loss function 153
Fig. 6.4 Basic idea of the error sparsification approach 155
Fig. 7.1 A one-dimensional PWA regression problem tackled via switching regression 171
Fig. 7.2 Learning a PWA target function from a noisy data set with local or global regularization 176
Fig. 7.3 Clustering local models for PWA regression. 180
Fig. 9.1 Illustration of the nonlinear error sparsification method for Example 9.1 217
Fig. 9.2 Iterations of the reweighting process in Example 9.2 218
Fig. 9.3 Example of PWS regression 223
Fig. 9.4 Clustering of the local nonlinear models for PWS regression. 223

List of Tables

Table 2.1 Common loss functions 23

Table 5.1 Computing time for the global optimization of a switching linear model 137

Table 5.2 Comparison of the computing time of the polynomial-time algorithm and the MIQP approach for PWA regression 139

Chapter 1
Introduction

This book is about *hybrid system identification*, which is a topic at the intersection of a number of fields, including hybrid dynamical systems, system identification, and machine learning. This first chapter briefly introduces these subjects and presents a number of applications, before giving an overview of the content of the book.

1.1 What are Hybrid Systems?

Hybrid dynamical systems—hybrid systems for short—are first and foremost dynamical systems.[1] A dynamical system can be broadly defined as a set of interacting components and a "rule" determining its evolution during time, with descriptive variables, internal (the "states") or linking the set with its environment (the "inputs" and "outputs").

1.1.1 Dynamical Systems

Dynamical systems can be broken down into two classes, according to the nature of the involved variables: continuous or discrete (dynamical) systems. For discrete systems, more precisely discrete-state systems, the state variables are countable; that is, each one takes a finite number of values, most often in a subset of $\mathbb{N}$. In continuous systems or analog systems, each variable can take an infinite number of values, most often in $\mathbb{R}$. The behavior of continuous systems is given by relationships, or "models," involving the derivatives (ordinary differential equations, in continuous time) or the delayed values (difference equations, in discrete time) of the variables.

[1] Note however that static phenomena can also be modeled by resorting to hybrid system identification methods as described in this book. In fact, due to the difficulty of hybrid system identification, the dynamical aspect of hybrid systems is mostly ignored in many dedicated approaches.

F. Lauer and G. Bloch, *Hybrid System Identification*,
Lecture Notes in Control and Information Sciences 478,
https://doi.org/10.1007/978-3-030-00193-3_1

Two general model forms of continuous systems can be chosen. The first one is the so-called state-space (SS) form, which can be written in continuous time $t \in \mathbb{R}$ as

$$\begin{cases} \dot{\mathbf{x}}(t) = \mathbf{f}(\mathbf{x}(t), \boldsymbol{u}(t)) \\ \mathbf{y}(t) = \mathbf{g}(\mathbf{x}(t), \boldsymbol{u}(t)), \end{cases}$$

or in discrete time $k \in \mathbb{N}$ as

$$\begin{cases} \mathbf{x}_{k+1} = \mathbf{f}(\mathbf{x}_k, \boldsymbol{u}_k) \\ \mathbf{y}_k = \mathbf{g}(\mathbf{x}_k, \boldsymbol{u}_k), \end{cases} \tag{1.1}$$

where $\mathbf{x}_k \in \mathbb{R}^{n_x}$, $\boldsymbol{u}_k \in \mathbb{R}^{n_u}$, and $\mathbf{y}_k \in \mathbb{R}^{n_y}$ are, respectively, the continuous (internal) "state," input and output vectors at time k. Note that in almost all of what follows, discrete-time systems are considered.

Another general form of discrete-time models is the input–output (I/O) form:

$$\mathbf{y}_k = \boldsymbol{f}(\boldsymbol{x}_k), \tag{1.2}$$

where $\mathbf{y}_k \in \mathbb{R}^{n_y}$ is the output vector, and the "regression vector" $\boldsymbol{x}_k \in \mathbb{R}^d$, although depending on the chosen subclass, typically includes lagged values of the input $\boldsymbol{u}_{k-i}$ and output $\mathbf{y}_{k-i}$. Note also that this regression vector $\boldsymbol{x}$ is not the state vector $\mathbf{x}$ and must not be confused with it, particularly in Chap. 8, where both I/O and SS models are dealt with.

In the preceding deterministic formulations, no noise is considered. We can take into account uncertainty in the models, for instance measurement noise or modeling errors, and this leads to the noisy variants of (1.1) and (1.2):

$$\begin{cases} \mathbf{x}_{k+1} = \mathbf{f}(\mathbf{x}_k, \boldsymbol{u}_k, \boldsymbol{w}_k) \\ \mathbf{y}_k = \mathbf{g}(\mathbf{x}_k, \boldsymbol{u}_k, \boldsymbol{v}_k), \end{cases} \tag{1.3}$$

where $\boldsymbol{w}_k \in \mathbb{R}^{n_x}$ and $\boldsymbol{v}_k \in \mathbb{R}^{n_y}$ are, respectively, the state and measurement noise terms, and

$$\mathbf{y}_k = \boldsymbol{f}(\boldsymbol{x}_k, \boldsymbol{\varepsilon}_k), \tag{1.4}$$

where $\boldsymbol{\varepsilon}_k \in \mathbb{R}^{n_y}$ is the noise term.

1.1.2 *Hybrid Systems*

Hybrid systems include both *interacting* continuous and discrete dynamical behaviors.[2] This results in systems that switch, according to the value of the discrete variables, between different operating modes with continuous dynamics modeled as in (1.3) or (1.4). Thus, a hybrid system can be formulated in SS form as

$$\begin{cases} \mathbf{x}_{k+1} = \mathbf{f}_{q_k}(\mathbf{x}_k, \boldsymbol{u}_k, \boldsymbol{w}_k) \\ \mathbf{y}_k = \mathbf{g}_{q_k}(\mathbf{x}_k, \boldsymbol{u}_k, \boldsymbol{v}_k), \end{cases} \tag{1.5}$$

where $\mathbf{x}_k, \boldsymbol{u}_k, \mathbf{y}_k, \boldsymbol{w}_k, \boldsymbol{v}_k$ are as in (1.3) and the new integer variable $q_k \in \{1, \dots, s\}$, known as the *discrete state* or "mode," selects which one among the s "submodels" $\{(\mathbf{f}_j, \mathbf{g}_j)\}_{j=1}^{s}$ is active at time k. Note that we restrain the formulation here to a single discrete variable q_k. However, if the system involves multiple discrete variables, each combination of values for these variables leads to a different mode and can be encoded as a single value for q_k.

For I/O models, this gives

$$\mathbf{y}_k = \boldsymbol{f}_{q_k}(\boldsymbol{x}_k, \boldsymbol{\varepsilon}_k), \tag{1.6}$$

with $\mathbf{y}_k, \boldsymbol{x}_k, \boldsymbol{\varepsilon}_k$ as in (1.4) and the discrete state $q_k \in \{1, \dots, s\}$ selecting the active submodel among $\{\boldsymbol{f}_j\}_{j=1}^{s}$ at time k.

1.1.2.1 Switching Mechanism

We consider *only two ways* by which the value of the discrete state q_k can change during time. In the first one, q_k is *arbitrarily* switched, both in its value and in the change instant. This corresponds to a discrete state generated externally, for instance by a supervisor, which forces the system to change its continuous behavior.

In the second way, for SS models, the discrete state q_k depends on the continuous state $\mathbf{x}_k$ and input $\boldsymbol{u}_k$ vectors, i.e., $q_k = j$ if $\begin{bmatrix} \mathbf{x}_k \\ \boldsymbol{u}_k \end{bmatrix} \in \Omega_j$, where the Ω_j's are regions that form a partition of the state-input domain Ω. Similarly, for I/O models, the discrete state q_k depends on the continuous regression vector $\boldsymbol{x}_k$, i.e., $q_k = j$ if $\boldsymbol{x}_k \in \mathcal{X}_j$, where the $\mathcal{X}_j$'s are regions that form a partition of the whole regression space $\mathcal{X}$. Such *piecewise*-defined models are often used to approximate a single nonlinear continuous behavior by a collection of linear or affine submodels, each of which is valid only in a particular region. These piecewise affine models, commonly abbreviated as PWA, have the capability of universally approximating any nonlinear dynamics, with arbitrary precision.

[2]Note that the keywords "hybrid systems" can have different meanings in different contexts. Our motivations for using this terminology are detailed in the Preface.

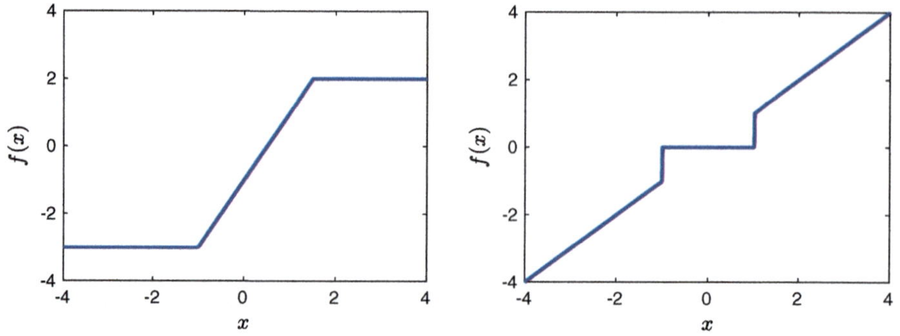

Fig. 1.1 Two examples of PWA function: a saturated affine function (left) and one with a dead zone (right)

Figure 1.1 shows two examples of PWA functions of a scalar variable x with a partition of $\mathcal{X} = [-4, 4]$ into three regions. The first PWA function,

$$f(x) = \max\{-3, \min\{2, 2x - 1\}\} = \begin{cases} -3, & \text{if } x < -1 \\ 2x - 1, & \text{if } x \in \left[-1, \frac{3}{2}\right] \\ 2, & \text{if } x > \frac{3}{2}, \end{cases}$$

is a saturated affine function. Such saturations are encountered in many applications. The second PWA function,

$$f(x) = \begin{cases} x, & \text{if } |x| > 1 \\ 0, & \text{otherwise,} \end{cases}$$

shows a dead zone in the interval $[-1, 1]$ which introduces discontinuities. Such a phenomenon occurs for instance in DC motors.

1.1.2.2 Other Hybrid Models

Many models of hybrid systems have been proposed in the literature, with a rather confused terminology.

Linear complementarity systems are discussed in [1–3], mixed logical dynamical (MDL) systems in [4], max-min-plus-scaling (MMPS) systems in [5]. Their equivalence with PWA systems and the transformation from one model into the other are exposed in [6]. Reference [7] regroups them in the class of discrete hybrid automata (DHA) with an associated high-level modeling language and a set of tools for translating DHA into any of the former hybrid models.

Besides, stochastic hybrid systems (SHSs), where, particularly, the mode transitions are triggered randomly or the new mode after transition is selected randomly, are not treated in this book. One can refer to [8] for an introduction. Reference [9] provides a comprehensive review of Markovian jump systems, where the transitions between the submodels are determined by a Markov chain, and the references therein include real-world applications, such as economic systems, flight systems, power systems, communication systems, and networked control systems.

1.1.2.3 Connections with Other Models

In what precedes, hard switchings are considered between the submodels. There are other possible ways to define transitions between submodels, as in linear parameter-varying (LPV) systems, where the parameters of a linear model evolve (most often) smoothly over time [10, 11]. In that sense, hybrid systems, as described here, can be considered as particular LPV systems with abrupt and nonsmooth parameter transitions.

There is also an important literature on multiple models. As in piecewise-defined hybrid systems, in multiple models, the operating space of the system is split into a finite number of regions, but these latter are possibly overlapping. Moreover, the current model is interpolated from the set of the submodels associated to the regions. The aggregation mechanism is based on a weighting function which quantifies the proximity of the current data to the submodel regions. This includes various approaches, such as fuzzy models, including Takagi–Sugeno models, or neuro-fuzzy models [12–15]. In the machine learning community, the mixture of experts [16] provides a model in which a *gating network* decides how to compute the output given those of an ensemble of submodels. This gating network can be made to yield either hard switchings between the submodels or combinations of their outputs.

1.2 What is System Identification?

System identification aims at building models of dynamical systems mainly from experimental data. Some prior knowledge, particularly from first principles, can be of help in selecting the input variables, choosing the model class and structure, or adding constraints linked to the system physical behavior, to yield parametric models. Then, system identification amounts to the estimation of a finite number of parameters in the model. Alternatively, when lacking precise knowledge on the physical phenomena involved, nonparametric models can be used to estimate unknown nonlinearities. In this case, system identification includes the estimation of both the model structure and its parameters from the data, which is a standard setting for machine learning techniques.

In the parametric case, the general SS model (1.3) can be clarified as

$$\begin{cases} \mathbf{x}_{k+1} = \mathbf{f}(\mathbf{x}_k, \boldsymbol{u}_k, \boldsymbol{w}_k, \boldsymbol{\theta}) \\ \mathbf{y}_k = \mathbf{g}(\mathbf{x}_k, \boldsymbol{u}_k, \boldsymbol{v}_k, \boldsymbol{\theta}), \end{cases} \tag{1.7}$$

exhibiting the parameters of the model grouped in the vector $\boldsymbol{\theta}$. For the I/O form (1.4), this reads:

$$\mathbf{y}_k = \boldsymbol{f}(\boldsymbol{x}_k[\boldsymbol{\theta}], \boldsymbol{\varepsilon}_k, \boldsymbol{\theta}), \tag{1.8}$$

where the regression vector $\boldsymbol{x}_k$ can itself depend on the parameter vector $\boldsymbol{\theta}$.

System identification then means selecting a particular structure for the model, which includes the determination of the dimension $n_{\mathbf{x}}$ of the state vector for SS models, or the components and the size of the regression vector for I/O models, and the estimation of the parameter vector $\boldsymbol{\theta}$ from recorded input-output data. This estimation is often performed by minimizing the prediction error, i.e., the difference between the measurements $\mathbf{y}_k$ and the values $\hat{\mathbf{y}}_k$ predicted by the model. We see here that system identification is tightly connected to the field of optimization and optimization problems will often be encountered throughout the book.

In the case of hybrid systems, the SS model (1.7) becomes

$$\begin{cases} \mathbf{x}_{k+1} = \mathbf{f}_{q_k}(\mathbf{x}_k, \boldsymbol{u}_k, \boldsymbol{w}_{k,}\boldsymbol{\theta}_{q_k}) \\ \mathbf{y}_k = \mathbf{g}_{q_k}(\mathbf{x}_k, \boldsymbol{u}_k, \boldsymbol{v}_{k,}\boldsymbol{\theta}_{q_k}), \end{cases} \tag{1.9}$$

while the I/O model (1.8) becomes

$$\mathbf{y}_k = \boldsymbol{f}_{q_k}(\boldsymbol{x}_k[\boldsymbol{\theta}_{q_k}], \boldsymbol{\varepsilon}_k, \boldsymbol{\theta}_{q_k}), \tag{1.10}$$

and we have to determine the number s of submodels in addition to all the parameter vectors $\boldsymbol{\theta}_j$. We see that to do the job, we also have to determine by which one of the submodels the different parts of the data are governed, which is a *classification* problem. Indeed, it is a fundamental feature of hybrid system identification to tightly combine data classification and parameter estimation.

Note that in order to avoid adding further complexity to the identification problem, we often assume a fixed structure for the submodels, that is, particularly, same state vector or parameter vector sizes.

1.3 Applications

For about two decades, there has been increased research interest in the study of hybrid systems [17], as they provide a suitable framework for modeling systems in a wide range of engineering applications [18, 19], and designing hybrid control laws [20]. In [21], a broader introduction to hybrid systems can be found. It is shown that such systems can be encountered at various levels of detail and accuracy, and in numerous application fields, such as chemical processes, process control systems, robotic manufacturing systems, interconnected power systems, transportation and intelligent highway systems, air traffic control, computer and communication networks. In the review of [22], a list of application domains for hybrid system identification is given, including computer vision, electromechanical/automotive systems, systems biology and environmental systems.

There are many physical processes involving different dynamical behaviors. For instance, threshold, dead zone, saturation, hysteresis phenomena are very common and the processes including them can be better described by a hybrid model than by a single dynamical model. Thus, most of the applications deal with piecewise affine (PWA) systems where the piecewise character expresses the different behaviors and the affine submodels are sufficient to describe each behavior.

In mechanical engineering, continuous motions may be interrupted by collisions. Other examples are backslash in gears or motion systems with friction models that distinguish between stick and slip modes. In [23, 24], the identification of the electronic component placement process in pick-and-place machines by PWA models is described. In such a process, unilateral contact and saturation phenomena characterize the hybrid dynamics of the system. In [25, 26], a rotary actuator, equipped with a speed reducer, which is commonly used in industrial robots and has strong nonlinear friction, is modeled as a PWA system. Examples can also be easily given for electromechanical systems, as, in [27] or [28], a DC motor with saturation and dead zone. An electronic throttle, that regulates air inflow into the engine of a car, is considered in [29]. The model predictive control (MPC) design of this highly nonlinear component (transmission friction and "limp-home" return spring) from a state-space PWA model is described in [30]. PWA modeling of the nonlinear displacement from the input voltage is also applied for a radio-frequency microelectromechanical system (MEMS) switch in [31]. In fluid mechanics, the level in a tank with several sections can be identified with respect to the flow rate with a PWA model, as in [32].

In electrical circuits, continuous phenomena, such as the charging of capacitors, are interrupted by switches or diodes, or involve saturation and hysteresis, as for the PWA modeling of an industrial current transformer in [33]. Applications in energy management are described in [34], from DC-DC boost converter, represented by a pair of affine continuous-time state-space equations, to a much more complicated case, the control of an electrical power generation and distribution system. In [35], the fault model of a track circuit in high-speed railway is identified to detect the failure of a capacitor. In [36], an electromagnetic actuator is modeled by a state-space piecewise linear model for fault detection and isolation by a piecewise linear observer.

Basically, the operation of internal combustion engines is also hybrid, with the well-known four-stroke cycle. Early, [37] described the modeling of an engine turbocharger by hinging hyperplanes, which form a subclass of PWA models. In gear shift control in automatic transmissions of cars, both continuous (throttle position, car velocity) and discrete (gear ratio) variables are involved. Reference [38] introduce various modeling and control challenges in power-train control, focusing on fast force-transient control, cutoff control, and idle speed control. From that time, numerous problems have been tackled in this field, as, for instance, in [7], the hybrid modeling of a car with robotized manual gear shift and cruise control, in [39], the control of magnetic actuators or homogeneous charge compression ignition engine, or, in [40], the piecewise affine modeling of NOx emission produced by a diesel engine, or yet the driving behavior for automatic driving systems [41]. In [42], no less than six PWA models are integrated in the complete model of a hybrid electric vehicle (HEV) powertrain. The resulting high-fidelity model is then used for prediction in a hybrid MPC control scheme. A recent comprehensive treatment of hybrid system control in automotive engineering is given in [43].

Applications can also be found in the environmental field. For instance, in [44], a sewer network is modeled by a mixed logical dynamical (MLD) model, a set of linear dynamic state-space equations subject to linear mixed-integer inequalities, and controlled by a model integer predictive control approach. In [45], an open-channel system is modeled through a piecewise affine or switched model.

More generally, almost all control systems can also be considered as hybrid systems. In chemical processes, the continuous evolution of chemical reactions is controlled by valves and pumps. In thermal processes, a thermostat controlling the temperature switches heating or cooling choices on or off, or links together different temperature profiles, such as in [46], for the PWA modeling of an olive oil esterification reactor. Reference [47] describes the identification by hinging hyperplanes and model predictive control of a laboratory water heater. The industrial control systems combine continuous control and logic, discrete-event controllers, that supervise high-level signals, trigger start-ups or shutdowns, alarms, or switch of operating mode. Hybrid systems are natural models for computer-controlled systems since they involve physical processes and computers. Besides, the communication between systems become more and more important and widespread. In the networked control systems, the behavior of systems is influenced or controlled by events communicated over the network. Such systems can be considered as complex hybrid systems.

In the slightly different but promising domain of computational biology, hybrid systems are proposed as a framework for modeling biological networks [48], particularly gene regulatory networks [49–51], or segmenting and modeling bacterial growth [52, 53]. Several biomedical applications of PWA system identification are described in [54]. Note also the review [55] of applications of hybrid systems to biological and medical systems. Besides, in computer vision, hybrid system identification techniques can be used for segmenting sequences of images of moving scenes (dynamic textures) and human gaits, as described in [56–58].

1.4 Outline of the Book

The content of this book can be divided into four major parts in an attempt to provide (i) sufficient background knowledge on topics related to hybrid system identification, (ii) a sound theoretical analysis of the main issues involved, (iii) an overview of the relevant methods to tackle these issues in practice, and (iv) a broader view including some advanced topics and open issues. However, it shall be clear that this does not constitute an exhaustive review of the literature on this rapidly growing field.

Figure 1.2 gives a graphical representation of the connections between the chapters of this book, whose contents are detailed below. Each chapter ends with bibliographic *Notes*. Throughout the book, technical sections that can be skipped upon first reading are marked with ($\star$).

Introductory material is given in Chaps. 2–4 with the first ones exposing the necessary background in related fields and Chap. 4 actually presenting hybrid system identification. Computational complexity will be introduced only in Chap. 5. Indeed, the involved technicalities can seem far from standard concerns of a researcher in control theory, and the general idea of computational complexity issues related to hybrid system identification can be grasped without all the details.

Chapter 2 starts with an overview of classical system identification, common models (state-space, input–output, linear, nonlinear, etc.) and standard estimation methods for parametric and nonparametric models. As most estimation methods considered in this book heavily rely on optimization, this field is also briefly introduced in Sect. 2.1.3.

Then, Chap. 3 introduces classification, which constitutes the second pillar of hybrid system identification. Two settings are distinguished depending on whether the learning problem is supervised or not. The first case corresponds to the estimation of the partition of the regression or state space used for determining the (current) mode for piecewise-defined systems, while the second case is tightly connected to hybrid system identification itself, as it amounts to assign data to the different modes.

Finally, Chap. 4 presents hybrid system identification. This chapter details the various formulations of the hybrid system models. In particular, the first distinction, which has been introduced above, is the one between state-space models and input–output models, these latter being considered in the majority of the works. The second concerns the nature of the submodels. The preceding formulations were very general, but linear models have been extensively investigated with strong theoretical results. So, linear submodels are mainly considered in the hybrid system literature. They are presented versus nonlinear ones. Finally, the third opposition, piecewise systems/arbitrarily switched systems, is introduced. The second part of the chapter poses the hybrid system identification problem and the inherent trade-off between the number of modes and the error of the model. This leads to different variants of the problems, depending on the assumptions and on which part of the trade-off is focused on. The chapter ends with a brief discussion on other problems studied in other fields that are related to hybrid system identification.

The following three chapters focus on linear hybrid systems in I/O form.

Theoretical material is exposed in Chap. 5, which focuses on algorithmic complexity and global optimization issues. The chapter starts with a brief introduction to computational complexity. Then, we analyze the opportunity of solving exactly the optimization problems set up in the previous chapter. In particular, it is shown that these problems are $\mathcal{NP}$-hard, i.e., that we cannot hope to solve them in all circumstances. Yet, the chapter also details algorithms that can be guaranteed to yield exact (or sufficiently close to exact) solutions in reasonable time for low-dimensional

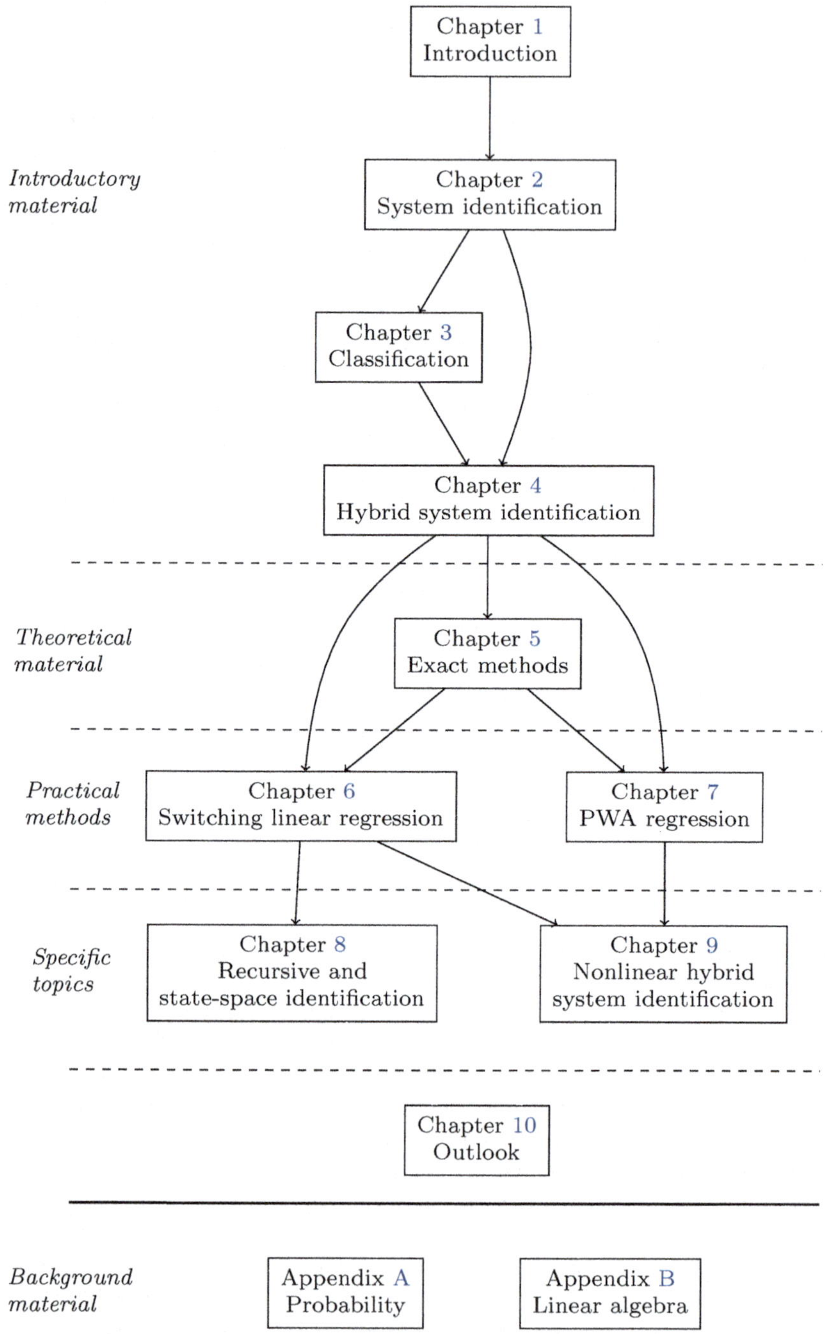

Fig. 1.2 Roadmap of the book

problems, i.e., when the number of regressors is small. The chapter ends with a few numerical results that highlight the limitations of such approaches and the need for the heuristics/approximations developed in the next chapters.

Practical methods that can deal with more challenging problems are detailed in Chaps. 6 and 7. More precisely, Chap. 6 focuses on switching regression for the identification of arbitrarily switched linear systems, while Chap. 7 deals with PWA regression, where the mode is a deterministic function of the regressors. For switching regression, we distinguish between methods dedicated to the minimization of the error for a fixed number of modes and those that aim at estimating the smallest number modes required to satisfy a given bound on the error. For PWA regression, the presentation is divided into three main approaches, each one considering a different starting point: either switching linear regression methods, nonlinear regression ones or a collection of local models. In that order, these approaches offer an increasing level of dedication to PWA models.

Specific topics are discussed in the following two chapters. Chap. 8 focuses on dynamical system identification issues with a more control-oriented perspective. As such, it considers recursive methods that apply on-line by taking one data point into account at a time and the direct identification of linear hybrid systems in state-space form. Finally, Chap. 9 introduces methods for nonlinear hybrid system identification with unknown nonlinearities estimated by nonparametric submodels, as introduced in Chap. 4, Sect. 2.4.2.

An *Outook* is given in Chap. 10, which summarizes the key points of the book and exposes a few open issues. These include topics related to the form of the model, the computational complexity of the identification problems, the statistical analysis of the identified models and model selection approaches.

Background material on probability and linear algebra is given in Appendices A and B. These recall the basic mathematical definitions and facts one should keep in mind while reading the book. More information can be found in dedicated sources and standard textbooks, such as [59] for probability and [60, 61] for linear algebra.

References

1. van der Schaft, A.J., Schumacher, J.M.: Complementarity modeling of hybrid systems. IEEE Trans. Autom. Control **43**(4), 483–490 (1998)
2. de Schutter, B., de Moor, B.: The extended linear complementarity problem and the modeling and analysis of hybrid systems. Hybrid Systems V, Proceedings of the 5th International Hybrid Systems Workshop, Notre Dame, IN, USA. Lecture Notes in Computer Science, vol. 1567, pp. 70–85. Springer, Berlin (1999)
3. Heemels, W.P.M.H., Schumacher, J.M., Weiland, S.: Linear complementarity systems. SIAM J. Appl. Math. **60**(4), 1234–1269 (2000)
4. Bemporad, A., Morari, M.: Control of systems integrating logic, dynamics, and constraints. Automatica **35**(3), 407–427 (1999)
5. de Schutter, B., van den Boom T.J.J.: Model predictive control for max-min-plus-scaling systems. In: Proceedings of the 2001 American Control Conference (ACC), Arlington, VA, USA, pp. 319–324 (2001)

6. Heemels, W.P.M.H., de Schutter, B., Bemporad, A.: Equivalence of hybrid dynamical models. Automatica **37**(7), 1085–1091 (2001)
7. Torrisi, F.D., Bemporad, A.: HYSDEL-A tool for generating computational hybrid models for analysis and synthesis problems. IEEE Trans. Control Syst. Technol. **12**(2), 235–249 (2004)
8. Cassandras, C.G., Lygeros, J.: Stochastic hybrid systems: research issues and areas. In: Cassandras, C.G., Lygeros, J. (eds.) Stochastic Hybrid Systems, pp. 1–14. CRC Press/Taylor & Francis, Boca Raton (2007)
9. Shi, P., Li, F.: Survey on Markovian jump systems: modeling and design. Int. J. Control Autom. Syst. **13**(1), 1–16 (2015)
10. Toth, R.: Modeling and Identification of Linear Parameter-Varying Systems. Lecture Notes in Control and Information Sciences, vol. 403. Springer, Berlin (2010)
11. Briat, C.: Introduction to LPV systems. Linear Parameter-Varying and Time-Delay Systems. Advances in Delays and Dynamics, vol. 3, pp. 3–36. Springer, Berlin (2015)
12. Babuška, R.: Fuzzy Modeling for Control. International Series in Intelligent Technologies, vol. 12. Springer, Berlin (1998)
13. Babuška R, Verbruggen H, : Neuro-fuzzy methods for nonlinear system identification. Ann. Rev. Control **27**(1), 73–85 (2003)
14. Nelles, O.: Nonlinear System Identification: From Classical Approaches to Neural Networks and Fuzzy Models. Springer, Berlin (2001)
15. Orjuela, R., Marx, B., Ragot, J., Maquin, D.: Nonlinear system identification using heterogeneous multiple models. J. Appl. Math. Comput. Sci. **23**(1), 103–115 (2013)
16. Jacobs, R.A., Jordan, M.I., Nowlan, S.J., Hinton, G.E.: Adaptive mixtures of local experts. Neural Comput. **3**(1), 79–87 (1991)
17. van der Schaft, A.J., Schumacher, H.: An Introduction to Hybrid Dynamical Systems. Lecture Notes in Control and Information Sciences, vol. 251. Springer, Berlin (2000)
18. Branicky, M.S.: Introduction to hybrid systems. In: Hristu-Varsakelis, D., Levine, W.S. (eds.) Handbook of Networked and Embedded Control Systems, pp. 91–116. Birkhäuser, Basel (2005)
19. Lunze, J., Lamnabhi-Lagarrigue, F. (eds.): Handbook of Hybrid Systems Control: Theory, Tools, Applications. Cambridge University Press, Cambridge (2009)
20. Sun, Z., Ge, S.S.: Switched Linear Systems - Control and Design. Communications and Control Engineering. Springer, Berlin (2005)
21. Antsaklis, P.J.: A brief introduction to the theory and applications of hybrid systems. Proc. IEEE, Spec. Issue Hybrid Syst. Theory Appl. **88**(7), 879–887 (2000)
22. Garulli, A., Paoletti, S., Vicino, A.: A survey on switched and piecewise affine system identification. In: Proceedings of the 16th IFAC Symposium on System Identification (SYSID), Brussels, Belgium, IFAC Proceedings Volumes, vol. 45(16), pp. 344–355 (2012)
23. Juloski, A.LJ., Heemels, W.P.M.H., Ferrari-Trecate, G.: Data-based hybrid modelling of the component placement process in pick-and-place machines. Control Eng. Pract. **12**(10), 1241–1252 (2004)
24. Juloski, A.Lj., Heemels, W.P.M.H., Ferrari-Trecate, G., Vidal, R., Paoletti, S., Niessen, J.H.G.: Comparison of four procedures for the identification of hybrid systems. In: Proceedings of the 8th International Workshop on Hybrid Systems: Computation and Control (HSCC), Zürich, Switzerland, Lecture Notes in Computer Science, vol. 3414, pp. 354–369 (2005)
25. Maruta, I., Sugie, T.: Identification of PWA models via optimal data compression. In: Proceedings of the 16th Symposium on System Identification, Brussels, Belgium, IFAC Proceedings Volumes, vol. 45(16), pp. 368–373 (2012)
26. Maruta, I., Ohlsson, H.: Compression based identification of PWA systems. In: Proceedings of the 19th IFAC World Congress, Cape Town, South Africa, IFAC Proceedings Volumes, vol. 47(3), pp. 4985–4992 (2014)
27. Canty, N., O'Mahony, T.: Design considerations for piecewise affine system identification of nonlinear systems. In: Proceedings of the 17th Mediterranean Conference on Control and Automation (MED), Thessaloniki, Greece, pp. 157–162 (2009)
28. Baptista, R.D.S., Ishihara, J.Y., Borges, G.A.: A split and merge algorithm for identification of piecewise affine systems. In: Proceedings of the 2011 American Control Conference, San Francisco, CA, USA, pp. 2018–2023 (2011)

29. Vašak, M., Mladenović, L., Perić, N.: Clustering-based identification of a piecewise affine electronic throttle model. In: Proceedings of the 31st Annual Conference of the IEEE Industrial Electronics Society (IECON), Raleigh, NC, USA, pp. 177–182 (2005)
30. Vašak, M., Baotić, M., Petrović, I., Perić, N.: Hybrid theory-based time-optimal control of an electronic throttle. IEEE Trans. Ind. Electron. **54**(3), 1483–1494 (2007)
31. Canty, N., O'Mahony, T., Cychowski, M.T. : An output error algorithm for piecewise affine system identification. Control Eng. Pract. **20**(4), 444–452 (2012)
32. Gegundez, M.E., Aroba, J., Bravo, J.M.: Identification of piecewise affine systems by means of fuzzy clustering and competitive learning. Eng. Appl. Artif. Intell. **21**(8), 1321–1329 (2008)
33. Ferrari-Trecate, G., Muselli, M., Liberati, D., Morari, M.: A clustering technique for the identification of piecewise affine systems. Automatica **39**(2), 205–217 (2003)
34. Morari, M., Beccuti, A.G., Mariéthoz, S., Papafotiou, G.: Energy management. In: Lunze, J., Lamnabhi-Lagarrigue, F. (eds.) Handbook of Hybrid Systems Control: Theory, Tools, Applications, pp. 377–404. Cambridge University Press, Cambridge (2009)
35. Li, L., Dong, W., Ji, Y., Zhang, Z.: An improved parameter identification approach for piecewise affine model. Control Eng. Pract. **21**(1), 54–64 (2013)
36. Forrai, A.: System identification and fault diagnosis of an electromagnetic actuator. IEEE Trans. Control Syst. Technol. **25**(3), 1028–1035 (2017)
37. Ernst, S.: Hinging hyperplane trees for approximation and identification. In: Proceedings of the 37th IEEE Conference on Decision and Control (CDC), Tampa, FL, USA, pp. 1266–1271 (1998)
38. Balluchi, A., Benvenuti, L., Di Benedetto, M.D., Pinello, C., Sangiovanni-Vincentelli, A.L.: Automotive engine control and hybrid systems: challenges and opportunities. Proc. IEEE Spec. Issue Hybrid Syst. Theory Appl. **88**(7), 888–910 (2000)
39. Benvenuti, L., Balluchi, A., Bemporad, A., Di Cairano, S., Johansson, B., Johansson, R., Sangiovanni Vincentelli, A.L., Tunestål, P.: Automotive control. In: Lunze, J., Lamnabhi-Lagarrigue, F. (eds.) Handbook of Hybrid Systems Control: Theory, Tools, Applications, pp. 439–470. Cambridge University Press, Cambridge (2009)
40. Vereshchaga, Y., Stadlbauer, S., Bako, L., del Re, L.: Piecewise affine modeling of NOx emission produced by a diesel engine. In: Proceedings of the 2013 European Control Conference (ECC), Zürich, Switzerland, pp. 167–172 (2013)
41. Taguchi, S., Suzuki, T., Hayakawa, S., Inagaki, S.: Identification of probability weighted multiple ARX models and its application to behavior analysis. In: Proceedings of the 48th IEEE Conference on Decision and Control (CDC), Shanghai, China, pp. 3952–3957 (2009)
42. Ripaccioli, G., Bemporad, A., Assadian, F., Dextreit, C., Di Cairano, S., Kolmanovsky, I.V.: Hybrid modeling, identification, and predictive control: an application to hybrid electric vehicle energy management. In: Proceedings of the 12th International Conference on Hybrid Systems: Computation and Control (HSCC), San Francisco, CA, USA. Lecture Notes in Computer Science, vol. 5469, pp. 321–335 (2009)
43. Böhme, T.J., Frank, B.: Hybrid Systems, Optimal Control and Hybrid Vehicles: Theory, Methods and Applications. Advances in Industrial Control. Springer, Berlin (2017)
44. Ocampo-Martinez, C., Bemporad, A., Ingimundarson, A., Puig Cayuela, V.: On hybrid model predictive control of sewer networks. In: Sánchez Peña, R.S., Quevedo Casín, J., Puig Cayuela, V. (eds.) Identification and Control - The Gap between Theory and Practice, pp. 87–116. Springer, Berlin (2007)
45. Bako, L., Boukharouba, K., Duviella, E., Lecoeuche, S.: A recursive identification algorithm for switched linear/affine models. Nonlinear Anal. Hybrid Syst. **5**(2), 242–253 (2011)
46. Lassoued, Z., Abderrahim, K.: An experimental validation of a novel clustering approach to PWARX identification. Eng. Appl. Artif. Intell. **28**, 201–209 (2014)
47. Kenesei, T., Abonyi, J.: Hinging hyperplane based regression tree identified by fuzzy clustering and its application. Appl. Soft Comput. **13**(2), 782–792 (2013)
48. Tenazinha, N., Vinga, S.: A survey on methods for modeling and analyzing integrated biological networks. IEEE/ACM Trans. Comput. Biol. Bioinform. **8**(4), 943–958 (2011)

49. de Jong, H.: Modeling and simulation of genetic regulatory systems: a literature review. J. Comput. Biol. **9**(1), 67–103 (2002)
50. Porreca, R., Ferrari-Trecate, G.: Identification of piecewise affine models of genetic regulatory networks: the data classification problem. In: Proceedings of the 17th IFAC World Congress on Automatic Control, Seoul, Korea, IFAC Proceedings Volumes, vol. 41(2), pp. 307–312 (2008)
51. Bernot, G., Comet, J.P., Richard, A., Chaves, M., Gouzé, J.L., Dayan, F.: Modeling and analysis of gene regulatory networks. In: Cazals, F., Kornprobst, P. (eds.) Modeling in Computational Biology and Biomedicine, pp. 47–80. Springer, Berlin (2013)
52. Hartmann, A., Lemos, J.M., Costa, R.S., Xavier, J., Vinga, S.: Identification of switched ARX models via convex optimization and expectation maximization. J. Process Control **28**, 9–16 (2015)
53. Hartmann, A., Neves, A.R., Lemos, J.M., Vinga, S.: Identification and automatic segmentation of multiphasic cell growth using a linear hybrid model. Math. Biosci. **279**, 83–89 (2016)
54. Liberati, D.: Biomedical applications of piece-wise affine identification for hybrid systems. Ann. Biomed. Eng. **37**(9), 1871–1876 (2009)
55. Aihara, K., Suzuki, H.: Theory of hybrid dynamical systems and its applications to biological and medical systems. Philos. Trans. R. Soc. A **368**(1930), 4893–4914 (2010)
56. Vidal, R., Soatto, S., Chiuso, A.: Application of hybrid system identification in computer vision. In: Proceedings of the European Control Conference (ECC), Kos, Greece, pp. 27–34 (2007)
57. Vidal, R.: Recursive identification of switched ARX systems. Automatica **44**(9), 2274–2287 (2008)
58. Ozay, N., Sznaier, M., Lagoa, C., Camps, O.: A sparsification approach to set membership identification of a class of affine hybrid systems. In: Proceedings of the 47th IEEE Conference on Decision and Control (CDC), Cancun, Mexico, pp. 123–130 (2008)
59. Bertsekas, D.P., Tsitsiklis, J.N.: Introduction to Probability, 2nd edn. Athena Scientific, Belmont (2008)
60. Golub, G.H., Van Loan, C.F.: Matrix Computations, 4th edn. John Hopkins University Press, Baltimore (2013)
61. Horn, R.A., Johnson, C.R.: Matrix Analysis, 2nd edn. Cambridge University Press, Cambridge (2013)

Chapter 2
System Identification

System identification aims at building models of continuous dynamical systems mainly from experimental data. This can be decomposed in several steps, specified in Procedure 1 below.

Procedure 1 Identification procedure

1. Record a data set of input–output pairs (u_k, y_k).
2. Choose the class of models or the model structure.
3. Estimate the model in the class, guided by the data with a criterion of fit.
4. Assess the obtained model.

In this chapter, we deal, for regular, nonhybrid, systems, with the last three steps, leaving aside the choice of input-output signals used to form the data set, which enters into the topic of experimental design. Clearly, these signals must be at least sufficiently exciting to reveal the dynamical behavior of the considered system.

In the most common case, Step 2 in the procedure above refers to the difficult task of expressing our prior knowledge on the system as a model within which a number of parameters remain unknown. Then, Step 3 amounts to estimating these parameters on the basis of the data collected in Step 1. In a more complex situation, where prior knowledge is scarce, Step 2 must typically settle on the choice of a highly flexible model class able to sufficiently well approximate many system behaviors. Such model classes are usually nonparametric and, in this case, Step 3 becomes more involved as it now amounts to estimating both the structure and the parameters of the model.

The identification procedure ends in Step 4 with the evaluation of the model quality. Indeed, due to the presence of noise or the lack of prior knowledge, we cannot hope to obtain a perfect model and thus it is crucial to quantify the model accuracy before using its predictions. Yet, model assessment techniques can also be used in the middle of the procedure, for model selection purposes and the choice of the most appropriate model class and user-level parameters.

F. Lauer and G. Bloch, *Hybrid System Identification*,
Lecture Notes in Control and Information Sciences 478,
https://doi.org/10.1007/978-3-030-00193-3_2

Chapter Outline

The chapter is organized with respect to the model classes. For each class, we start with a presentation of the models before the description of standard methods for their estimation. As we will see, there are many model classes and subclasses, which can be first distinguished according to their linear or nonlinear nature. "Linear" models are described in Sects. 2.1 and 2.2. From the general form (1.8), Sect. 2.1 is devoted to linear Input–Output (I/O) models and includes also, by an abuse of language, pseudo-linear models, while Sect. 2.2 presents the linear state-space (SS) models with their properties, from the general SS form (1.7). For I/O models, the parameter estimation based on the well-known prediction error method (PEM) is introduced in Sect. 2.1.2, with background material on optimization given in Sect. 2.1.3. For SS models, Sect. 2.2.2 briefly describes subspace methods in addition to PEM.

Procedure 1 is a batch one as a number of sampled data, inputs $\{u_k\}$ and outputs $\{y_k\}$, are used together. The corresponding recursive approaches, where the data samples (u_k, y_k) are used one by one, are presented in Sect. 2.3.

Next, nonlinear models are presented in Sect. 2.4 with a focus on nonparametric models, for which both the structure and the parameters must be estimated. We present these models and some related estimation methods within a machine learning framework with an emphasis on some fundamental concepts such as overfitting, regularization, and kernel methods.

Finally, the approaches and criteria used to assess the quality of the obtained models and to select the "best" one are introduced in Sect. 2.5.

We recall that the reader is referred to Appendix A for the definition of probabilistic terms and to Appendix B for the definitions and tools from linear algebra, used in the sequel.

2.1 Input–Output (I/O) Models

We start with the definition of standard Input–Output (I/O) models and their prediction errors, on which will be based the parameter estimation described in Sect. 2.1.2.

2.1.1 Models, Predictor, and Prediction Error

Basic model structures are able to describe many dynamical system behaviors. They make it possible to introduce useful ideas, while avoiding complications of more advanced cases. Here, we deal with systems which include only one input and one output, and that are time-invariant, i.e., shortly, the output does not depend *directly* on the time. We will see below that time invariance is expressed by transfer functions constant in time.

A linear time-invariant dynamical Single-Input, Single-Output (SISO) system can be described in discrete-time k, with the output y_k and input u_k, from its impulse response $\{\tilde{g}_i\}_{i=0}^{\infty}$ as

$$y_k = \sum_{i=0}^{\infty} \tilde{g}_i u_{k-i} + v_k. \tag{2.1}$$

To account for various types of disturbances (measurement noise, incontrollable inputs, unmeasurable disturbances, modeling errors, i.e., misfit between noise-free system outputs and model outputs), the disturbance v_k is written as

$$v_k = \sum_{i=0}^{\infty} \tilde{h}_i \varepsilon_{k-i}, \tag{2.2}$$

where $\{\varepsilon_k\}$ is white noise, i.e., a sequence of independent and identically distributed random variables (see Appendix A.3.1.1), with zero mean and variance σ_ε^2, or as

$$v_k = \varepsilon_k + \sum_{i=1}^{\infty} \tilde{h}_i \varepsilon_{k-i}, \tag{2.3}$$

without loss of generality as the variance σ_ε^2 can be fixed arbitrarily.

Defining q as the forward shift operator, such that $\mathrm{q}x_k = x_{k+1}$, and q^{-1} as the backward shift operator, such that $\mathrm{q}^{-1}x_k = x_{k-1}$, this can be rewritten as

$$y_k = G(\mathrm{q})u_k + v_k = G(\mathrm{q})u_k + H(\mathrm{q})\varepsilon_k, \tag{2.4}$$

where $G(\mathrm{q}) = \sum_{i=0}^{\infty} \tilde{g}_i \mathrm{q}^{-i}$ is the transfer function, and $H(\mathrm{q}) = 1 + \sum_{i=1}^{\infty} \tilde{h}_i \mathrm{q}^{-i}$ is monic ($\tilde{h}_0 = 1$). Moreover, $H(\mathrm{q})$ is assumed to be stable and invertible, that is $\varepsilon_k = H^{-1}(\mathrm{q})v_k$.

There are different ways of representing the sequences $\{\tilde{g}_i\}$ and $\{\tilde{h}_i\}$ which will be reviewed in the sequel. We now define the prediction error, also called innovation, that is the best we can say about the disturbance v_k.

Let us consider the one-step-ahead prediction of the disturbance term (2.3), $v_k = \varepsilon_k + \sum_{i=1}^{\infty} \tilde{h}_i \varepsilon_{k-i} = H(\mathrm{q})\varepsilon_k$, given measurements of $v_l, l \leq k-1$. Note that $\varepsilon_l, l \leq k-1$, are then also given, because of the invertible relation between v_k and ε_k. As ε_k has zero mean, we have

$$\begin{aligned}\hat{v}(k \mid k-1) &= \mathbb{E}\left[v(k \mid k-1)\right] = \mathbb{E}\left[\varepsilon_k\right] + \mathbb{E}\left[\sum_{i=1}^{\infty} \tilde{h}_i \varepsilon_{k-i}\right] \\ &= \sum_{i=1}^{\infty} \tilde{h}_i \varepsilon_{k-i} = \left(H(\mathrm{q}) - 1\right)\varepsilon_k = \left(1 - H^{-1}(\mathrm{q})\right) v_k.\end{aligned}$$

Then, the one-step-ahead prediction of model (2.4), given u_l and y_l, and thus v_l, $l \leq k-1$, can be written as

$$\begin{aligned}\hat{y}(k \mid k-1) &= G(\mathrm{q})u_k + \hat{v}(k \mid k-1)\\ &= G(\mathrm{q})u_k + (1 - H^{-1}(\mathrm{q}))v_k\\ &= G(\mathrm{q})u_k + (1 - H^{-1}(\mathrm{q}))(y_k - G(\mathrm{q})u_k),\end{aligned}$$

or

$$\hat{y}(k \mid k-1) = H^{-1}(\mathrm{q})G(\mathrm{q})u_k + (1 - H^{-1}(\mathrm{q}))y_k. \tag{2.5}$$

Thus, the part of the output y_k which cannot be predicted from the data, the *prediction error* or *innovation*, is given by

$$e_k = y_k - \hat{y}(k \mid k-1) = -H^{-1}(\mathrm{q})G(\mathrm{q})u_k + H^{-1}(\mathrm{q})y_k. \tag{2.6}$$

Moreover,

$$e_k = H^{-1}(\mathrm{q})(y_k - G(\mathrm{q})u_k) = H^{-1}(\mathrm{q})v_k = \varepsilon_k.$$

The transfer functions $G(\mathrm{q})$ and $H(\mathrm{q})$ in (2.4) can be represented as rational fractions, with the coefficients at the numerator and denominator as parameters. This leads to various model structures, briefly recalled here, which can be distinguished as equation error or output error structures. The corresponding predictors will be expressed by including the different choices for $G(\mathrm{q})$ and $H(\mathrm{q})$ in (2.5).

2.1.1.1 Equation Error Models

In the equation error models, the noise term is directly present in the model equation.

Simply considering $G(\mathrm{q})$ as a polynomial of degree n_b in q^{-1}: $B(\mathrm{q}) = b_1\mathrm{q}^{-1} + \cdots + b_{n_b}\mathrm{q}^{-n_b}$, which corresponds to a (pure) time delay of one sampling period, and taking $H(\mathrm{q}) = 1$ in (2.4) leads to the finite impulse response (FIR) model:

$$y_k = B(\mathrm{q})u_k + \varepsilon_k. \tag{2.7}$$

The associated predictor (2.5) reads

$$\hat{y}_k = f(\boldsymbol{x}_k, \boldsymbol{\theta}) = B(\mathrm{q})u_k = \boldsymbol{x}_k^\top \boldsymbol{\theta}, \tag{2.8}$$

with the regression vector $\boldsymbol{x}_k = \begin{bmatrix} u_{k-1} & \dots & u_{k-n_b} \end{bmatrix}^\top$ and the parameter vector $\boldsymbol{\theta} = \begin{bmatrix} b_1 & \dots & b_{n_b} \end{bmatrix}^\top$, of dimension $d = n_b$.

Considering now in (2.4) $G(\mathrm{q}) = B(\mathrm{q})/A(\mathrm{q})$ and $H(\mathrm{q}) = 1/A(\mathrm{q})$, with $A(\mathrm{q}) = 1 + a_1\mathrm{q}^{-1} + \cdots + a_{n_a}\mathrm{q}^{-n_a}$ and $B(\mathrm{q})$ as above, gives the AutoRegressive with eXogenous input (ARX) model, with the model orders n_a and n_b:

$$A(\mathrm{q})y_k = B(\mathrm{q})u_k + \varepsilon_k, \tag{2.9}$$

or

$$y_k = -\sum_{i=1}^{n_a} a_i y_{k-i} + \sum_{i=1}^{n_b} b_i u_{k-i} + \varepsilon_k.$$

The associated predictor (2.5) becomes

$$\hat{y}_k = f(\boldsymbol{x}_k, \boldsymbol{\theta}) = (1 - A(\mathrm{q}))y_k + B(\mathrm{q})u_k = \boldsymbol{x}_k^\top \boldsymbol{\theta}, \tag{2.10}$$

with the regression vector

$$\boldsymbol{x}_k = \left[-y_{k-1} \, \dots \, -y_{k-n_a} \; u_{k-1} \, \dots \, u_{k-n_b}\right]^\top, \tag{2.11}$$

and the parameter vector, of dimension $d = n_a + n_b$,

$$\boldsymbol{\theta} = \left[a_1 \, \dots \, a_{n_a} \; b_1 \, \dots \, b_{n_b}\right]^\top. \tag{2.12}$$

If the system has a pure delay $n_\mathrm{d} > 1$, the ARX model (2.9) reads:

$$A(\mathrm{q})y_k = \mathrm{q}^{-n_\mathrm{d}} B(\mathrm{q})u_k + \varepsilon_k,$$

or

$$y_k = -\sum_{i=1}^{n_a} a_i y_{k-i} + \sum_{i=1}^{n_b} b_i u_{k-n_\mathrm{d}-i+1} + \varepsilon_k. \tag{2.13}$$

Again, the corresponding predictor is simply

$$\hat{y}_k = f(\boldsymbol{x}_k, \boldsymbol{\theta}) = \boldsymbol{x}_k^\top \boldsymbol{\theta}. \tag{2.14}$$

with the same parameter vector $\boldsymbol{\theta} \in \mathbb{R}^{n_a+n_b}$ but with the regression vector

$$\boldsymbol{x}_k = \left[-y_{k-1} \, \dots \, -y_{k-n_a} \; u_{k-n_\mathrm{d}} \, \dots \, u_{k-n_\mathrm{d}-n_b+1}\right]^\top. \tag{2.15}$$

These two FIR and ARX predictors, $\hat{y}_k = \boldsymbol{x}_k^\top \boldsymbol{\theta}$, are the only true linear ones, as the regression vector $\boldsymbol{x}_k$ depends only on measured data. On the other hand, the ARX model needs to be calculated from both the input and output of the system, hence the name of series–parallel model.

If we take in (2.4) $G(\mathrm{q}) = B(\mathrm{q})/A(\mathrm{q})$ and $H(\mathrm{q}) = C(\mathrm{q})/A(\mathrm{q})$, with $C(\mathrm{q}) = 1 + c_1\mathrm{q}^{-1} + \cdots + c_{n_c}\mathrm{q}^{-n_c}$, the AutoRegressive Moving Average with eXogenous input (ARMAX) model is obtained as

$$A(\mathrm{q})y_k = B(\mathrm{q})u_k + C(\mathrm{q})\varepsilon_k. \tag{2.16}$$

The associated predictor (2.5) is now

$$\hat{y}_k = \frac{B(\mathrm{q})}{C(\mathrm{q})}u_k + \left(1 - \frac{A(\mathrm{q})}{C(\mathrm{q})}\right) y_k.$$

Multiplying by $C(\mathrm{q})$, then adding $(1 - C(\mathrm{q}))\hat{y}_k$ in both sides of this expression allows us to write the predictor as

$$\hat{y}_k = B(\mathrm{q})u_k + (1 - A(\mathrm{q}))y_k + (C(\mathrm{q}) - 1)e_k,$$

where $e_k = y_k - \hat{y}_k$ is the prediction error, or

$$\hat{y}_k = f(\boldsymbol{x}_k, \boldsymbol{\theta}) = \boldsymbol{x}_k(\boldsymbol{\theta})^\top \boldsymbol{\theta}, \tag{2.17}$$

with the regression vector

$$\boldsymbol{x}_k(\boldsymbol{\theta}) = \begin{bmatrix} -y_{k-1} & \ldots & -y_{k-n_a} & u_{k-1} & \ldots & u_{k-n_b} & e_{k-1} & \ldots & e_{k-n_c} \end{bmatrix}^\top$$

and the parameter vector $\boldsymbol{\theta} = \begin{bmatrix} a_1 & \ldots & a_{n_a} & b_1 & \ldots & b_{n_b} & c_1 & \ldots & c_{n_c} \end{bmatrix}^\top$, of dimension $d = n_a + n_b + n_c$. As for the ARX model, a pure delay $n_\mathrm{d} > 1$ can be introduced. This ARMAX model is pseudo-linear, as the regression vector $\boldsymbol{x}_k(\boldsymbol{\theta})$ includes measured data but also unknown terms depending on the parameters $\boldsymbol{\theta}$.

The equation error models share a feature: They have a common polynomial $A(\mathrm{q})$ in the denominators.

Remark 2.1 We will see below that, due to its linearity, the ARX model is easy to handle for parameter estimation and, as a matter of fact, it is the most common I/O model for hybrid system identification. However, ARMAX models can naturally arise when parameter estimation is done from noisy measurements of the output. To see this, recall the model of the output (2.4): $y_k = G(\mathrm{q})u_k + H(\mathrm{q})\varepsilon_k$. Whenever the outputs $-y_{k-i}$ are directly corrupted by additive white noise and included in the regression vector of the ARX model (2.11), the transfer function associated with the noise is $H(\mathrm{q}) = 1$ and the resulting model is actually an ARMAX one, with $C(\mathrm{q}) = A(\mathrm{q})$ in (2.16). In this case, the basic assumption $H(\mathrm{q}) = 1/A(\mathrm{q})$ of the ARX model (2.9) does not hold and the ARX predictor (2.10) yields a prediction error (2.6) of $e_k = A(\mathrm{q})\varepsilon_k$, which is not a white noise anymore. Hence, the name of colored noise.

2.1.1.2 Output Error Models

Conversely, in output error models, the linear difference equation is error-free, but a noise term is added to the output measurement. Writing the model error-free equation as $z_k + f_1 z_{k-1} + \cdots + f_{n_f} z_{k-n_f} = b_1 u_{k-1} + \cdots + b_{n_b} u_{k-n_b}$ and the measurement equation as $y_k = z_k + \varepsilon_k$ corresponds to the choice $G(\mathrm{q}) = B(\mathrm{q})/F(\mathrm{q})$ and $H(\mathrm{q}) = 1$ in (2.4) and leads to the output error (OE) model

$$y_k = \frac{B(\mathrm{q})}{F(\mathrm{q})} u_k + \varepsilon_k, \tag{2.18}$$

which can be also written as

$$\begin{cases} z_k = \boldsymbol{r}_k^\top \boldsymbol{\theta} \\ y_k = z_k + \varepsilon_k, \end{cases} \tag{2.19}$$

with $\boldsymbol{r}_k = \left[u_{k-1} \; \ldots \; u_{k-n_b} \; -z_{k-1} \; \ldots \; -z_{k-n_f}\right]^\top$ and $\boldsymbol{\theta} = \left[b_1 \; \ldots \; b_{n_b} \; f_1 \; \ldots \; f_{n_f}\right]^\top$, of dimension $d = n_b + n_f$. The associated predictor (2.5) is simply

$$\hat{y}_k = \frac{B(\mathrm{q})}{F(\mathrm{q})} u_k, \tag{2.20}$$

or

$$\hat{y}_k = f(\boldsymbol{x}_k, \boldsymbol{\theta}) = \boldsymbol{x}_k(\boldsymbol{\theta})^\top \boldsymbol{\theta}, \tag{2.21}$$

where $\boldsymbol{x}_k(\boldsymbol{\theta}) = \left[u_{k-1} \; \ldots \; u_{k-n_b} \; -\hat{y}_{k-1} \; \ldots \; -\hat{y}_{k-n_f}\right]^\top$.

Finally, the Box–Jenkins (BJ) model is obtained by choosing $G(\mathrm{q}) = B(\mathrm{q})/F(\mathrm{q})$ and $H(\mathrm{q}) = C(\mathrm{q})/D(\mathrm{q})$ in (2.4):

$$y_k = \frac{B(\mathrm{q})}{F(\mathrm{q})} u_k + \frac{C(\mathrm{q})}{D(\mathrm{q})} \varepsilon_k. \tag{2.22}$$

In this case, the predictor (2.5) is more involved:

$$\hat{y}_k = \frac{D(\mathrm{q})B(\mathrm{q})}{C(\mathrm{q})F(\mathrm{q})} u(k) + \left(1 - \frac{D(\mathrm{q})}{C(\mathrm{q})}\right) y(k),$$

or

$$\hat{y}_k = B(\mathrm{q})u_k + (1 - F(\mathrm{q}))\hat{y}_{(u)k} + (C(\mathrm{q}) - 1)e_k + (D(\mathrm{q}) - 1)e_{(u)k}, \tag{2.23}$$

where

$$\hat{y}_{(u)k} = \frac{B(\mathrm{q})}{F(\mathrm{q})} u_k$$

is the output predicted only from the past inputs u (see (2.20)), $e_k = y_k - \hat{y}_k$ the prediction error, and $e_{(u)k} = y_k - \hat{y}_{(u)k}$ the simulation error. With

$$\boldsymbol{x}_k(\boldsymbol{\theta}) = \begin{bmatrix} u_{k-1} & \dots & u_{k-n_b} & -\hat{y}_{(u)k-1} & \dots & -\hat{y}_{(u)k-n_f} & e_{k-1} & \dots & e_{k-n_c} & e_{(u)k-1} & \dots & e_{(u)k-n_d} \end{bmatrix}^\top$$

and $\boldsymbol{\theta} = \begin{bmatrix} b_1 & \dots & b_{n_b} & f_1 & \dots & f_{n_f} & c_1 & \dots & c_{n_c} & d_1 & \dots & d_{n_d} \end{bmatrix}^\top$, of dimension $d = n_b + n_c + n_d + n_f$, the predictor can be rewritten as (2.21).

These output error (OE and BJ) models need only the system input to be calculated and are also termed simulation models.

2.1.2 Parameter Estimation

For parameter identification, if the model class is fixed, for instance, given orders n_a and n_b, and pure delay n_d in the case of ARX models, the data set considered in Procedure 1 is

$$\mathcal{D} = \{(\boldsymbol{x}_k, y_k)\}_{k=1}^N. \tag{2.24}$$

Then, recalling the prediction error (2.6) as the difference between the system output y_k and the model prediction $\hat{y}_k = f(\boldsymbol{x}_k, \boldsymbol{\theta})$ (2.8) (2.10) (2.14) (2.17) (2.21) (2.23),

$$e_k = y_k - \hat{y}_k = y_k - f(\boldsymbol{x}_k, \boldsymbol{\theta}), \tag{2.25}$$

the most common approach to estimate the parameter vector $\boldsymbol{\theta} \in \mathbb{R}^d$ is the *prediction error method* (PEM). This amounts to solving

$$\min_{\boldsymbol{\theta} \in \mathbb{R}^d} \sum_{k=1}^N \ell(e_k), \tag{2.26}$$

where the loss function

$$\ell : \mathbb{R} \longrightarrow \mathbb{R}^+ \tag{2.27}$$

provides a positive measure of the error and is such that $\ell(0) = 0$.

The loss functions considered in this book are recalled in Table 2.1. Among these, the squared loss

$$\ell_2(e) = e^2 \tag{2.28}$$

is probably the most common. However, it is also the most sensitive to outliers, i.e., points that do not coincide with the general distribution of the data and that can result from temporary sensor or data transmission failures. Indeed, such outliers typically lead to large errors whose influence on the objective function of (2.26) is amplified by the squared loss. However, it remains largely used in practice for its computational benefits detailed below. By penalizing large errors in a similar manner

Table 2.1 Common loss functions (plotted with $\epsilon = 1$). The last three plots are reprinted from [1], Copyright (2018), with permission from Elsevier

Loss function	$\ell(e)$	Plot	Properties
Squared loss	$\ell_2(e) = e^2$	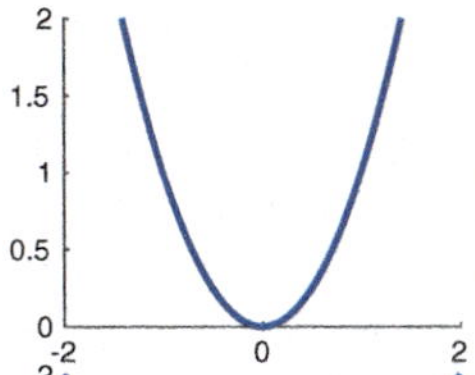	Convex, smooth, non-robust
Absolute loss	$\ell_1(e) = \|e\|$	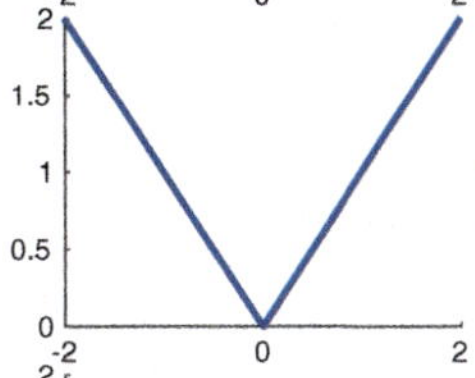	Convex, non-smooth, slightly robust
Huber's loss	$\begin{cases} \frac{e^2}{2}, & \text{if } \|e\| \le \epsilon \\ \epsilon(\|e\| - \frac{\epsilon}{2}), & \text{otherwise} \end{cases}$	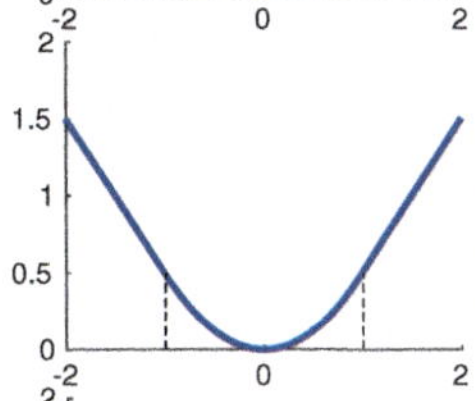	Convex, smooth, slightly robust
Hampel's loss	$\begin{cases} \frac{\epsilon^2}{\pi}\left(1 - \cos \frac{\pi e}{\epsilon}\right), & \text{if } \|e\| \le \epsilon \\ \frac{2\epsilon^2}{\pi}, & \text{otherwise} \end{cases}$	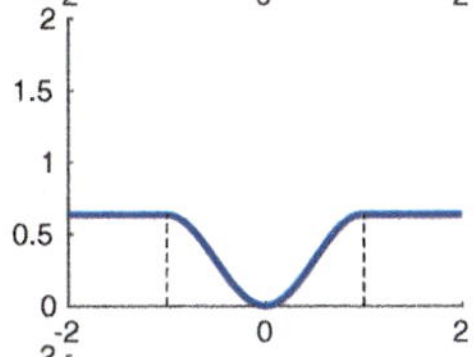	Non-convex, smooth, robust
Saturated squared loss	$\ell_{2,\epsilon}(e) = \min\{e^2, \epsilon^2\}$	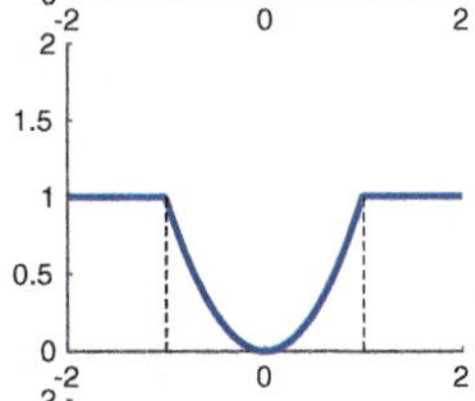	Non-convex, non-smooth, robust
Saturated absolute loss	$\ell_{1,\epsilon}(e) = \min\{\|e\|, \epsilon\}$	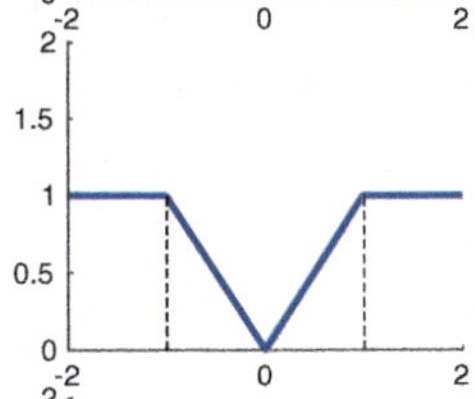	Non-convex, non-smooth, robust
Bounded-error loss	$\ell_{0,\epsilon}(e) = 1_{\|e\|>\epsilon}$	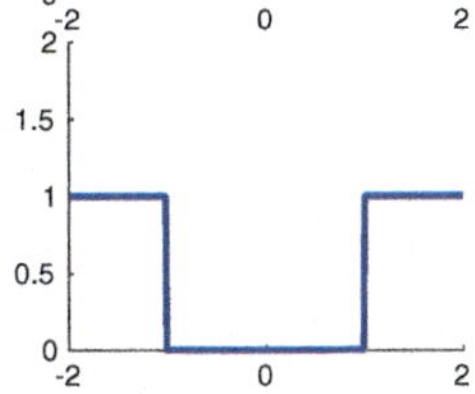	Non-convex, non-smooth, robust

as small ones, the absolute loss provides a partial answer to this issue. Though nonsmooth, its minimization remains tractable since (2.26) amounts in this case to a linear programming problem, for which efficient solvers exist (see below). A smooth version is provided by Huber's loss, which also benefits from a quadratic penalization of small errors and thus a more pronounced importance given to inliers. By annealing the effect of the increase of large errors, the remaining saturated loss functions become truly robust to outliers. However, this cannot be obtained without losing convexity, which makes solving (2.26) much more challenging with such loss functions. Among these, Hampel's loss was designed to be smooth, while the others are obtained by plain saturation of standard losses.

2.1.3 Optimization

The identification scheme depicted so far relies on our ability to solve the optimization problem (2.26). Two kinds of optimal solutions must be distinguished. Given a general minimization problem of the form

$$\min_{\boldsymbol{\theta}\in\mathbb{R}^d} J(\boldsymbol{\theta}), \tag{2.29}$$

a *local* solution $\boldsymbol{\theta}^*$ is a point for which there is some $\epsilon > 0$ such that

$$\forall\boldsymbol{\theta} \in B_\epsilon(\boldsymbol{\theta}^*), \quad J(\boldsymbol{\theta}^*) \leq J(\boldsymbol{\theta}),$$

where $B_\epsilon(\boldsymbol{\theta}^*)$ denotes a ball of radius ϵ around $\boldsymbol{\theta}^*$. If this holds for all $\epsilon > 0$, then $\boldsymbol{\theta}^*$ is also a *global* solution and we write[1]

$$\boldsymbol{\theta}^* = \operatorname*{argmin}_{\boldsymbol{\theta}\in\mathbb{R}^d} J(\boldsymbol{\theta}).$$

Optimality conditions can be formulated in terms of the gradient (see (B.17))

$$\nabla J = \begin{bmatrix} \frac{\partial J}{\partial \theta_1} \\ \vdots \\ \frac{\partial J}{\partial \theta_d} \end{bmatrix}$$

and the Hessian (see (B.18))

[1] We also use the set notation $\boldsymbol{\theta}^* \in \operatorname{argmin}_{\boldsymbol{\theta}\in\mathbb{R}^d} J(\boldsymbol{\theta})$ for cases where (2.29) has multiple global solutions.

$$\nabla^2 J = \begin{bmatrix} \frac{\partial^2 J}{\partial \theta_1 \partial \theta_1} & \cdots & \frac{\partial^2 J}{\partial \theta_1 \partial \theta_d} \\ \vdots & \ddots & \\ \frac{\partial^2 J}{\partial \theta_d \partial \theta_1} & \cdots & \frac{\partial^2 J}{\partial \theta_d \partial \theta_d} \end{bmatrix}$$

of the cost function J. In particular, a necessary condition for $\boldsymbol{\theta}^*$ to be a local solution is

$$\nabla J(\boldsymbol{\theta}^*) = \mathbf{0}. \tag{2.30}$$

Together with the positive semi-definiteness (see (B.15)) of the Hessian at $\boldsymbol{\theta}^*$, this also becomes a sufficient condition for $\boldsymbol{\theta}^*$ to be a local minimizer.

Converse definitions and notations (with max replacing min) are used for maximization problems, for which the positive semi-definiteness of the *opposite* of the Hessian at $\boldsymbol{\theta}^*$ associated with (2.30) provides a sufficient condition for $\boldsymbol{\theta}^*$ to be a local maximizer.

2.1.3.1 Convex Optimization

A function $J : \mathbb{R}^d \to \mathbb{R}$ is convex if, for all $\boldsymbol{\theta}_1 \in \mathbb{R}^d$, $\boldsymbol{\theta}_2 \in \mathbb{R}^d$ and $\beta \in [0, 1]$,

$$J(\beta\boldsymbol{\theta}_1 + (1-\beta)\boldsymbol{\theta}_2) \leq \beta J(\boldsymbol{\theta}_1) + (1-\beta)J(\boldsymbol{\theta}_2).$$

This means that the graph of J lies below the line segment between $(\boldsymbol{\theta}_1, J(\boldsymbol{\theta}_1))$ and $(\boldsymbol{\theta}_2, J(\boldsymbol{\theta}_2))$, as illustrated in Fig. 2.1. An unconstrained optimization problem of the form (2.29) is said to be convex when the cost function J is convex. For such convex optimization problems, any local solution is also a global solution; and if J is differentiable, then (2.30) becomes a necessary and sufficient condition for global optimality.

We can extend the convexity condition when considering constraints on the search domain. The constrained optimization problem

$$\min_{\boldsymbol{\theta} \in \Theta \subseteq \mathbb{R}^d} J(\boldsymbol{\theta}),$$

is convex, if J is convex and, in addition, the domain Θ is a convex set, that is if, for all $\boldsymbol{\theta}_1 \in \Theta, \boldsymbol{\theta}_2 \in \Theta$ and $\beta \in [0, 1]$, $\beta\boldsymbol{\theta}_1 + (1-\beta)\boldsymbol{\theta}_2 \in \Theta$, meaning that every point on the line segment connecting $\boldsymbol{\theta}_1$ and $\boldsymbol{\theta}_2$ is in Θ. If the constrained optimization problem is formulated as

$$\begin{aligned} &\min_{\boldsymbol{\theta} \in \mathbb{R}^d} J(\boldsymbol{\theta}) \\ &\text{s.t. } c_i(\boldsymbol{\theta}) \leq 0, \ i = 1, \ldots, m \\ &\quad\;\; c'_i(\boldsymbol{\theta}) = 0, \ i = 1, \ldots, m', \end{aligned}$$

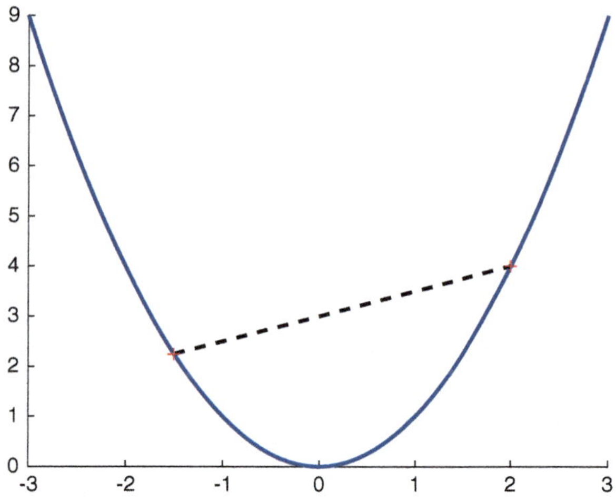

Fig. 2.1 A one-dimensional convex function $J(\theta)$ with a graph lying below the line segment between any points of the graph, as illustrated with the points $(-1.5, J(-1.5))$ and $(2, J(2))$

all the functions J, c_i, and c_i' must be convex for the problem to be convex. Particularly easy-to-handle convex problems are linear and quadratic programming problems, in which the constraint functions c_i and c_i' are all linear and the cost function J is either linear or quadratic. An important use case for linear programming includes the minimization of a sum of absolute values of linear functions of $\boldsymbol{\theta}$. Consider, for instance, the problem (2.26) with the absolute loss:

$$\min_{\boldsymbol{\theta}\in\mathbb{R}^d} \sum_{k=1}^{N} |y_k - \boldsymbol{x}_k^\top \boldsymbol{\theta}|. \tag{2.31}$$

Though the cost function is here nonlinear, (2.31) can be reformulated as a linear program by introducing additional slack variables $\zeta_k \geq 0$ upper bounding the absolute values. Indeed, $\zeta_k \geq |y_k - \boldsymbol{x}_k^\top \boldsymbol{\theta}|$ can be equivalently written as two linear inequalities. Therefore, we can minimize ζ_k under these constraints instead of the absolute value and ensure that, at the optimum, $\zeta_k^* = |y_k - \boldsymbol{x}_k^\top \boldsymbol{\theta}^*|$. This means that (2.31) is equivalent to the linear program

$$\begin{aligned} \min_{\boldsymbol{\theta}\in\mathbb{R}^d, \boldsymbol{\zeta}\in\mathbb{R}^N} \ & \sum_{k=1}^{N} \zeta_k \\ \text{s.t.} \ & -\zeta_k \leq y_k - \boldsymbol{x}_k^\top \boldsymbol{\theta} \leq \zeta_k. \end{aligned} \tag{2.32}$$

Another class of optimization problems for which dedicated solvers can also be found is the class of second-order cone programming problems, in which J is linear and some inequalities define second-order cones with $c_i(\boldsymbol{\theta}) = \|\boldsymbol{A}_i\boldsymbol{\theta} + \boldsymbol{b}_i\|_2 + \boldsymbol{c}_i^\top \boldsymbol{\theta} + d_i$.

2.1.3.2 Least Squares

As previously mentioned, since its associated regression vector (2.11) or (2.15) contains only measured variables, the ARX predictor (2.10) or (2.14) is truly linear, and, for a squared loss (2.28) in (2.26), the parameter vector can be estimated in a computationally efficient manner by standard techniques described here. However, recall that in some situations, as discussed in Remark 2.1, the prediction error can be a colored noise, which introduces a bias in the least squares estimates of the parameters.

From the optimization viewpoint, with the squared loss (2.28) and a linear model $f(\boldsymbol{x}_k, \boldsymbol{\theta}) = \boldsymbol{x}_k^\top \boldsymbol{\theta}$, (2.26) is a least squares problem that can be solved explicitly by applying directly (2.30), i.e., by setting $\boldsymbol{\theta}$ to the vector at which the gradient of the cost function vanishes. Indeed, rewriting this *convex* cost function in matrix form, with $\boldsymbol{y} = \begin{bmatrix} y_1 \dots y_N \end{bmatrix}^\top$ and $\boldsymbol{X} = \begin{bmatrix} \boldsymbol{x}_1 \dots \boldsymbol{x}_N \end{bmatrix}^\top$, as

$$J^{LS}(\boldsymbol{\theta}) = \|\boldsymbol{y} - \boldsymbol{X}\boldsymbol{\theta}\|_2^2,$$

with gradient

$$\nabla J^{LS}(\boldsymbol{\theta}) = -2\boldsymbol{X}^\top \boldsymbol{y} + 2\boldsymbol{X}^\top \boldsymbol{X}\boldsymbol{\theta},$$

leads to[2]

$$\boldsymbol{\theta}^* = (\boldsymbol{X}^\top \boldsymbol{X})^{-1} \boldsymbol{X}^\top \boldsymbol{y}. \tag{2.33}$$

2.1.3.3 Local Search

The other models are only pseudo-linear, as the regression vector depends on unknown parameters. For the parameter estimation, one must resort to iterative local search, which may in this case get stuck in local minima. Note nevertheless, that, even for linear models, but with nonsquared losses, iterative search must be employed.

Local optimization methods minimize an unconstrained cost function $J(\boldsymbol{\theta})$ by iteratively updating the variables as

$$\boldsymbol{\theta} \leftarrow \boldsymbol{\theta} + \mu\boldsymbol{\delta}$$

along a direction of descent given by $\boldsymbol{\delta}$ and with a step size $\mu > 0$. In particular, first-order methods compute μ and $\boldsymbol{\delta}$ on the basis of information on the gradient $\nabla J(\boldsymbol{\theta})$ only. The most straightforward of these methods, known as *steepest gradient descent*, uses a direction opposite to that of the gradient, in which we know the function J is decreasing:

$$\boldsymbol{\theta} \leftarrow \boldsymbol{\theta} - \mu\nabla J(\boldsymbol{\theta}).$$

---

[2]Formula (2.33) only applies when $\boldsymbol{X}$ is full rank. Otherwise, the singular value decomposition can be used to obtain the Moore–Penrose pseudo-inverse (B.27) of $\boldsymbol{X}$ (see Appendix B.2.8)

In this case, the step size μ can be set to a constant in (0, 1] or via a predefined decreasing sequence. Alternatively, one can search for the optimal step size at each iteration by solving $\min_{\mu \geq 0} J(\boldsymbol{\theta} - \mu \nabla J(\boldsymbol{\theta}))$.

Second-order methods improve the convergence by taking more information into account to modify the descent direction. A typical choice leading to the well-known Newton method is based on the inverse of the Hessian:

$$\boldsymbol{\theta} \leftarrow \boldsymbol{\theta} - \mu \left(\nabla^2 J(\boldsymbol{\theta})\right)^{-1} \nabla J(\boldsymbol{\theta}).$$

2.2 State-Space (SS) Models

Unlike input–output models, state-space (SS) models often allow to express directly physical equations, with meaningful parameters. They are particularly suitable for multivariable systems (Multi-Input Multi-Output (MIMO) systems), which can be represented elegantly in a simple and compact form.

2.2.1 *Models and Properties*

A linear system in discrete-time, without noise, can be represented in state-space form as

$$\begin{cases} \mathbf{x}_{k+1} = \boldsymbol{A}\mathbf{x}_k + \boldsymbol{B}\boldsymbol{u}_k \\ \mathbf{y}_k = \boldsymbol{C}\mathbf{x}_k + \boldsymbol{D}\boldsymbol{u}_k, \end{cases} \tag{2.34}$$

where $\mathbf{x}_k \in \mathbb{R}^{n_x}$, $\boldsymbol{u}_k \in \mathbb{R}^{n_u}$, and $\mathbf{y}_k \in \mathbb{R}^{n_y}$ are, respectively, the continuous state, input and output vectors, and $\boldsymbol{A} \in \mathbb{R}^{n_x \times n_x}$, $\boldsymbol{B} \in \mathbb{R}^{n_x \times n_u}$, $\boldsymbol{C} \in \mathbb{R}^{n_y \times n_x}$, and $\boldsymbol{D} \in \mathbb{R}^{n_y \times n_u}$ are the system, input, output, and direct feed-through matrices, respectively. The number of state variables, i.e., the dimension of $\mathbf{x}$, is called the system order.

There are two issues related to the state-space representation. The first is that the state vector is generally unmeasured, and thus unavailable for parameter identification of the matrices $\boldsymbol{A}, \boldsymbol{B}, \boldsymbol{C}, \boldsymbol{D}$. The second is the nonuniqueness of the state vector.

2.2.1.1 Similarity Transformation

The state vector $\mathbf{x}_k$ is defined up to a so-called *similarity transformation*. The simplest example is a permutation of its components. More generally, a similarity transformation $\boldsymbol{T}$ is a one-to-one linear change of coordinates of the state vector $\mathbf{x}_k$: $\mathbf{x}_k = \boldsymbol{T}\tilde{\mathbf{x}}_k$, where $\boldsymbol{T}$ is any square nonsingular matrix, or $\tilde{\mathbf{x}}_k = \boldsymbol{T}^{-1}\mathbf{x}_k$. Substituting $\mathbf{x}_k = \boldsymbol{T}\tilde{\mathbf{x}}_k$ in model (2.34) gives

$$\begin{cases} \boldsymbol{T}\tilde{\mathbf{x}}_{k+1} = \boldsymbol{A}\boldsymbol{T}\tilde{\mathbf{x}}_k + \boldsymbol{B}\boldsymbol{u}_k \\ \mathbf{y}_k = \boldsymbol{C}\boldsymbol{T}\tilde{\mathbf{x}}_k + \boldsymbol{D}\boldsymbol{u}_k, \end{cases}$$

or, with $\tilde{\boldsymbol{A}} = \boldsymbol{T}^{-1}\boldsymbol{A}\boldsymbol{T}$, $\tilde{\boldsymbol{B}} = \boldsymbol{T}^{-1}\boldsymbol{B}$, $\tilde{\boldsymbol{C}} = \boldsymbol{C}\boldsymbol{T}$,

$$\begin{cases} \tilde{\mathbf{x}}_{k+1} = \tilde{\boldsymbol{A}}\tilde{\mathbf{x}}_k + \tilde{\boldsymbol{B}}\boldsymbol{u}_k \\ \mathbf{y}_k = \tilde{\boldsymbol{C}}\tilde{\mathbf{x}}_k + \boldsymbol{D}\boldsymbol{u}_k. \end{cases}$$

It can be verified that the two "realizations" share the same transfer function $G(\mathrm{q}) = y_k/u_k$ (see (2.4)) or transfer function matrix $\boldsymbol{G}(\mathrm{q})$, in the multivariable case $\mathbf{y}_k = \boldsymbol{G}(\mathrm{q})\boldsymbol{u}_k$:

$$\begin{aligned} \tilde{\boldsymbol{G}}(\mathrm{q}) &= \tilde{\boldsymbol{C}}\left(\mathrm{q}\boldsymbol{I} - \tilde{\boldsymbol{A}}\right)^{-1}\tilde{\boldsymbol{B}} + \boldsymbol{D} \\ &= \boldsymbol{C}\boldsymbol{T}\left(\mathrm{q}\boldsymbol{I} - \boldsymbol{T}^{-1}\boldsymbol{A}\boldsymbol{T}\right)^{-1}\boldsymbol{T}^{-1}\boldsymbol{B} + \boldsymbol{D} \\ &= \boldsymbol{C}\left[\boldsymbol{T}\left(\mathrm{q}\boldsymbol{I} - \boldsymbol{T}^{-1}\boldsymbol{A}\boldsymbol{T}\right)\boldsymbol{T}^{-1}\right]^{-1}\boldsymbol{B} + \boldsymbol{D} \\ &= \boldsymbol{C}\left(\mathrm{q}\boldsymbol{I} - \boldsymbol{A}\right)^{-1}\boldsymbol{B} + \boldsymbol{D} = \boldsymbol{G}(\mathrm{q}), \end{aligned}$$

where we used (B.24). Thus, different choices of the state vector lead to different internal descriptions of the model, but to the same I/O model.

Besides, two important control properties can be defined from the state-space representation (2.34).

2.2.1.2 Controllability and Observability

A system is said to be *completely controllable* if, using the input $\boldsymbol{u}$, any point $\mathbf{x}$ of the state space can be reached in finite time. From the origin $\mathbf{x}_0$, we have $\mathbf{x}_i = \boldsymbol{A}^i\mathbf{x}_0 + \boldsymbol{A}^{i-1}\boldsymbol{B}\boldsymbol{u}_0 + \boldsymbol{A}^{i-2}\boldsymbol{B}\boldsymbol{u}_1 + \cdots + \boldsymbol{A}\boldsymbol{B}\boldsymbol{u}_{i-2} + \boldsymbol{B}\boldsymbol{u}_{i-1}$, or

$$\left(\mathbf{x}_i - \boldsymbol{A}^i\mathbf{x}_0\right) = \boldsymbol{\mathcal{C}}_i\tilde{\boldsymbol{u}}_i, \tag{2.35}$$

where

$$\tilde{\boldsymbol{u}}_i = \left[\boldsymbol{u}_{i-1}^\top\ \boldsymbol{u}_{i-2}^\top\ \ldots\ \boldsymbol{u}_0^\top\right]^\top$$

and

$$\boldsymbol{\mathcal{C}}_i = \left[\boldsymbol{B}\ \boldsymbol{A}\boldsymbol{B}\ \ldots\ \boldsymbol{A}^{i-1}\boldsymbol{B}\right] \tag{2.36}$$

is the *extended controllability matrix*. System (2.34), of order $n_{\mathbf{x}}$, is completely controllable if and only if $\mathrm{rank}(\boldsymbol{\mathcal{C}}_{n_{\mathbf{x}}}) = n_{\mathbf{x}}$, in which case, for $i \geq n_{\mathbf{x}}$, (2.35) admits the solution $\tilde{\boldsymbol{u}}_i = \boldsymbol{\mathcal{C}}_i^\top(\boldsymbol{\mathcal{C}}_i\boldsymbol{\mathcal{C}}_i^\top)^{-1}\left(\mathbf{x}_i - \boldsymbol{A}^i\mathbf{x}_0\right)$ (see (B.26) with (B.28)).

A system is said to be *completely observable* if, from inputs $\boldsymbol{u}$ and outputs $\mathbf{y}$ observed on a finite time interval, the initial state can be determined, and then the state at any time of the interval, by integrating the state equation. The relation between the initial state $\mathbf{x}_0$, the input and the output over $k = 0, \ldots, i-1$ can be written as

$$\begin{bmatrix} \mathbf{y}_0 \\ \mathbf{y}_1 \\ \mathbf{y}_2 \\ \vdots \\ \mathbf{y}_{i-1} \end{bmatrix} = \begin{bmatrix} \boldsymbol{C} \\ \boldsymbol{CA} \\ \boldsymbol{CA}^2 \\ \vdots \\ \boldsymbol{CA}^{i-1} \end{bmatrix} \mathbf{x}_0 + \begin{bmatrix} \boldsymbol{D} & \mathbf{0} & \mathbf{0} & \cdots & \mathbf{0} \\ \boldsymbol{CB} & \boldsymbol{D} & \mathbf{0} & \cdots & \mathbf{0} \\ \boldsymbol{CAB} & \boldsymbol{CB} & \boldsymbol{D} & \ddots & \vdots \\ \vdots & & \ddots & \ddots & \mathbf{0} \\ \boldsymbol{CA}^{i-2}\boldsymbol{B} & \cdots & \cdots & \boldsymbol{CB} & \boldsymbol{D} \end{bmatrix} \begin{bmatrix} \boldsymbol{u}_0 \\ \boldsymbol{u}_1 \\ \boldsymbol{u}_2 \\ \vdots \\ \boldsymbol{u}_{i-1} \end{bmatrix}, \tag{2.37}$$

or

$$\tilde{\boldsymbol{y}}_i - \boldsymbol{H}\tilde{\boldsymbol{u}}_i = \mathcal{O}_i \mathbf{x}_0, \tag{2.38}$$

where $\tilde{\boldsymbol{y}}_i = \left[\mathbf{y}_0^\top \; \mathbf{y}_1^\top \ldots \mathbf{y}_{i-1}^\top\right]^\top$, $\tilde{\boldsymbol{u}}_i = \left[\boldsymbol{u}_0^\top \; \boldsymbol{u}_1^\top \ldots \boldsymbol{u}_{i-1}^\top\right]^\top$,

$$\boldsymbol{H} = \begin{bmatrix} \boldsymbol{D} & \mathbf{0} & \mathbf{0} & \cdots & \mathbf{0} \\ \boldsymbol{CB} & \boldsymbol{D} & \mathbf{0} & \cdots & \mathbf{0} \\ \boldsymbol{CAB} & \boldsymbol{CB} & \boldsymbol{D} & \ddots & \vdots \\ \vdots & & \ddots & \ddots & \mathbf{0} \\ \boldsymbol{CA}^{i-2}\boldsymbol{B} & \cdots & \cdots & \boldsymbol{CB} & \boldsymbol{D} \end{bmatrix} \tag{2.39}$$

is a lower block triangular Toeplitz matrix (see (B.8)), and where

$$\mathcal{O}_i = \begin{bmatrix} \boldsymbol{C} \\ \boldsymbol{CA} \\ \boldsymbol{CA}^2 \\ \vdots \\ \boldsymbol{CA}^{i-1} \end{bmatrix} \tag{2.40}$$

is the *extended observability matrix*. Particularly, we have

$$\mathbf{y}_{i-1} = \boldsymbol{CA}^{i-1}\mathbf{x}_0 + \left[\boldsymbol{D} \; \boldsymbol{CB} \ldots \boldsymbol{CA}^{i-2}\boldsymbol{B}\right]\left[\boldsymbol{u}_{i-1}^\top \; \boldsymbol{u}_{i-2}^\top \cdots \boldsymbol{u}_0^\top\right]^\top, \tag{2.41}$$

where the matrix terms

$$\boldsymbol{H}_0 = \boldsymbol{D}, \quad \boldsymbol{H}_i = \boldsymbol{CA}^{i-1}\boldsymbol{B}, \qquad i > 0, \tag{2.42}$$

the impulse response coefficients (see (2.1) for I/O models), are known as the Markov parameters. If the system is stable, the term $\boldsymbol{CA}^{i-1}$ vanishes for high values of i and the influence of the initial state $\boldsymbol{CA}^{i-1}\mathbf{x}_0$ can be neglected.

System (2.34), of order $n_\mathbf{x}$, is completely observable if and only if $\text{rank}(\mathcal{O}_{n_\mathbf{x}}) = n_\mathbf{x}$, in which case $\mathbf{x}_0 = (\mathcal{O}_i^\top \mathcal{O}_i)^{-1}\mathcal{O}_i^\top(\tilde{\boldsymbol{y}}_i - \boldsymbol{H}\tilde{\boldsymbol{u}}_i)$ is a solution to (2.38) (see (B.26) with (B.27)).

If $\mathcal{C}_{n_\mathbf{x}}$ and $\mathcal{O}_{n_\mathbf{x}}$ are both of full rank $n_\mathbf{x}$, system (2.34) is a *minimal* realization of order $n_\mathbf{x}$.

Remark 2.2 A necessary condition for minimality can be expressed from the characteristic polynomial of $\boldsymbol{A}$. For the model to be minimal, the characteristic polynomial $P(\boldsymbol{A}) = \det(\mathrm{q}\boldsymbol{I} - \boldsymbol{A})$ must be a minimal polynomial, i.e., of least degree such that $P(\boldsymbol{A}) = 0$. For SISO systems, $P(\boldsymbol{A})$ is the denominator of the transfer function and thus the numerator and denominator of the transfer function must be coprime, i.e., have no common factors. The order $n_\mathbf{x}$ is the degree of the characteristic polynomial, or, in the SISO case, the degree of the denominator.

2.2.1.3 Obtaining State-Space Models

The form of model (2.34) can be directly given or obtained, after discretization, from first principles modeling in continuous time. Alternatively, such models can be the result of translating in state-space form a model originally in I/O form.

Consider a physical model in continuous-time t:

$$\begin{cases} \dot{\mathbf{x}}(t) = \boldsymbol{F}\mathbf{x}(t) + \boldsymbol{G}\boldsymbol{u}(t) \\ \mathbf{y}(t) = \boldsymbol{H}\mathbf{x}(t) + \boldsymbol{J}\boldsymbol{u}(t). \end{cases} \tag{2.43}$$

This model can be discretized by several means to yield the discrete-time model (2.34). Integrating, from $\mathbf{x}(0)$, the equation of $\dot{\mathbf{x}}(t)$ in (2.43) to compute $\mathbf{x}(kT)$ and $\mathbf{x}((k+1)T)$, with T the sampling period, leads after some substitutions to the exact expressions $\boldsymbol{A} = \mathrm{e}^{\boldsymbol{F}T}$ (see (B.10)) and $\boldsymbol{B} = \int_{\tau=0}^{T} \mathrm{e}^{\boldsymbol{F}\tau}\mathrm{d}\tau\, \boldsymbol{G} = \boldsymbol{F}^{-1}(\boldsymbol{A} - \boldsymbol{I})\boldsymbol{G}$. Alternatively, the zero-order hold approximation of the first-order derivative

$$\dot{\mathbf{x}}(kT) \simeq \frac{\mathbf{x}((k+1)T) - \mathbf{x}(kT)}{T} = \frac{\mathbf{x}_{k+1} - \mathbf{x}_k}{T},$$

more directly gives $\boldsymbol{A} = \boldsymbol{I} - T\boldsymbol{F}$, $\boldsymbol{B} = T\boldsymbol{G}$. Besides, the matrices of the static measurement equation are not affected by sampling: $\boldsymbol{C} = \boldsymbol{H}$, $\boldsymbol{D} = \boldsymbol{J}$.

Different forms of state-space model can be derived from I/O models. Note that, for instance, for a SISO model with $n_a + n_b$ parameters ($n_b \le n_a$), there are possibly n_a^2 parameters for matrix $\boldsymbol{A}$, n_a parameters for each one of the vectors $\boldsymbol{B}$ and $\boldsymbol{C}$, and one parameter for $\boldsymbol{D}$. This leads to overparametrization[3] issues, and even identifiability[4] problems, particularly for multivariable systems. Thus, special state-space forms, so-called *canonical forms*, are often chosen.

[3] Overparametrization, i.e., the inclusion of too many parameters in the model, leads to larger variance of the parameters and overfitting (see Sect. 2.4.2.1).

[4] Among various more formal definitions, it can be said that a model is identifiable if and only if a *unique* value of the parameters can be obtained from the data.

2.2.1.4 Canonical Forms

Consider, for instance, the SISO I/O model without noise

$$y_k = \frac{B'(\mathrm{q})}{A(\mathrm{q})}u_k = \frac{b_0 + b'_1\mathrm{q}^{-1} + \cdots + b'_{n_\mathbf{x}}\mathrm{q}^{-n_\mathbf{x}}}{1 + a_1\mathrm{q}^{-1} + \cdots + a_{n_\mathbf{x}}\mathrm{q}^{-n_\mathbf{x}}}u_k,$$

where the degree n_a of the polynomial $A(\mathrm{q})$ is equal to the system order $n_\mathbf{x}$, and the degree n_b of the polynomial $B'(\mathrm{q})$ is such that $n_b \leq n_\mathbf{x}$, meaning that the coefficients $b'_{n_b+1}, \ldots, b'_{n_\mathbf{x}}$ are zero.

If there is no pure time delay, i.e., $b_0 \neq 0$, the model can be rewritten by polynomial long division as

$$y_k = \frac{B(\mathrm{q})}{A(\mathrm{q})}u_k + b_0 u_k = \frac{b_1\mathrm{q}^{-1} + \cdots + b_{n_\mathbf{x}}\mathrm{q}^{-n_\mathbf{x}}}{1 + a_1\mathrm{q}^{-1} + \cdots + a_{n_\mathbf{x}}\mathrm{q}^{-n_\mathbf{x}}}u_k + b_0 u_k,$$

where $b_i = b'_i - b_0\, a_i$. Then, in (2.34), $\boldsymbol{D} = [b_0]$, otherwise, for a pure delay of at least 1, $\boldsymbol{D} = [0]$.

The *companion forms* are directly derived from the coefficients of the transfer function $G(\mathrm{q}) = B(\mathrm{q})/A(\mathrm{q})$ in

$$y_k = G(\mathrm{q})u_k = \frac{B(\mathrm{q})}{A(\mathrm{q})}u_k. \tag{2.44}$$

Rewriting the model as $y_k = B(\mathrm{q})\tilde{u}_k$ with

$$\tilde{u}_k = \frac{u_k}{1 + a_1\mathrm{q}^{-1} + \cdots + a_{n_\mathbf{x}}\mathrm{q}^{-n_\mathbf{x}}} = -(a_1\mathrm{q}^{-1} + \cdots + a_{n_\mathbf{x}}\mathrm{q}^{-n_\mathbf{x}})\tilde{u}_k + u_k$$

and choosing the state vector as $\mathbf{x}_k = \begin{bmatrix}\tilde{u}_{k-1}\ \tilde{u}_{k-2}\ \ldots\ \tilde{u}_{k-n_\mathbf{x}}\end{bmatrix}^\mathsf{T}$ leads to the *controller canonical form*:

$$\boldsymbol{A}_c = \begin{bmatrix} -a_1 & \ldots & -a_{n_\mathbf{x}-1} & -a_{n_\mathbf{x}} \\ 1 & \ldots & 0 & 0 \\ \vdots & \ddots & \vdots & \vdots \\ 0 & \ldots & 1 & 0 \end{bmatrix}, \quad \boldsymbol{B}_c = \begin{bmatrix} 1 \\ 0 \\ \vdots \\ 0 \end{bmatrix}, \quad \boldsymbol{C}_c = \begin{bmatrix} b_1 \ \ldots \ b_{n_\mathbf{x}} \end{bmatrix}.$$

Also, the *controllability canonical form* can be obtained as

$$\boldsymbol{A}_{co} = \begin{bmatrix} 0 & \ldots & 0 & -a_{n_\mathbf{x}} \\ 1 & \ldots & 0 & -a_{n_\mathbf{x}-1} \\ \vdots & \ddots & \vdots & \vdots \\ 0 & \ldots & 1 & -a_1 \end{bmatrix}, \quad \boldsymbol{B}_{co} = \begin{bmatrix} 1 \\ 0 \\ \vdots \\ 0 \end{bmatrix}, \quad \boldsymbol{C}_{co} = \begin{bmatrix} \beta_1 \ \ldots \ \beta_{n_\mathbf{x}} \end{bmatrix}, \tag{2.45}$$

where $\beta_i = h_i, i = 1, \ldots, n_\mathrm{x}$, with the (scalar) impulse response/Markov parameters $h_i = \boldsymbol{C}\boldsymbol{A}^{i-1}\boldsymbol{B}$.

The model (2.44) can also be written as

$$y_k = -\left(a_1 \mathrm{q}^{-1} + \cdots + a_{n_\mathrm{x}} \mathrm{q}^{-n_\mathrm{x}}\right) y_k + \left(b_1 \mathrm{q}^{-1} + \cdots + b_{n_\mathrm{x}} \mathrm{q}^{-n_\mathrm{x}}\right) u_k,$$

which, according to Horner's rule, can be reformulated as

$$y_k = \mathrm{q}^{-1}\left(-a_1 y_k + b_1 u_k + \mathrm{q}^{-1}\left(-a_2 y_k + b_2 u_k + \cdots + \mathrm{q}^{-1}\left(-a_{n_\mathrm{x}} y_k + b_{n_\mathrm{x}} u_k\right)\cdots\right)\right).$$

This leads to the *observer canonical form*:

$$\boldsymbol{A}_o = \begin{bmatrix} -a_1 & 1 \ldots 0 \\ \vdots & \vdots \ddots \vdots \\ -a_{n_\mathrm{x}-1} & 0 \ldots 1 \\ -a_{n_\mathrm{x}} & 0 \ldots 0 \end{bmatrix}, \quad \boldsymbol{B}_o = \begin{bmatrix} b_1 \\ \vdots \\ b_{n_\mathrm{x}} \end{bmatrix}, \quad \boldsymbol{C}_o = \begin{bmatrix} 1 & 0 & \ldots & 0 \end{bmatrix}.$$

Similarly, we have the *observability canonical form*:

$$\boldsymbol{A}_{ob} = \begin{bmatrix} 0 & 1 & \ldots & 0 \\ \vdots & \vdots & \ddots & \vdots \\ 0 & 0 & \ldots & 1 \\ -a_{n_\mathrm{x}} & -a_{n_\mathrm{x}-1} & \ldots & -a_1 \end{bmatrix}, \quad \boldsymbol{B}_{ob} = \begin{bmatrix} \beta_1 \\ \vdots \\ \beta_{n_\mathrm{x}} \end{bmatrix}, \quad \boldsymbol{C}_{ob} = \begin{bmatrix} 1 & 0 & \ldots & 0 \end{bmatrix}. \tag{2.46}$$

Note the duality: $\boldsymbol{A}_o = \boldsymbol{A}_c^\top$, $\boldsymbol{B}_o = \boldsymbol{C}_c^\top$, $\boldsymbol{C}_o = \boldsymbol{B}_c^\top$, and $\boldsymbol{A}_{ob} = \boldsymbol{A}_{co}^\top$, $\boldsymbol{B}_{ob} = \boldsymbol{C}_{co}^\top$, $\boldsymbol{C}_{ob} = \boldsymbol{B}_{co}^\top$.[5]

Finally, a transfer function $G(\mathrm{q})$, with distinct poles, can be expressed by partial fraction decomposition in the weighted sum of its modes as $G(\mathrm{q}) = y_k/u_k = \sum_{i=1}^{n_\mathrm{x}} c_i/(\mathrm{q} - \lambda_i)$ or $y_k = \sum_{i=1}^{n_\mathrm{x}} c_i \tilde{u}_{ik}$, with $\tilde{u}_{ik} = u_k/(\mathrm{q} - \lambda_i)$. This leads to n_x decoupled equations $\tilde{u}_{i(k+1)} = \lambda_i \tilde{u}_{ik} + u_k$, and thus the *modal form*:

$$\boldsymbol{A}_m = \begin{bmatrix} \lambda_1 & \ldots & 0 \\ \vdots & \ddots & \vdots \\ 0 & \ldots & \lambda_{n_\mathrm{x}} \end{bmatrix}, \quad \boldsymbol{B}_m = \begin{bmatrix} 1 \\ \vdots \\ 1 \end{bmatrix}, \quad \boldsymbol{C}_m = \begin{bmatrix} c_1 & \ldots & c_{n_\mathrm{x}} \end{bmatrix}.$$

Multiple poles show Jordan blocks (see (B.7)) in $\boldsymbol{A}_m$.

It is useful to derive the similarity transformations (see Sect. 2.2.1.1) leading to the canonical forms above. To this end, we consider two realizations $(\boldsymbol{A}, \boldsymbol{B}, \boldsymbol{C}, \boldsymbol{D})$, with state vector $\mathbf{x}_k$ of dimension n_x, and $(\tilde{\boldsymbol{A}}, \tilde{\boldsymbol{B}}, \tilde{\boldsymbol{C}}, \boldsymbol{D})$, with state vector $\tilde{\mathbf{x}}_k = \boldsymbol{T}^{-1}\mathbf{x}_k$, and $\tilde{\boldsymbol{A}} = \boldsymbol{T}^{-1}\boldsymbol{A}\boldsymbol{T}$, $\tilde{\boldsymbol{B}} = \boldsymbol{T}^{-1}\boldsymbol{B}$, $\tilde{\boldsymbol{C}} = \boldsymbol{C}\boldsymbol{T}$. It is easy to show that the respective

[5] Other canonical forms can be found while keeping the sparse structure. Think, for instance, of reversing the order of the state components.

observability (2.40) and controllability (2.36) matrices are such as $\tilde{\mathcal{O}}_i = \mathcal{O}_i \boldsymbol{T}$ and $\tilde{\mathcal{C}}_i = \boldsymbol{T}^{-1}\mathcal{C}_i$.

Now, we consider *minimal* realizations with the respective full rank (not extended) observability matrices $\mathcal{O}_{n_x}, \tilde{\mathcal{O}}_{n_x} \in \mathbb{R}^{n_y.n_x \times n_x}$, and controllability matrices $\mathcal{C}_{n_x}, \tilde{\mathcal{C}}_{n_x} \in \mathbb{R}^{n_x \times n_u.n_x}$. Then, for MISO systems, we have $\boldsymbol{T} = \tilde{\mathcal{O}}_{n_x}\mathcal{O}_{n_x}^{-1}$ and $\boldsymbol{T}^{-1} = \tilde{\mathcal{C}}_{n_x}\mathcal{C}_{n_x}^\top \left(\mathcal{C}_{n_x}\mathcal{C}_{n_x}^\top\right)^{-1}$, yielding $\boldsymbol{T}^{-1} = \tilde{\mathcal{C}}_{n_x}\mathcal{C}_{n_x}^{-1}$ for SISO systems. Finally, the observability matrix of the observability canonical form (2.46) is $\tilde{\mathcal{O}}_{n_x} = \boldsymbol{I}_{n_x}$, and the similarity transformation is simply $\boldsymbol{T} = \mathcal{O}_{n_x}^{-1}$. Also, for SISO systems, the controllability matrix of the controllability canonical form (2.45) is $\tilde{\mathcal{C}}_{n_x} = \boldsymbol{I}_{n_x}$, and $\boldsymbol{T} = \mathcal{C}_{n_x}$.

The case of MIMO systems is more involved and will be partially dealt with in Sect. 8.2.2.4.

2.2.2 Parameter Estimation

For parameter estimation, we start from the noisy version of the state-space model (2.34):

$$\begin{cases} \mathbf{x}_{k+1} = \boldsymbol{A}\mathbf{x}_k + \boldsymbol{B}\boldsymbol{u}_k + \boldsymbol{w}_k \\ \mathbf{y}_k = \boldsymbol{C}\mathbf{x}_k + \boldsymbol{D}\boldsymbol{u}_k + \boldsymbol{v}_k, \end{cases} \tag{2.47}$$

with the process noise vector $\boldsymbol{w}_k \in \mathbb{R}^{n_x}$ and the measurement noise vector $\boldsymbol{v}_k \in \mathbb{R}^{n_y}$ such that

$$\mathbb{E}\left[\begin{bmatrix}\boldsymbol{w}_m \\ \boldsymbol{v}_m\end{bmatrix}\begin{bmatrix}\boldsymbol{w}_n^\top & \boldsymbol{v}_n^\top\end{bmatrix}\right] = \begin{bmatrix}\boldsymbol{Q} & \boldsymbol{S} \\ \boldsymbol{S}^\top & \boldsymbol{R}\end{bmatrix}\delta_{m,n} \geq 0, \tag{2.48}$$

where $\delta_{m,n}$ is the Kronecker delta, meaning that the noise sequences are uncorrelated. It is assumed that they are white noise vector sequences (see Appendix A.3.1.1) and uncorrelated with the inputs $\boldsymbol{u}_k$.

2.2.2.1 Prediction Error Approach

In the favorable (but uncommon) case where the state vector $\mathbf{x}$ is measured, the parameter identification is as follows. As all the variables $\mathbf{x}$, $\mathbf{y}$, and $\boldsymbol{u}$ are measured, model (2.47) can be written as a linear regression model:

$$\begin{bmatrix}\mathbf{x}_{k+1} \\ \mathbf{y}_k\end{bmatrix} = \boldsymbol{\Theta}\begin{bmatrix}\mathbf{x}_k \\ \boldsymbol{u}_k\end{bmatrix} + \begin{bmatrix}\boldsymbol{w}_k \\ \boldsymbol{v}_k\end{bmatrix},$$

with $\boldsymbol{\Theta} = \begin{bmatrix}\boldsymbol{A} & \boldsymbol{B} \\ \boldsymbol{C} & \boldsymbol{D}\end{bmatrix} \in \mathbb{R}^{(n_x+n_y)\times(n_x+n_u)}$, or $\begin{bmatrix}\mathbf{x}_{k+1}^\top & \mathbf{y}_k^\top\end{bmatrix} = \begin{bmatrix}\mathbf{x}_k^\top & \boldsymbol{u}_k^\top\end{bmatrix}\boldsymbol{\Theta}^\top + \begin{bmatrix}\boldsymbol{w}_k^\top & \boldsymbol{v}_k^\top\end{bmatrix}$. Stacking this model for different values of k yields:

$$\underbrace{\begin{bmatrix} \mathbf{x}_1^\top & \mathbf{y}_0^\top \\ \vdots & \vdots \\ \mathbf{x}_N^\top & \mathbf{y}_{N-1}^\top \end{bmatrix}}_{Y} = \underbrace{\begin{bmatrix} \mathbf{x}_0^\top & \boldsymbol{u}_0^\top \\ \vdots & \vdots \\ \mathbf{x}_{N-1}^\top & \boldsymbol{u}_{N-1}^\top \end{bmatrix}}_{Z} \boldsymbol{\Theta}^\top + \underbrace{\begin{bmatrix} \boldsymbol{w}_0^\top & \boldsymbol{v}_0^\top \\ \vdots & \vdots \\ \boldsymbol{w}_{N-1}^\top & \boldsymbol{v}_{N-1}^\top \end{bmatrix}}_{W},$$

or

$$\boldsymbol{Y} = \boldsymbol{Z}\boldsymbol{\Theta}^\top + \boldsymbol{W}.$$

From the prediction error,

$$\boldsymbol{e}_k = \begin{bmatrix} \mathbf{x}_k \\ \mathbf{y}_{k-1} \end{bmatrix} - \begin{bmatrix} \hat{\mathbf{x}}_k \\ \hat{\mathbf{y}}_{k-1} \end{bmatrix} = \begin{bmatrix} \mathbf{x}_k \\ \mathbf{y}_{k-1} \end{bmatrix} - \boldsymbol{\Theta} \begin{bmatrix} \mathbf{x}_{k-1} \\ \boldsymbol{u}_{k-1} \end{bmatrix},$$

as in (2.26) for I/O models, the unknown parameters of the matrices $\boldsymbol{A}$, $\boldsymbol{B}$, $\boldsymbol{C}$, $\boldsymbol{D}$ in $\boldsymbol{\Theta}$ are estimated by solving

$$\min_{\boldsymbol{\Theta} \in \mathbb{R}^{(n_\mathbf{x}+n_\mathbf{y}) \times (n_\mathbf{x}+n_u)}} \sum_{k=0}^{N-1} \ell(\|\boldsymbol{e}_k\|_p), \tag{2.49}$$

with ℓ a loss function as defined in (2.27). With $p = 2$ and the squared loss (2.28), this directly leads to the least squares solution which is the counterpart of (2.33) for multi-output systems (see (B.30)):

$$\hat{\boldsymbol{\Theta}}^\top = (\boldsymbol{Z}^\top \boldsymbol{Z})^{-1} \boldsymbol{Z}^\top \boldsymbol{Y}.$$

The noise covariance matrices $\boldsymbol{Q}$, $\boldsymbol{R}$, and $\boldsymbol{S}$ in (2.48) can be estimated from the model residuals.

In the more realistic case of unmeasured state vector, a Kalman filter can be applied to the noisy model (2.47) to compute the predictions:

$$\begin{cases} \hat{\mathbf{x}}_{k+1} = \boldsymbol{A}\hat{\mathbf{x}}_k + \boldsymbol{B}\boldsymbol{u}_k + \boldsymbol{K}\left[\mathbf{y}_k - \boldsymbol{C}\hat{\mathbf{x}}_k - \boldsymbol{D}\boldsymbol{u}_k\right] \\ \hat{\mathbf{y}}_k = \boldsymbol{C}\hat{\mathbf{x}}_k + \boldsymbol{D}\boldsymbol{u}_k, \end{cases}$$

where $\boldsymbol{K}$ is the steady state Kalman gain, and thus the part of the output that cannot be predicted from the past data, the prediction error or innovation

$$\boldsymbol{e}_k = \mathbf{y}_k - \hat{\mathbf{y}}_k = \boldsymbol{C}\mathbf{x}_k + \boldsymbol{D}\boldsymbol{u}_k + \boldsymbol{v}_k - \left[\boldsymbol{C}\hat{\mathbf{x}}_k + \boldsymbol{D}\boldsymbol{u}_k\right] = \boldsymbol{C}\left[\mathbf{x}_k - \hat{\mathbf{x}}_k\right] + \boldsymbol{v}_k. \tag{2.50}$$

This allows us to write model (2.47) in the *innovation form* as

$$\begin{cases} \hat{\mathbf{x}}_{k+1} = \boldsymbol{A}\hat{\mathbf{x}}_k + \boldsymbol{B}\boldsymbol{u}_k + \boldsymbol{K}\boldsymbol{e}_k \\ \mathbf{y}_k = \boldsymbol{C}\hat{\mathbf{x}}_k + \boldsymbol{D}\boldsymbol{u}_k + \boldsymbol{e}_k, \end{cases}$$

where the innovation vector $\boldsymbol{e}_k$ has zero mean and covariance matrix $\boldsymbol{R}$.

Then, as for (2.49), the unknown parameters of the matrices $\boldsymbol{A}, \boldsymbol{B}, \boldsymbol{C}, \boldsymbol{D}$ in $\boldsymbol{\Theta}$ and $\boldsymbol{K}$ are estimated by solving

$$\min_{\boldsymbol{\Theta}\in\mathbb{R}^{(n_{\mathrm{x}}+n_{\mathrm{y}})\times(n_{\mathrm{x}}+n_u)},\,\boldsymbol{K}\in\mathbb{R}^{n_{\mathrm{x}}\times n_{\mathrm{y}}}} \sum_{k=0}^{N-1} \ell(\|\boldsymbol{e}_k\|_p), \tag{2.51}$$

and $\boldsymbol{R}$ can be estimated from the residuals.

One can use various parametrizations for these matrices: full matrices, with all or only a subset of entries as free parameters, or sparse matrices as in the canonical forms previously described in Sect. 2.2.1. The estimation problem is a nonconvex optimization one and the iterative minimization for solving (2.51) needs a wise initialization of the parameters of the system matrices. This can be done from scratch, by bounding the possible values, by giving an initial state-space model, converted or not from an I/O model into a canonical form or obtained with a noniterative subspace approach, presented below.

2.2.2.2 Subspace Methods

Contrary to PEM, subspace methods do not optimize any error criterion and use tools of system theory, geometry, and numerical linear algebra to solve the following problem: given a large number of measurements of the input $\boldsymbol{u}_k$ and the output $\mathbf{y}_k$ generated by the unknown system (2.47)–(2.48), determine its order $n_{\mathbf{x}}$, the system matrices $\boldsymbol{A}, \boldsymbol{B}, \boldsymbol{C}, \boldsymbol{D}$ up to a similarity transformation and an estimate of the matrices $\boldsymbol{Q}, \boldsymbol{S}, \boldsymbol{R}$. Solving this problem is now presented.

Integrating model (2.47) over $k = i, \ldots, 2i-1$, similarly to (2.37) for model (2.34), leads to

$$\tilde{\mathbf{y}}_i = \mathcal{O}_i \mathbf{x}_i + \boldsymbol{H}\tilde{\boldsymbol{u}}_i + \boldsymbol{H}^s\tilde{\boldsymbol{w}}_i + \tilde{\boldsymbol{v}}_i, \tag{2.52}$$

with $\tilde{\mathbf{y}}_i \in \mathbb{R}^{i.n_{\mathrm{y}}}$, $\tilde{\boldsymbol{u}}_i \in \mathbb{R}^{i.n_u}$, $\tilde{\boldsymbol{w}}_i \in \mathbb{R}^{i.n_{\mathrm{x}}}$ and $\tilde{\boldsymbol{v}}_i \in \mathbb{R}^{i.n_{\mathrm{y}}}$ defined as

$$\tilde{\mathbf{y}}_i = \begin{bmatrix} \mathbf{y}_i \\ \mathbf{y}_{i+1} \\ \vdots \\ \mathbf{y}_{2i-1} \end{bmatrix}, \quad \tilde{\boldsymbol{u}}_i = \begin{bmatrix} \boldsymbol{u}_i \\ \boldsymbol{u}_{i+1} \\ \vdots \\ \boldsymbol{u}_{2i-1} \end{bmatrix}, \quad \tilde{\boldsymbol{w}}_i = \begin{bmatrix} \boldsymbol{w}_i \\ \boldsymbol{w}_{i+1} \\ \vdots \\ \boldsymbol{w}_{2i-1} \end{bmatrix}, \quad \tilde{\boldsymbol{v}}_i = \begin{bmatrix} \boldsymbol{v}_i \\ \boldsymbol{v}_{i+1} \\ \vdots \\ \boldsymbol{v}_{2i-1} \end{bmatrix},$$

and with the extended observability matrix $\mathcal{O}_i$, assumed to be of rank $n_{\mathbf{x}}$, given by (2.40), $\boldsymbol{H}$ by (2.39), and $\boldsymbol{H}^s$ defined as

$$H^s = \begin{bmatrix} 0 & 0 & 0 & \cdots & 0 \\ C & 0 & 0 & \cdots & 0 \\ CA & C & 0 & \ddots & \vdots \\ \vdots & & \ddots & \ddots & 0 \\ CA^{i-2} & \cdots & \cdots & C & 0 \end{bmatrix}.$$

Then, stacking the signals in data block Hankel matrices (see (B.9)) splitted into "past" (with subscript p) and "future" (with subscript f) parts as

$$\begin{bmatrix} & \bullet_0 & \bullet_1 & \cdots & \bullet_{n-1} \\ & \bullet_1 & \bullet_2 & \cdots & \bullet_n \\ \text{past} & \vdots & \vdots & \ddots & \vdots \\ & \bullet_{i-1} & \bullet_i & \cdots & \bullet_{i+n-2} \\ \hline & \bullet_i & \bullet_{i+1} & \cdots & \bullet_{i+n-1} \\ & \bullet_{i+1} & \bullet_{i+2} & \cdots & \bullet_{i+n} \\ \text{future} & \vdots & \vdots & \ddots & \vdots \\ & \bullet_{2i-1} & \bullet_{2i} & \cdots & \bullet_{2i+n-2} \end{bmatrix}, \tag{2.53}$$

yields, for the "future" horizon, to the basic equation of subspace methods:

$$Y_f = \mathcal{O}_i X_i + H U_f + H^s M_f + N_f, \tag{2.54}$$

with $Y_f \in \mathbb{R}^{n_y.i \times n}$ defined as

$$Y_f = \left[\tilde{\mathbf{y}}_i \ \tilde{\mathbf{y}}_{i+1} \cdots \tilde{\mathbf{y}}_{i+n-1}\right],$$

$U_f \in \mathbb{R}^{n_u.i \times n}$, $M_f \in \mathbb{R}^{n_x.i \times n}$, $N_f \in \mathbb{R}^{n_y.i \times n}$ defined accordingly, and with the state sequence $X_i \in \mathbb{R}^{n_x \times n}$ such as

$$X_i = \left[\mathbf{x}_i \ \mathbf{x}_{i+1} \ \ldots \ \mathbf{x}_{i+n-1}\right]. \tag{2.55}$$

The sizes i and n of the two horizons are quite arbitrary design variables, provided they are not too small, at least $i, n > n_x$. The use of the equation similar to (2.54) but for "past" data, or, at least, of the involved matrices, particularly Y_p and U_p will be made clear below.

The first step of subspace methods always performs a weighted projection of the row space of the Hankel matrices Y_f, U_f, M_f, N_f on the orthogonal complement of the row space of U_f (see (B.32)). From this projection, estimates $\hat{\mathcal{O}}_i$ of the extended observability matrix (2.40) and/or $\hat{X}_i$ of the state sequence (2.55) can be retrieved. With $\Pi_{U_f}^{\perp} = I - U_f^{\top}(U_f U_f^{\top})^{-1} U_f \in \mathbb{R}^{n \times n}$, we have $U_f \Pi_{U_f}^{\perp} = 0$ and U_f is eliminated from (2.54):

$$Y_f \Pi_{U_f}^{\perp} = \mathcal{O}_i X_i \Pi_{U_f}^{\perp} + H^s M_f \Pi_{U_f}^{\perp} + N_f \Pi_{U_f}^{\perp}. \tag{2.56}$$

The noise contribution is removed by multiplying by matrices $\boldsymbol{W}_1$ and $\boldsymbol{W}_2$ while maintaining the ranks, $\text{rank}(\boldsymbol{W}_1\mathcal{O}_i) = \text{rank}(\mathcal{O}_i)$, $\text{rank}(\boldsymbol{X}_i\boldsymbol{\Pi}^{\perp}_{U_f}\boldsymbol{W}_2) = \text{rank}(\boldsymbol{X}_i)$ and such that $(\boldsymbol{H}^s\boldsymbol{M}_f\boldsymbol{\Pi}^{\perp}_{U_f} + \boldsymbol{N}_f\boldsymbol{\Pi}^{\perp}_{U_f})\boldsymbol{W}_2 = \boldsymbol{0}$. To give some idea about the choice of the weighting matrices $\boldsymbol{W}_1$ and $\boldsymbol{W}_2$, in the basic variant IV-SID of the subspace methods, $\boldsymbol{W}_1 = \boldsymbol{I}$ and $\boldsymbol{W}_2 = \boldsymbol{Z}^\top$, with $\boldsymbol{Z} \in \mathbb{R}^{i(n_u+n_y)\times n}$ such that

$$\boldsymbol{Z} = \begin{bmatrix} \boldsymbol{U}_p \\ \boldsymbol{Y}_p \end{bmatrix},$$

including instrumental variables (IV), uncorrelated with the noise in the future data. This ensures that $\frac{1}{n}\boldsymbol{H}^s\boldsymbol{M}_f\boldsymbol{\Pi}^{\perp}_{U_f}\boldsymbol{W}_2 \xrightarrow{n\to\infty} \boldsymbol{0}$ and $\frac{1}{n}\boldsymbol{N}_f\boldsymbol{\Pi}^{\perp}_{U_f}\boldsymbol{W}_2 \xrightarrow{n\to\infty} \boldsymbol{0}$.

Then, from the singular value decomposition (SVD) (see Appendix B.2.8),

$$\boldsymbol{W}_1\boldsymbol{Y}_f\boldsymbol{\Pi}^{\perp}_{U_f}\boldsymbol{W}_2 = \boldsymbol{W}_1\mathcal{O}_i\boldsymbol{X}_i\boldsymbol{\Pi}^{\perp}_{U_f}\boldsymbol{W}_2 = \begin{bmatrix} \boldsymbol{U}_1 & \boldsymbol{U}_2 \end{bmatrix} \begin{bmatrix} \boldsymbol{S}_1 & \boldsymbol{0} \\ \boldsymbol{0} & \boldsymbol{S}_2 \end{bmatrix} \begin{bmatrix} \boldsymbol{V}_1^\top \\ \boldsymbol{V}_2^\top \end{bmatrix} \approx \boldsymbol{U}_1\boldsymbol{S}_1\boldsymbol{V}_1^\top,$$

the model order n_x can be retrieved as the number of diagonal elements in $\boldsymbol{S}_1$, i.e., the number of "nonzero" singular values,[6] and $\boldsymbol{W}_1\hat{\mathcal{O}}_i = \boldsymbol{U}_1\boldsymbol{S}_1^{1/2}$ and $\hat{\boldsymbol{X}}_i\boldsymbol{\Pi}^{\perp}_{U_f}\boldsymbol{W}_2 = \boldsymbol{S}_1^{1/2}\boldsymbol{V}_1^\top$. Thus $\hat{\mathcal{O}}_i = \boldsymbol{W}_1^{-1}\boldsymbol{U}_1\boldsymbol{S}_1^{1/2}$.

In the second step, the system matrices $\boldsymbol{A}$, $\boldsymbol{B}$, $\boldsymbol{C}$, $\boldsymbol{D}$ and $\boldsymbol{Q}$, $\boldsymbol{R}$, $\boldsymbol{S}$ are determined, either from $\hat{\mathcal{O}}_i$ or from $\hat{\boldsymbol{X}}_i$. From the state sequence $\hat{\boldsymbol{X}}_i$, the system matrices can be estimated by a PEM approach (see Sect. 2.2.2.1).

We start here from $\hat{\mathcal{O}}_i$. The matrix $\hat{\boldsymbol{C}} \in \mathbb{R}^{n_y\times n_x}$ is simply the first n_y rows of $\hat{\mathcal{O}}_i$. To obtain $\hat{\boldsymbol{A}}$, we use its *shift invariance property*. Extracting the matrices

$$\boldsymbol{\Gamma}_1 = \begin{bmatrix} \boldsymbol{C} \\ \boldsymbol{C}\boldsymbol{A} \\ \boldsymbol{C}\boldsymbol{A}^2 \\ \vdots \\ \boldsymbol{C}\boldsymbol{A}^{n_x-1} \end{bmatrix}, \quad \boldsymbol{\Gamma}_2 = \begin{bmatrix} \boldsymbol{C}\boldsymbol{A} \\ \boldsymbol{C}\boldsymbol{A}^2 \\ \vdots \\ \boldsymbol{C}\boldsymbol{A}^{n_x-1} \\ \boldsymbol{C}\boldsymbol{A}^{n_x} \end{bmatrix}$$

from the extended observability matrix $\hat{\mathcal{O}}_i$, it is easy to see that $\boldsymbol{\Gamma}_2 = \boldsymbol{\Gamma}_1\boldsymbol{A}$ and $\boldsymbol{A}$ can be estimated by least squares as $\hat{\boldsymbol{A}} = \left(\boldsymbol{\Gamma}_1^\top\boldsymbol{\Gamma}_1\right)^{-1}\boldsymbol{\Gamma}_1^\top\boldsymbol{\Gamma}_2$ (see (B.30)). The matrices $\hat{\boldsymbol{B}}$ and $\hat{\boldsymbol{D}}$ are usually obtained by solving a least squares problem using the data and the estimates of $\boldsymbol{A}$ and $\boldsymbol{C}$. Starting from (2.41), we obtain using (B.13)–(B.14):

[6]The determination of the model order by inspecting the singular values is not a trivial task when the system is corrupted by a high level of noise.

$$\begin{aligned}\mathbf{y}_k &= \hat{\boldsymbol{C}}\hat{\boldsymbol{A}}^k \mathbf{x}_0 + \sum_{i=0}^{k-1} \hat{\boldsymbol{C}}\hat{\boldsymbol{A}}^{k-i-1} \boldsymbol{B}\boldsymbol{u}_i + \boldsymbol{D}\boldsymbol{u}_k \\ &= \hat{\boldsymbol{C}}\hat{\boldsymbol{A}}^k \mathbf{x}_0 + \left[\sum_{i=0}^{k-1} \boldsymbol{u}_i^\top \otimes \hat{\boldsymbol{C}}\hat{\boldsymbol{A}}^{k-i-1}\right] \operatorname{vec}(\boldsymbol{B}) + \left[\boldsymbol{u}_k^\top \otimes \boldsymbol{I}_{n_y}\right] \operatorname{vec}(\boldsymbol{D}),\end{aligned}$$

or $\mathbf{y}_k = \boldsymbol{\varphi}_k^\top \boldsymbol{\theta}$, with

$$\boldsymbol{\varphi}_k^\top = \left[\hat{\boldsymbol{C}}\hat{\boldsymbol{A}}^k \left[\sum_{i=0}^{k-1} \boldsymbol{u}_i^\top \otimes \hat{\boldsymbol{C}}\hat{\boldsymbol{A}}^{k-i-1}\right] \left[\boldsymbol{u}_k^\top \otimes \boldsymbol{I}_{n_y}\right]\right], \quad \boldsymbol{\theta} = \begin{bmatrix} \mathbf{x}_0 \\ \operatorname{vec}(\boldsymbol{B}) \\ \operatorname{vec}(\boldsymbol{D}) \end{bmatrix}.$$

Then $\hat{\boldsymbol{C}}$ and $\hat{\boldsymbol{D}}$ are obtained by solving $\min_{\boldsymbol{\theta}} \sum_{k=0}^{N} \|\mathbf{y}_k - \boldsymbol{\varphi}_k^\top \boldsymbol{\theta}\|_2^2$.

There are some variants of the method, mainly in the choice of weighting matrices: numerical algorithm for 4SID (N4SID), past inputs / past outputs multivariable output-error state-space (PI-MOESP, PO-MOESP), canonical variate analysis (CVA).

Compared to PEM, subspace methods, although suboptimal, have several advantages: intrinsic determination of the system order, no particular complication for MIMO systems, no additional problems for nonzero initial state, same treatment of stable and unstable systems, better numerical properties, as they do not involve nonlinear optimization techniques and thus are fast (since noniterative) and accurate (since no problems with local minima occur). Overall, subspace methods and prediction error methods are complementary: A good initial model can be quickly obtained with subspace methods while a further optimization of the parameters can be done with PEM.

2.3 Recursive Identification

Contrary to the batch approach presented so far, but nevertheless based on that approach, recursive or adaptive procedures have been developed to adjust a model on-line by taking only into account the new current measurement. Such algorithms also provide an alternative to the batch ones, since they can be less computationally demanding.

Recursive algorithms for the identification of single-output models can be written as

$$\hat{\boldsymbol{\theta}}_k = \hat{\boldsymbol{\theta}}_{k-1} + \mu_k \boldsymbol{R}_k^{-1} \boldsymbol{\psi}_k e_k, \tag{2.57}$$

where $\hat{\boldsymbol{\theta}}_k$ is the parameter estimate at time k, the prediction error $e_k = y_k - \hat{y}_k(\hat{\boldsymbol{\theta}}_{k-1})$ at time k is calculated from the parameter vector at time $k-1$, the scalar μ_k and the matrix $\boldsymbol{R}_k$ affect the size and the direction of the update, respectively, and where the derivative $\psi_k = \frac{d\hat{y}_k}{d\theta}\Big|_{\hat{\theta}_{k-1}}$ is, for linear models such as (2.10), simply $\psi_k = \boldsymbol{x}_k$.

From that general form, several update rules have been derived depending on the model structure, the criterion to be minimized, and also the assumptions on the parameter variations.

In particular, assume that the true parameters $\boldsymbol{\theta}$ of linear models, $y_k = \boldsymbol{x}_k^\top \boldsymbol{\theta} + v_k$, where v_k is a white Gaussian noise (see Appendix A.3.1.1) with variance $\mathbb{E}[v_k^2] = R_{2k}$, evolve as a random walk:

$$\boldsymbol{\theta}_k = \boldsymbol{\theta}_{k-1} + \boldsymbol{\omega}_k,$$

where $\boldsymbol{\omega}_k$ is a white Gaussian noise with covariance matrix $\mathbb{E}[\boldsymbol{\omega}_k \boldsymbol{\omega}_k^\top] = \boldsymbol{R}_{1k}$. Then, by considering the linear regression model as the measurement equation in a state-space model, a Kalman filter can be derived as

$$\begin{cases} \hat{\boldsymbol{\theta}}_k = \hat{\boldsymbol{\theta}}_{k-1} + \boldsymbol{l}_k \left(y_k - \boldsymbol{x}_k^\top \hat{\boldsymbol{\theta}}_{k-1} \right) \\ \boldsymbol{l}_k = \dfrac{\boldsymbol{P}_{k-1} \boldsymbol{x}_k}{R_{2k} + \boldsymbol{x}_k^\top \boldsymbol{P}_{k-1} \boldsymbol{x}_k} \\ \boldsymbol{P}_k = \boldsymbol{P}_{k-1} - \dfrac{\boldsymbol{P}_{k-1} \boldsymbol{x}_k \boldsymbol{x}_k^\top \boldsymbol{P}_{k-1}}{R_{2k} + \boldsymbol{x}_k^\top \boldsymbol{P}_{k-1} \boldsymbol{x}_k} + \boldsymbol{R}_{1k} \end{cases} \tag{2.58}$$

with $\hat{\boldsymbol{\theta}}_0 = \hat{\boldsymbol{\theta}}_{init}$ and $\boldsymbol{P}_0 = \boldsymbol{P}_{init}$. This filter minimizes the conditional expectation, given past observations, of the covariance matrix of the parameters, $\mathbb{E}[(\hat{\boldsymbol{\theta}}_k - \boldsymbol{\theta}_k)(\hat{\boldsymbol{\theta}}_k - \boldsymbol{\theta}_k)^\top]$. Note that $\boldsymbol{R}_{1k}$ and R_{2k} can be fixed to $\boldsymbol{R}_1$ and R_2 if known, or estimated along the process.

By simply taking $\boldsymbol{R}_{1k} = \mu^2 \frac{\boldsymbol{x}_k \boldsymbol{x}_k^\top}{1+\mu\|\boldsymbol{x}_k\|_2^2}$, $R_{2k} = 1$, $\boldsymbol{P}_0 = \mu \boldsymbol{I}$ in (2.58), or $\boldsymbol{R}_k = \boldsymbol{I}$ and $\mu_k = \frac{\mu}{1+\mu\|\boldsymbol{x}_k\|_2^2}$ in (2.57), this reduces to the classical algorithm in adaptive signal processing, the (normalized) least mean squares (NLMS) algorithm:

$$\begin{cases} \hat{\boldsymbol{\theta}}_k = \hat{\boldsymbol{\theta}}_{k-1} + \boldsymbol{l}_k \left(y_k - \boldsymbol{x}_k^\top \hat{\boldsymbol{\theta}}_{k-1} \right) \\ \boldsymbol{l}_k = \dfrac{\mu \boldsymbol{x}_k}{1 + \mu \|\boldsymbol{x}_k\|_2^2}, \end{cases} \tag{2.59}$$

where $\mu > 0$ is a tuning parameter. This algorithm minimizes simply the sum of the squared errors.

Also, taking $\boldsymbol{R}_{1k} = \left(\frac{1}{\gamma} - 1\right) \left(\boldsymbol{P}_{k-1} - \frac{\boldsymbol{P}_{k-1} \boldsymbol{x}_k \boldsymbol{x}_k^\top \boldsymbol{P}_{k-1}}{R_{2k} + \boldsymbol{x}_k^\top \boldsymbol{P}_{k-1} \boldsymbol{x}_k} \right)$ and $R_{2k} = \gamma$ leads to the recursive least squares (RLS) algorithm given as

$$\begin{cases} \hat{\theta}_k = \hat{\theta}_{k-1} + l_k \left(y_k - x_k^\top \hat{\theta}_{k-1} \right) \\ l_k = \dfrac{P_{k-1} x_k}{\gamma + x_k^\top P_{k-1} x_k} \\ P_k = \dfrac{1}{\gamma} \left(P_{k-1} - \dfrac{P_{k-1} x_k x_k^\top P_{k-1}}{\gamma + x_k^\top P_{k-1} x_k} \right) = \dfrac{1}{\gamma} \left(I - l_k x_k^\top \right) P_{k-1}, \end{cases} \tag{2.60}$$

where γ is a forgetting factor, usually chosen in the range [0.9, 0.999]. For a constant γ, the RLS algorithm minimizes a weighted sum of squared errors, where the weight γ^{N-k} gives exponential forgetting:

$$\hat{\theta}_N = \underset{\theta \in \mathbb{R}^d}{\operatorname{argmin}} \sum_{k=1}^{N} \gamma^{N-k} \left(y_k - x_k^\top \theta \right)^2 .$$

2.4 Nonlinear System Identification (⋆)

For nonlinear systems, two model classes can be considered in Step 2 of Procedure 1: parametric models and nonparametric models. We describe them next in the context of I/O models.

2.4.1 *Parametric Models*

Parametric models are models with a fixed structure and a finite number of unknown parameters to be determined from the data. A useful example for nonlinear systems is the class of block-structured models. In this class, the Hammerstein model consists of a static nonlinear map f, for instance, a saturation, a dead zone (see Fig. 1.1) or a polynomial model of fixed order, followed by a linear dynamical model. For the SISO ARX model (2.13), it reads

$$y_k = -\sum_{i=1}^{n_a} a_i y_{k-i} + \sum_{i=1}^{n_b} b_i f(u_{k-n_d-i+1}) + \varepsilon_k .$$

The Wiener model is the reverse:

$$y_k = f \left(-\sum_{i=1}^{n_a} a_i y_{k-i} + \sum_{i=1}^{n_b} b_i u_{k-n_d-i+1} \right) + \varepsilon_k .$$

2.4.2 Nonparametric Models

Nonparametric models are models for which both the structure and the parameters are to be estimated from the data. In this respect, it appears that the distinction between parametric and nonparametric is not so clear. For instance, if the structure of the static functions in the Hammerstein or Wiener models is not fixed, these models can be called semi-parametric, in fact nonparametric.

Nonparametric models are particularly well-suited for the estimation of arbitrary and unknown nonlinearities. Indeed, the model classes often offer enough flexibility to approximate sufficiently well any smooth function. An example for a nonparametric model class (a subset of the space $\mathbb{R}^{\mathcal{X}}$ of functions from the regression space $\mathcal{X}$ into $\mathbb{R}$) is the linearly parameterized function class of function expansions,

$$\mathcal{F} = \left\{ f \in \mathbb{R}^{\mathcal{X}} : f = \sum_{i=1}^{M} \alpha_i f_i, \ \alpha_i \in \mathbb{R}, M \in \mathbb{N} \right\}, \tag{2.61}$$

where α_i are function weights and f_i are referred to as basis functions. The typical ones are radial basis functions (RBF), wavelet functions, kernel functions, etc. Given such a model class, the identification procedure must determine the parameters defining the model structure, M and the f_i's, and those parametrizing the linear combination, the α_i's.

In parametric discrete-time models, the parameters can sometimes have meaning, if they combine, in a more or less complicated way, the physical parameters included in the original continuous-time models. At the opposite, it should be clear from the example above that the parameters (here the α_i's) of nonparametric models are always meaningless. The purpose of these *black-box* models is solely to predict the output the system, not to infer knowledge on the system from their internal parameters.

As a direct consequence, the goal of nonparametric system identification cannot be set in terms of estimation theory, which focuses on the recovery of the true parameter values, as it can be done for the parametric models of Sect. 2.1.2 (see the NPE criterion (2.80) to assess the quality of a model). With nonparametric models, the focus is on the ability of the model to accurately predict the system output. A convenient framework for the analysis and design of models and methods in that context is that of statistical learning. More precisely, we here formalize the agnostic learning framework, in which the only source of information is the data itself, and consider it for regression, which corresponds to the estimation of real-valued functions (again, see Appendix A for the definition of probabilistic terms).

Let (X, Y) be a random pair, of unknown distribution P, taking values $(\boldsymbol{x}, y)$ in $(\mathcal{X} \times \mathcal{Y}) \subseteq (\mathbb{R}^d \times \mathbb{R})$. Then, the goal is to find the model f that leads to the minimal prediction error on average. Therefore, the relevant quantity is the so-called *(expected) risk* (or generalization error) defined as

$$L(f) = \mathbb{E}[\ell(Y - f(X))] = \int_{\mathcal{X}} \int_{\mathcal{Y}} \ell(y - f(\boldsymbol{x}))\, p(\boldsymbol{x}, y)\, d\boldsymbol{x} dy, \tag{2.62}$$

where p is the probability density function (pdf) of P and ℓ is a loss function as defined in Sect. 2.1.2. The optimal f minimizing the risk is known as the *regression function* $f_{reg} = \text{argmin}_{f:\mathcal{X}\to\mathcal{Y}} L(f)$ and is given for the squared loss function, $\ell(e) = e^2$, by

$$f_{reg}(\boldsymbol{x}) = \mathbb{E}\left[Y|X = \boldsymbol{x}\right].$$

However, both the risk and the regression function cannot be computed without knowledge of P and, in practice, one has only access to a realization $\mathcal{D} = \{(\boldsymbol{x}_k, y_k)\}_{k=1}^N$ (2.24) of the training sample made of N independent copies of (X, Y). A standard approach is thus to minimize an estimate of the risk, known as the empirical risk (or the training error) and computed as

$$V = \frac{1}{N}\sum_{k=1}^{N} \ell\left(y_k - f(\boldsymbol{x}_k)\right).$$

This approach is the prediction error minimization one, already depicted for parametric models in Sect. 2.1.2 and also known as empirical risk minimization in the statistical learning literature.

2.4.2.1 Overfitting

One issue which must be considered with care is overfitting/underfitting. The typical nonparametric model classes provide sufficient flexibility for the model to yield a perfect fit of the data. Thus, if we were to minimize the prediction error on a training data set, the model would learn the noise as well as the target function, i.e., overfit the training data. Therefore, a central issue in nonparametric model estimation is the control of the model complexity, in order to limit its ability to fit the noise. Conversely, this control should not be too pronounced so as not to underfit, i.e., yield a too simple model not able to accurately approximate the regression function. These issues are illustrated in Fig. 2.2.

In general, a regression algorithm searches for f within a particular model class $\mathcal{F}$ and a standard approach to control the complexity of the model is to limit the "size" of $\mathcal{F}$, where the "size" can be understood as the cardinality for finite sets and as other measures of capacity for infinite ones. To understand the effect of this size, let us define the best function in the class as[7]

$$f^* = \underset{f\in\mathcal{F}}{\text{argmin}}\ L(f).$$

Then, the following classical decomposition of the risk is obtained by considering the difference between the risk of f and the best possible risk:

---

[7] Note that this is again a theoretical function that cannot be computed in practice without knowledge of P.

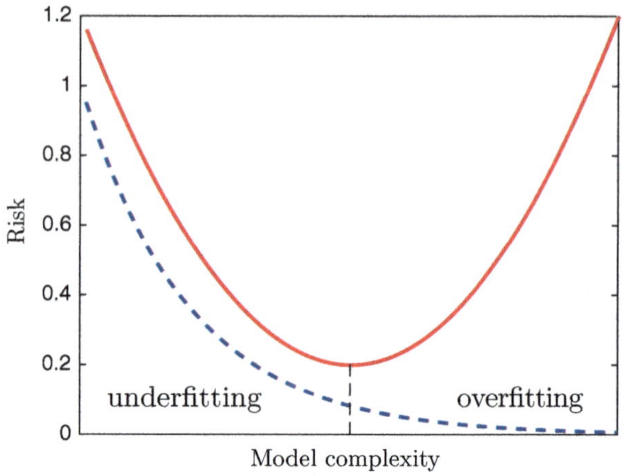

Fig. 2.2 Abstract view of the evolution of the empirical (−−) and expected (−) risks (or training and generalization errors) when the model complexity increases

$$L(f) - L(f_{reg}) = \underbrace{[L(f) - L(f^*)]}_{\text{approximation error}} + \underbrace{[L(f^*) - L(f_{reg})]}_{\text{estimation error}}.$$

On the one hand, the approximation error refers to the error implied by the choice of the model class $\mathcal{F}$, which limits the flexibility of the model f. On the other hand, the estimation error reflects the difficulty of finding the best function within a particular function class on the basis of the limited amount of information entailed in the training sample $\mathcal{D}$. This decomposition emphasizes the no-less classical bias–variance trade-off. More precisely, the approximation error can be decreased by increasing the capacity of $\mathcal{F}$ (think of adding functions to a finite function class). But doing so, it becomes less obvious for the algorithm to select the right function in $\mathcal{F}$ and the estimation error might increase. In other words, by decreasing the bias implied by the structure of the model, we increase the influence of the data on the estimation and thus the variance. Hence, and as already mentioned in the discussion on overfitting, a critical issue in nonlinear model estimation is the control of the function complexity or, equivalently, the choice of a model class with low capacity.

2.4.2.2 Regularized Learning

Regularized learning constitutes a classical and more practical approach to learn a nonlinear model while controlling its complexity. It amounts to adding a penalization term to the objective of the minimization of the training error. Thus, one minimizes a regularized functional representing a trade-off between the fit to the data and typically some regularity conditions of f:

$$\min_{f \in \mathcal{F}} \sum_{i=1}^{N} \ell(y_k - f(\boldsymbol{x}_k)) + \lambda R(f), \tag{2.63}$$

where $R(f)$ is a general regularization term and $\lambda \geq 0$ tunes the trade-off between the two terms. An illustration of the influence of the parameter λ will be given in Example 2.1, p. 48.

Though searching for f within a specific function class $\mathcal{F}$ can be related in some cases to a particular choice of structure for the nonlinear model f, this can also be more general. In particular, by assuming that f is an expansion over some functional basis, a single function $f \in \mathcal{F}$ can have multiple representations (and parametrizations) depending on the choice of the basis. In addition, we will see below that $\mathcal{F}$ can be an infinite dimensional function space with the universal approximation capability while still allowing for learning from a finite set of data. As a practical consequence, arbitrary target functions can be learned without prior knowledge on the nonlinearities and without introducing a bias due to an arbitrary choice of unsuitable or insufficiently flexible structure for f.[8]

Reviewing the huge literature on nonlinear model estimation and the various choices for the loss function ℓ, the regularizer R and the function class $\mathcal{F}$ goes far beyond the scope of this book. Instead, we focus the presentation on kernel machines. Note that the particular choice of kernel machines is in accordance with the current trend started twenty years ago in machine learning, with strong connections with support vector machines, presented for classification in Sect. 3.1.1.2.

2.4.2.3 Learning in Reproducing Kernel Hilbert Space

We start with a brief introduction to reproducing kernel functions before presenting their use in regularized learning.

In the following, we refer to real-valued positive definite functions according to the following definition as *kernel functions*.

Definition 2.1 (*Real-valued positive definite function*) A real-valued function K on $\mathcal{X}^2$ is called a positive definite function if it is symmetric and

$$\forall N \in \mathbb{N},\ \forall \{\boldsymbol{x}_k\}_{k=1}^N \in \mathcal{X}^N,\ \forall \{a_k\}_{k=1}^N \in \mathbb{R}^N,\quad \sum_{i=1}^N \sum_{k=1}^N a_i a_k K(\boldsymbol{x}_i, \boldsymbol{x}_k) \geq 0. \tag{2.64}$$

The condition in this definition can be equivalently formulated in terms of the positive semi-definiteness[9] (see (B.15)) of the *kernel matrix*, defined for a kernel function K and a data set, $\{\boldsymbol{x}_k\}_{k=1}^N$, as

---

[8]By arbitrary target functions, we refer to target functions with unknown components but that should nonetheless satisfy some basic smoothness assumption. Indeed, without such an assumption, the value of the target function at a point need not be related to its values over a neighborhood of that point and a predictive model could not generalize from a finite sample of values.

[9]Note the confusion in the dedicated terminologies, where a positive definite function yields a positive *semi*-definite matrix.

$$K = \begin{bmatrix} K(x_1, x_1) & \dots & K(x_1, x_N) \\ \vdots & \ddots & \vdots \\ K(x_N, x_1) & \dots & K(x_N, x_N) \end{bmatrix}. \tag{2.65}$$

Definition 2.2 (*Reproducing kernel Hilbert space*) Let $(\mathcal{H}, \langle\cdot,\cdot\rangle_{\mathcal{H}})$ be a Hilbert space of real-valued functions on $\mathcal{X}$ with inner product $\langle\cdot,\cdot\rangle_{\mathcal{H}}$ (see Appendix B.2.3). A real-valued function K on $\mathcal{X}^2$ is a reproducing kernel of $\mathcal{H}$ if and only if

1. $\forall x \in \mathcal{X},\ K(x, \cdot) \in \mathcal{H}$;
2. $\forall x \in \mathcal{X},\ \forall f \in \mathcal{H},\ \langle f, K(x, \cdot)\rangle_{\mathcal{H}} = f(x)$ (reproducing property).

A Hilbert space of real-valued functions which possesses a reproducing kernel is called a reproducing kernel Hilbert space (RKHS).

Note that the reproducing property of K implies in particular that

$$\langle K(x, \cdot), K(x', \cdot)\rangle_{\mathcal{H}} = K(x, x'). \tag{2.66}$$

The Moore–Aronszajn theorem states that for any kernel function K satisfying Definition 2.1, there is one and only one RKHS with K as reproducing kernel.

Let K be a kernel function as in Definition 2.1 and $(\mathcal{H}, \langle\cdot,\cdot\rangle_{\mathcal{H}})$ the associated RKHS. Then, the class of functions $\mathcal{H}$ can be written as

$$\mathcal{H} = \left\{ f \in \mathbb{R}^{\mathcal{X}} : f = \sum_{i=1}^{m} \alpha_i K(x_i, \cdot),\ m \in \mathbb{N}, \alpha_i \in \mathbb{R}, x_i \in \mathcal{X}, \|f\|_{\mathcal{H}} < +\infty \right\}, \tag{2.67}$$

where $\|f\|_{\mathcal{H}} = \sqrt{\langle f, f\rangle_{\mathcal{H}}}$ is the norm in $\mathcal{H}$ induced by the inner product defined for two functions, $f = \sum_{i=1}^{m} \alpha_i K(x_i, \cdot)$ and $f' = \sum_{i=1}^{m'} \alpha'_i K(x'_i, \cdot)$, as $\langle f, f'\rangle_{\mathcal{H}} = \sum_{i=1}^{m} \sum_{l=1}^{m'} \alpha_i \alpha'_l K(x_i, x_l)$, i.e.,

$$\|f\|_{\mathcal{H}}^2 = \langle f, f\rangle_{\mathcal{H}} = \sum_{i=1}^{m} \sum_{l=1}^{m} \alpha_i \alpha_l K(x_i, x_l). \tag{2.68}$$

A typical kernel function is the Gaussian RBF kernel (or Gaussian kernel for short),

$$K(x, x') = \exp\left(\frac{-\|x - x'\|_2^2}{2\sigma^2} \right), \quad \sigma > 0, \tag{2.69}$$

for which $\mathcal{H}$ consists of all infinitely differentiable (i.e., smooth) functions of $\mathcal{X} \to \mathbb{R}$. With such a kernel, $\mathcal{H}$ enjoys the so-called universal approximation capability, i.e., any continuous function can be arbitrarily well approximated by a function in $\mathcal{H}$. Other examples of kernel functions include the polynomial kernel of degree γ, $K(x, x') = (x^\top x' + 1)^\gamma$, and the linear kernel, $K(x, x') = x^\top x'$.

When learning in an RKHS, i.e., when $\mathcal{F} = \mathcal{H}$ in (2.63), a natural choice for $R(f)$ is based on the RKHS norm:

$$R(f) = \frac{1}{2}\|f\|_{\mathcal{H}}^2. \tag{2.70}$$

Such a regularizer is a measure of the function smoothness and is particularly suitable for cases without prior information on the shape of the target function. Indeed, the RKHS norm acts as an upper bound on the magnitude of derivatives. For instance with the Gaussian kernel (2.69), we have, for all $\boldsymbol{x} \in \mathcal{X}$ and $i, l \in [d]$,

$$\begin{aligned} |f(\boldsymbol{x})| &\leq \|f\|_{\mathcal{H}}, \\ \left|\frac{\partial f(\boldsymbol{x})}{\partial x_i}\right| &\leq \frac{1}{\sigma}\|f\|_{\mathcal{H}}, \\ \left|\frac{\partial^2 f(\boldsymbol{x})}{\partial x_i \partial x_l}\right| &\leq \frac{1}{\sigma^2}\|f\|_{\mathcal{H}} \end{aligned} \tag{2.71}$$

and similar bounds for higher-order derivatives. Thus, the smaller $\|f\|_{\mathcal{H}}$ is, the smoother f is. In addition, with (2.70), the representer theorem provides an explicit structure for the solution to (2.63).

Theorem 2.1 (Representer Theorem) *For any loss function ℓ, the solution f^* to* (2.63), *with $\mathcal{F} = \mathcal{H}$ and $\mathcal{H}$ defined as in* (2.67), $R(f) = r(\|f\|_{\mathcal{H}})$ *and a monotonically increasing function $r : \mathbb{R}^+ \to \mathbb{R}^+$, is a kernel expansion over the training set, i.e., f^* is in the span of $\{K(\boldsymbol{x}_k, \cdot)\}_{k=1}^N$.*

This result shows that minimizing any regularized functional of the form (2.63) over an RKHS leads to finite linear combinations of kernel functions computed at the training points:

$$f(\boldsymbol{x}) = \sum_{k=1}^N \alpha_k K(\boldsymbol{x}_k, \boldsymbol{x}). \tag{2.72}$$

Note that a semi-parametric version of Theorem 2.1 can also be obtained to allow for a bias term in the model. This is done by considering a model $\tilde{f} = f + b$, with $f \in \mathcal{H}$ and $b \in \mathbb{R}$, regularized only in f. Thus, the model can take the form of (2.72) or of

$$f(\boldsymbol{x}) = \sum_{k=1}^N \alpha_k K(\boldsymbol{x}_k, \boldsymbol{x}) + b. \tag{2.73}$$

Kernel Ridge Regression

A popular regularized learning algorithm working in RKHS for regression is *kernel ridge regression*. It is based on the learning problem (2.63) regularized with the squared RKHS norm and measuring the error with the squared loss $\ell_2(e) = e^2$.

Recalling Theorem 2.1 and the definition of the kernel matrix $\boldsymbol{K}$ in (2.65), we can write the vector of predictions on the training set as

$$\begin{bmatrix} f(\boldsymbol{x}_1) \\ \vdots \\ f(\boldsymbol{x}_N) \end{bmatrix} = \boldsymbol{K}\boldsymbol{\alpha},$$

where $\boldsymbol{\alpha} = \begin{bmatrix}\alpha_1 & \dots & \alpha_N\end{bmatrix}^\top$ is the weight vector of the linear combination (2.72). With these notations, the squared RKHS norm (2.68) of f can be computed as

$$\|f\|_{\mathcal{H}}^2 = \boldsymbol{\alpha}^\top \boldsymbol{K}\boldsymbol{\alpha}.$$

Then, problem (2.63) can be rewritten as the finite-dimensional convex and quadratic optimization problem

$$\min_{\boldsymbol{\alpha}\in\mathbb{R}^N} \|\boldsymbol{y} - \boldsymbol{K}\boldsymbol{\alpha}\|_2^2 + \lambda\boldsymbol{\alpha}^\top \boldsymbol{K}\boldsymbol{\alpha},$$

where $\boldsymbol{y} = \begin{bmatrix}y_1 & \dots & y_N\end{bmatrix}^\top$. The solution can be explicitly computed in closed form[10] with a formula obtained by setting the gradient of the cost function to zero:

$$2(\boldsymbol{K}\boldsymbol{K} + \lambda\boldsymbol{K})\boldsymbol{\alpha} - 2\boldsymbol{\alpha}^\top \boldsymbol{y} = \boldsymbol{0},$$

where we used the symmetry of $\boldsymbol{K}$. This leads to

$$\boldsymbol{\alpha} = (\boldsymbol{K} + \lambda\boldsymbol{I})^{-1}\boldsymbol{y}. \tag{2.74}$$

Example 2.1 (Kernel ridge regression and the influence of the regularization parameter) Figure 2.3 illustrates the influence of the regularization parameter λ in kernel ridge regression (the general idea remains the same for other regularized algorithms). A one-dimensional data set of 50 points is generated with x_k uniformly distributed in $[-3, 3]$ and $y_k = \mathrm{sinc}(x_k) + v_k$ with a Gaussian noise term v_k. Then, various values of λ are tested in the range $[0.01, 100]$. Figure 2.3 clearly shows that by increasing λ the variations of the model f are more and more penalized and smoother models are obtained. Thus, the small values of λ lead to overfitting and a model that fits the noise as well as the target function, whereas too large values of λ yield too simple models that cannot accurately approximate the target function.

Feature Space

Kernel functions can also be interpreted as inner products in some feature space $\mathcal{X}_\phi$. In fact, learning algorithms based on kernel functions were first popularized with this definition, while the RKHS point of view is now more and more considered. The feature space point of view is presented here and will be used more particularly by the methods described in Sect. 9.1.

[10] More precisely, the solution can *in principle* be explicitly computed in closed form. In practice, our ability to compute the explicit solution depends on the size $N \times N$ of $\boldsymbol{K}$ and the amount of available memory. Of course, when $\boldsymbol{K}$ fits in memory, (2.74) is computed with efficient linear system solving techniques rather than via the inverse of an $N \times N$ matrix.

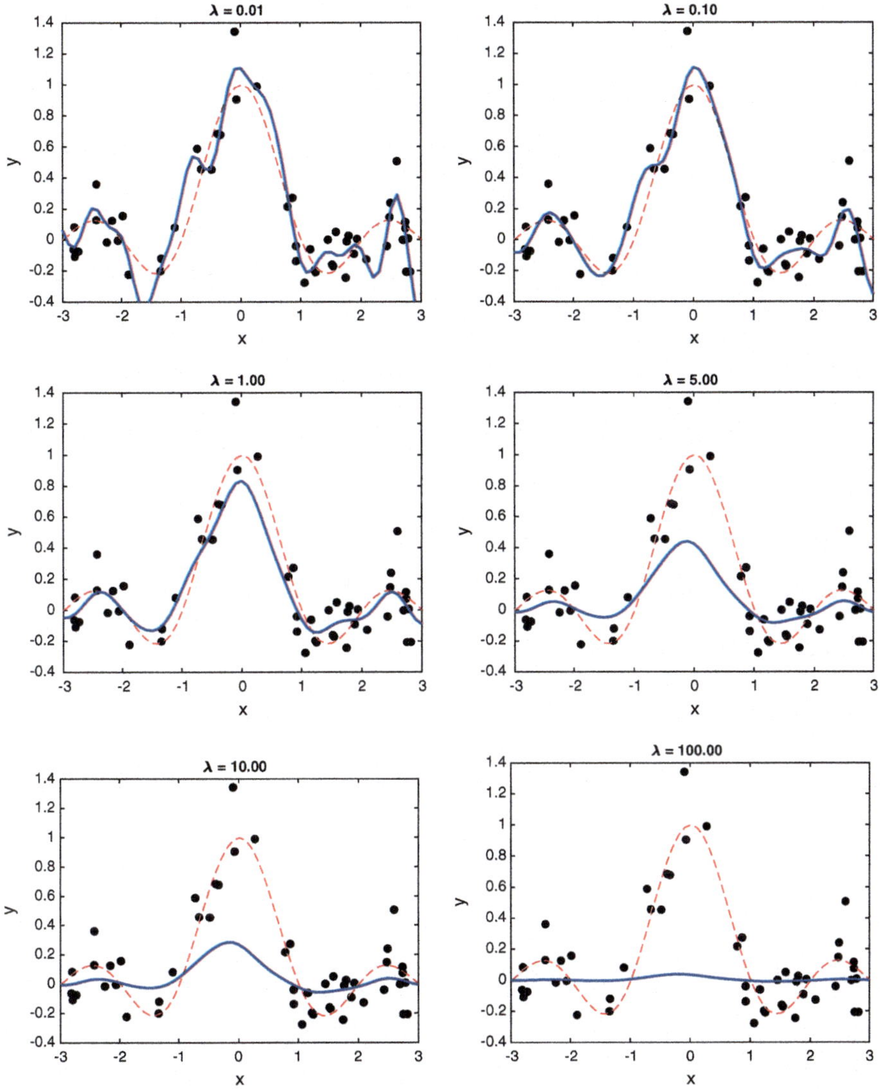

Fig. 2.3 Effect of the regularization parameter λ on kernel ridge regression in Example 2.1: the best value of $\lambda = 1$ yields the model (plain blue line) that best approximates the target function (dashed red line)

Consider the nonlinear mapping

$$\phi : \mathcal{X} \rightarrow \mathcal{X}_\phi \tag{2.75}$$

and the class of linear functions in $\mathcal{X}_\phi$, i.e., the class of functions f written as

$$f(\boldsymbol{x}) = \langle \boldsymbol{\theta}, \phi(\boldsymbol{x}) \rangle_{\mathcal{X}_\phi},$$

with parameters $\boldsymbol{\theta} \in \mathcal{X}_\phi$. Note that these functions are nonlinear with respect to $\boldsymbol{x}$ and can constitute function classes of high capacity. For instance, consider that $\phi(\boldsymbol{x})$ is a vector containing all the monomials of a polynomial of a certain degree. Then, $f(\boldsymbol{x})$ is a polynomial model and by increasing the dimension of $\phi(\boldsymbol{x})$, we increase the degree of the polynomial.

It can be shown that every positive definite kernel function K corresponds to an inner product in some feature space, i.e., that for some $\mathcal{X}_\phi$,

$$\forall(\boldsymbol{x}, \boldsymbol{x}') \in \mathcal{X}^2, \quad K(\boldsymbol{x}, \boldsymbol{x}') = \langle \phi(\boldsymbol{x}), \phi(\boldsymbol{x}') \rangle_{\mathcal{X}_\phi}. \tag{2.76}$$

For particular algorithms that only use the data through their inner products, this so-called *kernel trick* alleviates the computational difficulties of working with high dimensional vectors $\phi(\boldsymbol{x})$ and thus allows one to optimize (2.63) over classes of very high capacity.

The relationship with the RKHS framework can be seen by choosing ϕ as the most natural feature map for a kernel function, i.e., the one that maps $\mathcal{X}$ to the corresponding RKHS: $\phi(\boldsymbol{x}) = K(\boldsymbol{x}, \cdot)$ and $\mathcal{X}_\phi = \mathcal{H}$. This yields $f = \boldsymbol{\theta}$ and, by the reproducing property,

$$f(\boldsymbol{x}) = \langle f, K(\boldsymbol{x}, \cdot) \rangle_{\mathcal{H}} = \langle \boldsymbol{\theta}, \phi(x) \rangle_{\mathcal{X}_\phi}.$$

2.4.2.4 Learning with ℓ_1-norm Regularization

Another typical regularization scheme for models based on kernel functions is to penalize the ℓ_1-norm of the parameters, $\|\boldsymbol{\alpha}\|_1$. However, this scheme cannot apply to (2.63) where $\boldsymbol{\alpha}$ need not be uniquely defined. Therefore, this scheme is usually applied with the structure of f fixed a priori. With f chosen as in (2.72), this leads to

$$\min_{\alpha \in \mathbb{R}^N} \sum_{k=1}^{N} \ell(y_k - f(\boldsymbol{x}_k)) + \lambda \|\boldsymbol{\alpha}\|_1 \tag{2.77}$$

$$\text{s.t.} \quad f(\boldsymbol{x}) = \sum_{k=1}^{N} \alpha_k K(\boldsymbol{x}_k, \boldsymbol{x}).$$

This learning strategy is often chosen in order to favor sparse solutions with few nonzero α_k. Indeed, the sparsity of $\boldsymbol{\alpha}$ directly determines the sparsity of the model as defined below.

2.4.2.5 Sparse Models

A nonlinear model is said to be sparse when it corresponds to an expansion over a small number of basis functions. Equivalently, the models in the form (2.72) or (2.73) are sparse when the number of nonzero α_k, i.e., $\|\boldsymbol{\alpha}\|_0$ (see (B.2)), is small. Nonlinear estimation methods often favor sparse models, for which the sparsity allows for faster computations of their output for a given $\boldsymbol{x}$, due to a small number of terms involved in the sum in, e.g., (2.72).

Support Vectors

For models in the forms (2.72) and (2.73), the set of training data points, $\boldsymbol{x}_k$, which are retained in the model after the estimation of the parameters α_k, i.e., for which $\alpha_k \neq 0$, are the so-called *support vectors* (SVs) of the model. Indeed, these vectors support the model as they entirely determine the set of basis functions $\{K(\boldsymbol{x}_k, \cdot) : \alpha_k \neq 0,\ k = 1, \dots, N\}$ used to compute $f(\boldsymbol{x})$.

Obtaining Sparse Models

Various strategies can be used to obtain sparse models, and some that are particularly well suited for hybrid system identification will be reviewed in Sect. 9.1.2. In a classical regression setting, the sparsity of the model typically results from an appropriate choice of regularization term, as with the ℓ_1-norm above, or of loss function. In particular, support vector regression (SVR) is based on the ϵ-insensitive loss defined as

$$\ell(e) = \max\{0, |e| - \epsilon\} \tag{2.78}$$

and plotted in Fig. 2.4. This loss function builds a tube of insensitivity around the model so that errors below the threshold ϵ do not affect it. This insensitivity implies the model sparsity: With the regularization term (2.70) and the loss (2.78), the solution (2.73) to (2.63) can be shown to satisfy

$$\alpha_k \neq 0 \quad \Rightarrow \quad |y_k - f(\boldsymbol{x}_k)| \geq \epsilon \tag{2.79}$$

and the support vectors correspond only to data points lying outside the tube of insensitivity or exactly on its boundary.

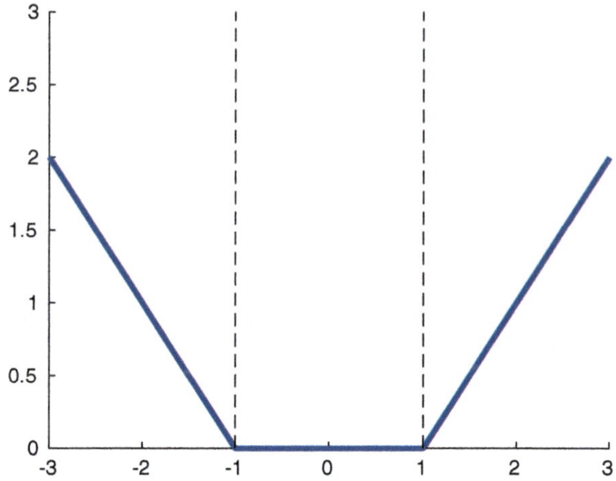

Fig. 2.4 ϵ-insensitive loss function (2.78) for $\epsilon = 1$

2.5 Model Selection and Assessment

The last Step 4 of Procedure 1 allows us to assess the quality of the model and to compare it with other models, obtained for different choices of model architecture, dynamics representation, model orders, in Step 2, or with other algorithms and criteria to determine the parameter values, in Step 3. The final goal is to retain the "best" model (model class and parameter values) among the others. The overall Procedure 1 comprises not only successive steps, but also steps backward and even complete procedures, such as cross-validation, testing together numerous models, to determine the final model. For the model selection, several criteria and procedures can be used. At the end, these criteria are also used to quantify the performance of the final model.

2.5.1 *Model Assessment*

Depending on the context, different tools can be used to assess a model. These tools are presented here in the case of I/O SISO models, but can be reformulated quite easily for MIMO I/O or state-space models.

When testing an identification procedure in simulation, i.e., when the true parameters of linear or nonlinear parametric models are known, the models are typically evaluated in terms of the normalized parametric error (NPE), defined as

$$NPE = \frac{\|\boldsymbol{\theta} - \hat{\boldsymbol{\theta}}\|_2}{\|\boldsymbol{\theta}\|_2}, \tag{2.80}$$

where $\boldsymbol{\theta}$ and $\hat{\boldsymbol{\theta}}$ are the true and estimated parameter vectors.

However, in practice or when the parametric structure of the model is not fixed, the quality of a model must be measured on the basis of data. Various criteria are usually used for this purpose.

The first and basic criterion is the residual criterion obtained after the estimation of model f (minimization (2.26) in the nonregularized linear case or (2.63) for the regularized nonlinear case):

$$V = \frac{1}{N}\sum_{k=1}^{N} \ell(e_k), \tag{2.81}$$

with the prediction error $e_k = y_k - \hat{y}_k$ (2.25), i.e., the difference between the system output y_k and the model prediction $\hat{y}_k$. With the squared loss (2.28), this yields to the common mean squared error (MSE) criterion:

$$MSE = \frac{1}{N}\sum_{k=1}^{N} \left(y_k - \hat{y}_k\right)^2, \tag{2.82}$$

also called empirical error variance.

Another error criterion used in practice is FIT which is calculated with normalized errors as

$$FIT = \left(1 - \frac{\sum_{k=1}^{N}(y_k - \hat{y}_k)^2}{\sum_{k=1}^{N}(y_k - \bar{y})^2}\right) \times 100\% \tag{2.83}$$

with the output mean $\bar{y} = \frac{1}{N}\sum_{k=1}^{N} y_k$.

However, simply computing the MSE or FIT on the data used for the estimation does not yield good estimates of the model accuracy. Indeed, this introduces a bias in the estimates which leads to an overestimation of the model quality (recall, for instance, the overfitting phenomenon discussed in Sect. 2.4.2). Instead, unbiased estimates of the model accuracy are obtained by computing such criteria on a data set independent of the one used for the estimation of the models. Thus, it is a common practice to divide the available data in two different sets: one for training and one for testing.

2.5.2 *Model Selection*

Model selection refers to the determination of the "best model" among several ones estimated from the same data set. Two major families of methods are available for this task: those based on information theory and data-oriented ones. For both approaches, the general scheme is to estimate a series of models for different choices of architecture, dynamics representation, orders, and other "hyperparameters." Then, their accuracy is measured by one of the criteria above and the retained model is the one leading to the smallest value of the criterion.

The term *hyperparameter* is particularly used in the context of machine learning. Hyperparameters are parameters which are fixed *a priori* and not automatically determined by the method under consideration. In a broad sense, they can be of various types. Some of them reflect the precise (sub)class of the model included in the optimization problem. In linear combinations of kernel functions (2.72), they can be the type of kernel (not strictly a numerical parameter), and, for instance, in Gaussian kernels (2.69), the "width" σ. In parametric I/O models, they can be the dynamics representation (ARX, ARMAX, etc.) (also not strictly a numerical parameter) and the orders of the included polynomials. Others influence the definition of the optimization problem itself, e.g., the type of regularizer $R(f)$ and the trade-off parameter λ in regularized learning (2.63). Yet others are specific to a particular optimization algorithm: the constant step size in the steepest gradient descent or its initial value, the parameters of the stopping criterion to name a few. For several models, they are typically tuned by a grid or random search, on the basis of the criteria described below.

2.5.2.1 Model Selection based on Information Theory

Based on information theory, a number of statistical measures have been derived for model selection, originally with models linear in the parameters. These include the final prediction error (FPE) criterion, the Akaike information criterion (AIC), the AIC corrected for small sample size (AICc), and the Bayesian information criterion (BIC). These are trade-offs between the goodness of fit expressed by the residual criterion (MSE) and a complexity penalty growing with the number d of parameters:

$$FPE = N\log(MSE) + N\log\left(\frac{N+d}{N-d}\right),$$

$$AIC = N\log(MSE) + 2d,$$

$$AICc = AIC + \frac{2d(d+1)}{N-d-1},$$

$$BIC = N\log(MSE) + d\log(N),$$

where N is the size of the data set used to estimate the models.

2.5.2.2 Data-Oriented Model Selection

Another strategy for model selection determines the "best" model in terms of the model accuracy, as computed by the MSE or FIT. However, in order not to bias these estimates or the ones to be computed in the final model assessment, the data set should be split once more. Thus, it is a common practice to randomly divide all the available data into three different sets: training, validation, and test sets. The training set is used only to estimate the model within a fixed model class and with fixed hyperparameters. The validation set is used to compare the quality of the different models and select the "seemingly" best one. Finally, the test set is held out of the procedure and kept only for the final assessment of the model. The training and validation sets should include the regimes representative of the whole system behavior. Note that, historically, for linear models, the validation set was often erroneously merged with the test set.

If a validation set cannot be obtained, or to avoid the waste of data, *cross-validation* approaches can be employed. The K-fold cross-validation technique divides the training data set into K parts. Each part is used once as a validation set while the other $K - 1$ parts are used as training set. The MSE averaged over the K different validation sets provides an (almost) unbiased estimate of the expected MSE of a model trained on the whole data. Setting $K = N$ leads to the leave-one-out (LOO) cross-validation, where only one data point is used for validation and training uses as many as $N - 1$ points.

The resulting LOO estimate can be computationally demanding. But for linearly parametrized models, the LOO estimate of the expected MSE,

$$MSE_{loo} = \frac{1}{N} \sum_{k=1}^{N} (y_k - \tilde{y}_k)^2 ,$$

where $\tilde{y}_k$ is the value predicted by the model estimated without the kth observation, can be calculated analytically.

Notes

This chapter provided a brief introduction to system identification, inspired by famous textbooks on this topic [2–4]. Particularly, the first two sections follow closely the definitive reference of [3] for linear models. For a more recent book, we can refer to [5]. State-space models and their properties are described in the classical book of [6]. Subspace methods are detailed in [7] among many others and are summarized and unified in [8], from which the above brief presentation is inspired.

Robust estimation in the presence of outliers is almost a research field in its own. A popular textbook covering all the basics and the analysis of robust loss functions is [9].

Recursive estimation is covered in [3] and the more focused reviews [10, 11].

Recent approaches for block-oriented nonlinear system identification are covered in [12]. Nonparametric models were popularized for system identification by [13, 14]. The statistical learning perspective developed in Sect. 2.4.2 is described in more details in [15–17]. Support vector regression based on kernel functions and the ϵ-

insensitive loss was proposed by [18] (see also [19] for a nice tutorial overview) and linear programming versions based on ℓ_1-norm regularization by [20]. Kernel ridge regression is due to [21]. Mathematical details of reproducing kernel Hilbert spaces can be found in [22] and [17]. For instance, (2.71) is derived from Corollary 4.36 in the latter. The representer theorem is originally due to [23] and in the generalized form of Theorem 2.1 to [24]. Other well-known models include neural networks, fully described in [25]. The universal approximation capability of neural networks was first investigated in [26–28] and in [29] for kernel machines.

The criteria for model selection based on information theory are described in detail in [30, 31]. FPE and AIC were proposed in [32] and [33], AICc in [34], BIC in [35]. Model selection and cross-validation procedures in the nonparametric case are an active field of research in statistical learning, see e.g., [36], and the references therein.

Optimization remains one of the main tools for system identification, which often amounts to minimize a cost function made of a data fit term and a regularization term. More details on this topic can be found in many textbooks, such as [37]. We also recommend [38] for a focus on convex optimization.

Software

All the basic system identification tools are available in the popular system identification toolbox for MATLAB. For nonparametric regression, see the neural network toolbox, or LibSVM [39] (http://www.csie.ntu.edu.tw/~cjlin/libsvm/), which provides an efficient implementation of support vector regression with the ϵ-insensitive loss.

Convex optimization problems can be efficiently modeled with the CVX toolbox [40, 41] available at http://cvxr.com/cvx. Efficient solvers such as MOSEK [42] http://www.mosek.com are usually based on interior-point algorithms.

References

1. Lauer, F.: Global optimization for low-dimensional switching linear regression and bounded-error estimation. Automatica **89**, 73–82 (2018)
2. Söderström, T., Stoica, P.: System Identification. Prentice-Hall, Upper Saddle (1989)
3. Ljung, L.: System Identification – Theory for the User, 2nd edn. Prentice-Hall, Upper Saddle (1999)
4. Nelles, O.: Nonlinear System Identification: From Classical Approaches to Neural Networks and Fuzzy Models. Springer, Berlin (2001)
5. Keesman, K.J.: System Identification - An Introduction. Advanced Textbooks in Control and Signal Processing. Springer, Berlin (2011)
6. Kailath, T.: Linear Systems. Prentice-Hall, Upper Saddle (1980)
7. van Overschee, P., De Moor, B.: Subspace Identification for Linear Systems: Theory - Implementation - Applications. Kluwer, Netherlands (1996)
8. Favoreel, W., De Moor, B., Van Overschee, P.: Subspace state space system identification for industrial processes. J. Process Control **10**, 149–155 (2000)
9. Rousseeuw, P.J., Leroy, A.M.: Robust Regression and Outlier Detection. Wiley, New York (2003)

10. Ljung, L., Gunnarsson, S.: Adaptation and tracking in system identification - a survey. Automatica **26**(1), 7–21 (1990)
11. Ljung, L.: Recursive identification algorithms. Circuits Syst. Signal Process. **21**(1), 57–68 (2002)
12. Giri, F., Bai, E.W. (eds.): Block-oriented Nonlinear System Identification. Lecture Notes in Control and Information Sciences, vol. 404. Springer, Berlin (2010)
13. Narendra, K.S., Parthasarathy, K.: Identification and control of dynamical systems using neural networks. IEEE Trans. Neural Netw. **1**(1), 4–27 (1990)
14. Sjöberg, J., Zhang, Q., Ljung, L., Benveniste, A., Delyon, B., Glorennec, P.Y., Hjalmarsson, H., Juditsky, A.: Nonlinear black-box modeling in system identification: a unified overview. Automatica **31**(12), 1691–1724 (1995)
15. Vapnik, V.N.: Statistical Learning Theory. Wiley, New York (1998)
16. Hastie, T., Tibshirani, R., Friedman, J.: The Elements of Statistical Learning: Data Mining, Inference, and Prediction, 2nd edn. Springer, Berlin (2009)
17. Steinwart, I., Christmann, A.: Support Vector Machines. Springer, Berlin (2008)
18. Vapnik, V.N.: The Nature of Statistical Learning Theory. Springer, Berlin (1995)
19. Smola, A.J., Schölkopf, B.: A tutorial on support vector regression. Stat. Comput. **14**(3), 199–222 (2004)
20. Mangasarian, O.L., Musicant, D.R.: Large scale kernel regression via linear programming. Mach. Learn. **46**(1–3), 255–269 (2002)
21. Saunders, C., Gammerman, A., Vovk, V.: Ridge regression learning algorithm in dual variables. In: Proceedings of the 15th International Conference on Machine Learning (ICML), Madison, WI, USA, pp. 515–521 (1998)
22. Berlinet, A., Thomas-Agnan, C.: Reproducing Kernel Hilbert Spaces in Probability and Statistics. Springer/Kluwer, Berlin (2004)
23. Kimeldorf, G., Wahba, G.: Some results on Tchebycheffian spline functions. J. Math. Anal. Appl. **33**(1), 82–95 (1971)
24. Schölkopf, B., Herbrich, R., Smola, A.J.: A generalized representer theorem. In: Proceedings of the 14th Annual Conference on Computational Learning Theory (COLT), Amsterdam, The Netherlands. Lecture Notes in Computer Science, vol. 2111, pp. 416–426 (2001)
25. Haykin, S.: Neural Networks and Learning Machines, 3rd edn. Prentice Hall, Upper Saddle (2009)
26. Cybenko, G.: Approximation by superpositions of a sigmoidal function. Math. Control Signals Syst. **2**(4), 303–314 (1989)
27. Hornik, K., Stinchcombe, M., White, H.: Multilayer feedforward networks are universal approximators. Neural Netw. **2**(5), 359–366 (1989)
28. Park, J., Sandberg, I.W.: Universal approximation using radial-basis-function networks. Neural Comput. **3**(2), 246–257 (1991)
29. Hammer, B., Gersmann, K.: A note on the universal approximation capability of support vector machines. Neural Process. Lett. **17**(1), 43–53 (2003)
30. McQuarrie, A.D.R., Tsai, C.L.: Regression and Time Series Model Selection. World Scientific, Singapore (1998)
31. Burnham, K.P., Anderson, D.R.: Model Selection and Multimodel Inference. A Practical Information-Theoretic Approach, 2nd edn. Springer, Berlin (2002)
32. Akaike, H.: Statistical predictor identification. Ann. Inst. Stat. Math. **22**(2), 203–217 (1970)
33. Akaike, H.: A new look at the statistical model identification. IEEE Trans. Autom. Control **19**(6), 716–723 (1974)
34. Hurvich, C.M., Tsai, C.L.: Regression and time series model selection in small samples. Biometrica **76**(2), 297–307 (1989)
35. Schwarz, G.: Estimating the dimension of a model. Ann. Stat. **6**(2), 461–464 (1978)
36. Arlot, S., Celisse, A.: A survey of cross-validation procedures for model selection. Stat. Surv. **4**, 40–79 (2010)
37. Bertsekas, D.P.: Nonlinear Programming. Athena Scientific, Belmont (1999)

38. Boyd, S., Vandenberghe, L.: Convex Optimization. Cambridge University Press, Cambridge (2004)
39. Chang, C.C., Chih-Jen, L.: LIBSVM: a library for support vector machines. ACM Trans. Intell. Syst. Technol. **2**(3), 27:1–27:27 (2001). http://www.csie.ntu.edu.tw/~cjlin/libsvm/
40. Grant, M., Boyd, S.: CVX: Matlab software for disciplined convex programming, version 2.1 (2017). http://cvxr.com/cvx
41. Grant, M.C., Boyd, S.P.: Graph implementations for nonsmooth convex programs. In: Blondel, V., Boyd, S.P., Kimura, H. (eds.) Recent Advances in Learning and Control. Lecture Notes in Control and Information Sciences, vol. 2, pp. 95–110. Springer, Berlin (2008). http://stanford.edu/~boyd/papers/pdf/graph_dcp.pdf
42. Andersen, E.D., Andersen, K.D.: The MOSEK interior point optimizer for linear programming: an implementation of the homogeneous algorithm. High Performance Optimization, Applied Optimization, vol. 33, pp. 197–232. Springer, Berlin (2000)

Chapter 3
Classification

This chapter presents an introduction to classification and more precisely to two fundamental problems in machine learning and pattern recognition, stated informally as follows.

Problem 3.1 (*Discrimination*) Given a set of N labeled data instances, i.e., data pairs $(\boldsymbol{x}_k, q_k)$ made of an input/pattern vector $\boldsymbol{x}_k$ from an input space $\mathcal{X}$ and a label q_k from a finite set of categories $\mathcal{Q}$, estimate a classifier g that can predict the label $q \in \mathcal{Q}$ of any $\boldsymbol{x} \in \mathcal{X}$ with minimum probability of misclassification.

Problem 3.2 (*Clustering*) From a set of N unlabeled data instances, i.e., data points $\boldsymbol{x}_k \in \mathcal{X}$, estimate their labels q_k so as to create homogeneous groups of points.

These problems are encountered in the context of hybrid system identification, where the labels q_k correspond to the active mode or discrete state at time k.

In the machine learning literature, the first problem above is studied in the framework of *supervised learning* and the second one belongs to the family of *unsupervised learning* problems. In supervised learning, the labels q_k of the training data $\boldsymbol{x}_k$, i.e., their correctly associated categories, are given and the machine can learn under the guidance of a teacher, whereas without this information, there is no knowing teacher in unsupervised learning.

This difference in the available information translates to different goals. While Problem 3.1 focuses on learning a classification function (or decision function) g, Problem 3.2 is solely concerned with the estimation of the labels for the given data points. This is a fundamental difference in the sense that the aim for a discrimination algorithm is to find a classifier which will be able to generalize and correctly predict the labels of new test data points, whereas a clustering algorithm need not. While the goal of discrimination seems higher and more difficult, we will see that it is indeed more prone to analysis and that algorithms with guaranteed performance can be derived. On the other hand, the clustering problem, though stated with a less ambitious goal, does not enjoy such ease of analysis.

F. Lauer and G. Bloch, *Hybrid System Identification*,
Lecture Notes in Control and Information Sciences 478,
https://doi.org/10.1007/978-3-030-00193-3_3

Chapter Outline

Section 3.1 starts with the exposition of the discrimination problem. Then, it focuses on linear classifiers (Sect. 3.1.1) and presents the popular support vector machine in Sect. 3.1.1.2. The section ends with a discussion on multi-class problems with more than two categories (Sect. 3.1.2) and nonlinear classifiers (Sect. 3.1.3). The second part of the chapter, Sect. 3.2, is dedicated to the clustering problem and details in particular the well-known k-means algorithm.

3.1 Discrimination

Discrimination aims at predicting the category of previously unseen objects based on previously acquired data on similar objects whose categories are known. A typical (and very successful) application of discrimination is handwritten digit recognition: By learning from tens of thousands of labeled images of handwritten digits, a system can nowadays easily recognize new digits with human-level accuracy.

Formally, we aim at predicting the label $q \in \mathcal{Q}$ of a description $\boldsymbol{x} \in \mathcal{X}$ (typically a real vector) of an object. Discrimination is usually described as a nonparametric learning problem, under the formalism depicted in Sect. 2.4.2. In this framework, we consider the training data set, $\{(\boldsymbol{x}_k, q_k)\}_{k=1}^N$, as a realization of N independent copies of the random pair $(X, Q) \in \mathcal{X} \times \mathcal{Q}$ of unknown distribution P (think, for instance, of $\boldsymbol{x}_k$ as a vector representation of an image and of q_k as the number written in this image). The ultimate goal of discrimination is to find a classifier, i.e., a function

$$g : \mathcal{X} \to \mathcal{Q}, \tag{3.1}$$

which minimizes the probability of misclassification. With the assumptions above, for a given g, this probability corresponds to the *(expected) risk* of g (or its generalization error)

$$L(g) = \mathbb{E}[\ell_C(Q - g(X))] = \mathbb{P}(g(X) \neq Q),$$

when using the 0-1 loss function, $\ell_C(e) = \mathbb{1}_{e \neq 0}$, which gives

$$\ell_C(Q - g(X)) = \mathbb{1}_{g(X) \neq Q} = \begin{cases} 1, & \text{if } g(X) \neq Q \\ 0, & \text{otherwise.} \end{cases}$$

The optimal classifier, i.e., the one with minimal risk, is known as the *Bayes classifier* and is given for all $\boldsymbol{x} \in \mathcal{X}$ by

$$g_{Bayes}(\boldsymbol{x}) = \operatorname*{argmax}_{q \in \mathcal{Q}} \mathbb{P}(Q = q \mid X = \boldsymbol{x}), \tag{3.2}$$

i.e., as the label with maximal conditional probability given the observation $\boldsymbol{x}$ of X.

However, without knowledge of the distribution P, neither the risk nor the Bayes classifier can be computed. The classical approach is then to minimize an estimate of the risk computed as the average loss over the training data:

$$\hat{g} = \operatorname*{argmin}_{g \in \mathcal{G}} \frac{1}{N} \sum_{k=1}^{N} \mathbb{1}_{g(\boldsymbol{x}_k) \neq q_k}. \tag{3.3}$$

Note that in (3.3), the function class $\mathcal{G}$, within which the classifier g is searched for, has to be specified prior to the optimization. Depending on the capacity of this class of functions, i.e., the flexibility and complexity of the classifiers it contains, this direct approach can be prone to overfitting. As discussed in Sect. 2.4.2, this is typically dealt with by penalizing complex functions g in a regularized learning framework as

$$\hat{g} = \operatorname*{argmin}_{g \in \mathcal{G}} \frac{1}{N} \sum_{k=1}^{N} \mathbb{1}_{g(\boldsymbol{x}_k) \neq q_k} + \lambda R(g), \tag{3.4}$$

where $R(g)$ is a regularizer increasing with the complexity of g and $\lambda \geq 0$ is a trade-off parameter which controls the complexity of the classifier $\hat{g}$.

3.1.1 Binary Linear Classification

We first concentrate on binary classification, which refers to discrimination problems with two categories, usually with labels in $\mathcal{Q} = \{-1, +1\}$. Though the analysis of this particular case may seem restrictive, many works in this field are dedicated to it, both theoretically and algorithmically. This is in part due to the fact that, as will be seen in Sect. 3.1.2, problems with more than two categories can be dealt with by solving a combination of binary subproblems.

Due to the discrete nature of its codomain $\mathcal{Q}$, the classifier (3.1) cannot be a linear function in general. However, with a slight abuse of terminology, a linear classifier is one that is based on a linear function, i.e., for $\mathcal{X} = \mathbb{R}^d$:

$$g(\boldsymbol{x}) = \operatorname{sign}(g_0(\boldsymbol{x})),$$

with a real-valued function

$$g_0(\boldsymbol{x}) = \boldsymbol{g}^\top \boldsymbol{x} + b,$$

of parameters $\boldsymbol{g} \in \mathbb{R}^d$ and $b \in \mathbb{R}$. Note that it is customary to define linear classifiers on the basis of *affine* functions (yet another abuse of terminology). Therefore, in the binary case, we write the set of linear classifiers of $\mathcal{X} = \mathbb{R}^d$ as

$$\mathcal{G} = \left\{ g \in \{-1, +1\}^{\mathbb{R}^d} : g(\boldsymbol{x}) = \text{sign}\left(\boldsymbol{g}^\top \boldsymbol{x} + b\right), \ \boldsymbol{g} \in \mathbb{R}^d, \ b \in \mathbb{R} \right\}. \tag{3.5}$$

A classifier from this family implements a separating hyperplane, $H = \{\boldsymbol{x} \in \mathbb{R}^d : g_0(\boldsymbol{x}) = 0\}$, that divides $\mathbb{R}^d$ in two halfspaces: one in which $g_0(\boldsymbol{x}) > 0$ and one with $g_0(\boldsymbol{x}) < 0$. Therefore, a linear classifier that perfectly classifies a data set $\{(\boldsymbol{x}_k, q_k)\}_{k=1}^N \subset \mathcal{X} \times \{-1, +1\}$ without error can be found by solving the system of linear inequalities

$$q_k(\boldsymbol{g}^\top \boldsymbol{x}_k + b) > 0, \quad k = 1, \dots, N, \tag{3.6}$$

which ensures that for all $k \in [N]$,

$$g(\boldsymbol{x}_k) = \text{sign}\left(\boldsymbol{g}^\top \boldsymbol{x}_k + b\right) = \text{sign}(q_k) = q_k.$$

When such a classifier exists, the data set is said to be *linearly separable*.

3.1.1.1 Combinatorial Results (⋆)

Statistical learning theory is concerned with the analysis of learning problems and algorithms and in particular in the derivation of guarantees, typically taking the form of upper bounds on the risk. A large part of such an analysis relies on the determination of the number of possible classifications of N points that can be produced by a set of classifiers, as measured by the growth function. We now give a glimpse at these combinatorial results which will find their use for hybrid system identification in Chap. 5.

Definition 3.1 (*Trace of a function class*) The trace $\mathcal{G}_\mathcal{S}$ of a set of classifiers $\mathcal{G} \subset \mathcal{Q}^\mathcal{X}$ onto a set of N points $\mathcal{S} = \{\boldsymbol{x}_k\}_{k=1}^N \subset \mathcal{X}$ is the set of all labelings of $\mathcal{S}$ that can be produced by a classifier in $\mathcal{G}$:

$$\mathcal{G}_\mathcal{S} = \{(g(\boldsymbol{x}_1), \dots, g(\boldsymbol{x}_N)) \ : \ g \in \mathcal{G}\} \subset \mathcal{Q}^N.$$

Definition 3.2 (*Growth function*) The *growth function* $\Pi_\mathcal{G}(N)$ of a set of classifiers $\mathcal{G}$ at N is the maximal number of labelings of N points that can be produced by classifiers from $\mathcal{G}$:

$$\Pi_\mathcal{G}(N) = \sup_{\mathcal{S} \in \mathcal{X}^N} |\mathcal{G}_\mathcal{S}|.$$

For binary classifiers, $\mathcal{Q} = \{-1, +1\}$ and the growth function is trivially upper bounded by $|\mathcal{Q}|^N = 2^N$. But for the family of linear classifiers, $\mathcal{G}$ in (3.5), a major result in statistical learning states that the growth function can be more tightly bounded by a polynomial function of N. This is obtained via two fundamental combinatorial arguments: Sauer's lemma and the computation of the Vapnik–Chervonenkis (VC) dimension of the set of linear classifiers (3.5).

Definition 3.3 (*VC dimension*) The *VC dimension*, d_{VC}, of a set of binary classifiers $\mathcal{G}$ is the largest integer N such that $\Pi_{\mathcal{G}}(N) = 2^N$. If this equation holds for all $N > 0$, then the VC dimension is infinite.

In other words, the VC dimension is the cardinality of the largest set of points that can be classified by functions of $\mathcal{G}$ in any arbitrary manner.

Lemma 3.1 (Sauer–Shelah lemma) *Let $\mathcal{G}$ be a class of binary classifiers of VC dimension d_{VC}. Then, for any $N > d_{VC}$, the growth function is bounded as*[1]

$$\Pi_{\mathcal{G}}(N) \leq \sum_{k=0}^{d_{VC}} \binom{N}{k} \leq \left(\frac{eN}{d_{VC}}\right)^{d_{VC}}.$$

The VC dimension of the class of linear (actually, affine) classifiers (3.5) can be shown to be $d_{VC} = d + 1$. Thus, for linear classification, Lemma 3.1 leads to the polynomial bound

$$\Pi_{\mathcal{G}}(N) \leq \left(\frac{eN}{d+1}\right)^{d+1}. \tag{3.7}$$

3.1.1.2 Support Vector Machines

In the linearly separable case, among all the feasible solutions to (3.6), the so-called *optimal separating hyperplane* is the one with maximum *margin* γ defined as the distance from the nearest point to the hyperplane:

$$\gamma = \min_{k \in [N]} \frac{|\boldsymbol{g}^\top \boldsymbol{x}_k + b|}{\|\boldsymbol{g}\|_2}.$$

By maximizing this margin, we ensure that new data points randomly drawn with the same distribution as those in the data set $\{(\boldsymbol{x}_k, q_k)\}_{k=1}^N$ will likely fall on the correct side of the hyperplane. However, there are infinitely many scaled versions of the parameters $(\boldsymbol{g}, b)$ for that hyperplane. In order to formulate a well-posed problem, we arbitrarily fix the scale by searching for the canonical form of the hyperplane, i.e., the one with parameters that satisfy

$$\min_{k \in [N]} |\boldsymbol{g}^\top \boldsymbol{x}_k + b| = 1.$$

In this case, the margin becomes $\gamma = 1/\|\boldsymbol{g}\|_2$ and its maximization can be written as the minimization of $\|\boldsymbol{g}\|_2$. Therefore, we see that the (canonical form of the) optimal

[1] Here, $\binom{N}{k} = N!/k!(N-k)!$ is a binomial coefficient and $e = \exp(1)$ is Euler's number.

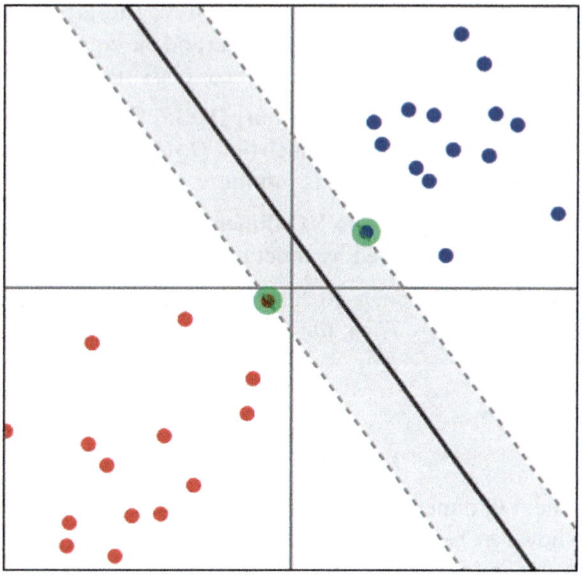

Fig. 3.1 Optimal separating hyperplane between the sets of blue and red points (plain line) with maximum margin (gray-shaded area). The points lying on the border of the margin (dashed lines) are highlighted in green

separating hyperplane is given by the solution to

$$\min_{\boldsymbol{g}\in\mathbb{R}^d, b\in\mathbb{R}} \frac{1}{2}\|\boldsymbol{g}\|_2^2 \quad \text{s.t. } q_k(\boldsymbol{g}^\top \boldsymbol{x}_k + b) \geq 1, \quad k = 1, \dots, N. \tag{3.8}$$

Figure 3.1 shows an example in the plane. This plot illustrates an interesting feature: The optimal separating hyperplane is entirely defined by a small subset of points lying on the border of the margin. Indeed, adding or removing points outside of the margin does not change the solution to (3.8). This is why these points are known as the *support vectors*.

In the nonseparable case, where (3.8) has no feasible solution, slack variables $\boldsymbol{\xi} = [\xi_1 \dots \xi_N]^\top$ are introduced to relax the constraints. This leads to the training algorithm for a linear support vector machine (SVM):

$$\min_{\boldsymbol{g}\in\mathbb{R}^d, b\in\mathbb{R}, \boldsymbol{\xi}\geq\boldsymbol{0}} \frac{1}{2}\|\boldsymbol{g}\|_2^2 + C\sum_{k=1}^{N}\xi_k \quad \text{s.t. } q_k(\boldsymbol{g}^\top \boldsymbol{x}_k + b) \geq 1 - \xi_k, \quad k = 1, \dots, N, \tag{3.9}$$

where the hyperparameter C is used to tune the trade-off between the maximization of the margin and the satisfaction of the constraints enforcing the points to lie outside of this margin. The effect of the relaxation of the constraints and the value of C is illustrated in Fig. 3.2. Such a trade-off was already discussed in Sect. 2.4.2, and

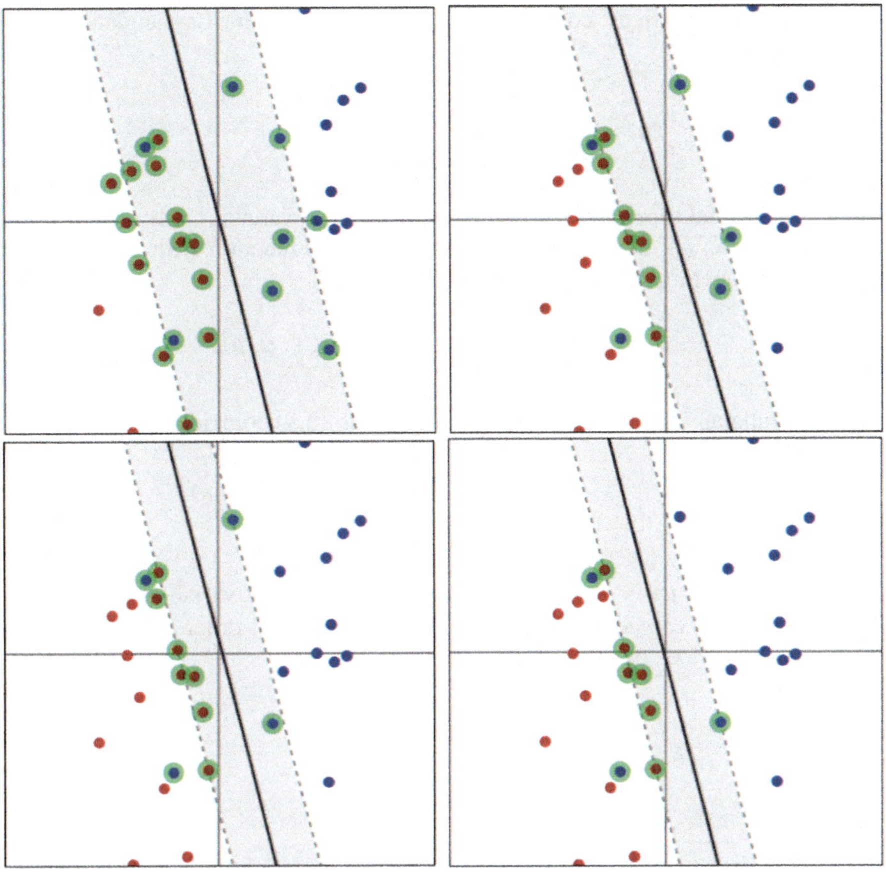

Fig. 3.2 Support vector machine classifier (plain line) trained on a nonseparable data set and the corresponding margin (gray-shaded area) for different values of C (from top left to bottom right: $C = 0.01, 0.1, 1$, and 10). When C increases, more weight is given to the satisfaction of the constraints and the margin becomes thinner. Note that for other data distributions this can imply a more pronounced change in the separating hyperplane

indeed, this convex quadratic problem can be reformulated as a regularized learning one:

$$\min_{\boldsymbol{g}\in\mathbb{R}^d, b\in\mathbb{R}} \frac{1}{2}\|\boldsymbol{g}\|_2^2 + C\sum_{k=1}^{N} \max\{0, 1 - q_k(\boldsymbol{g}^\top \boldsymbol{x}_k + b)\}, \tag{3.10}$$

where $\max\{0, 1 - q_k(\boldsymbol{g}^\top \boldsymbol{x}_k + b)\}$ is a loss function known as the hinge loss.

However, the training problem for an SVM is usually considered in its constrained form in order to apply Lagrangian duality. Let us introduce, for (3.9), the Lagrange multipliers, or dual variables, $\boldsymbol{\alpha} \in \mathbb{R}^N$ associated with the margin constraints and

$\boldsymbol{\mu} \in \mathbb{R}^N$ for the positivity constraints $\boldsymbol{\xi} \geq \mathbf{0}$, and define the Lagrangian function of (3.9) as

$$J(g, b, \boldsymbol{\xi}, \boldsymbol{\alpha}, \boldsymbol{\mu}) = \frac{1}{2}\|g\|_2^2 + C\sum_{k=1}^{N}\xi_k - \sum_{k=1}^{N}\alpha_k\left[q_k(g^\top \boldsymbol{x}_k + b) - 1 + \xi_k\right] - \sum_{k=1}^{N}\mu_k\xi_k.$$

The Lagrange dual function J_D is a function of the dual variables only, which yields, for any fixed $(\boldsymbol{\alpha}, \boldsymbol{\mu})$, the infimum of the Lagrangian function with respect to the primal variables (those of the original problem (3.9)):

$$J_D(\boldsymbol{\alpha}, \boldsymbol{\mu}) = \inf_{g\in\mathbb{R}^d, b\in\mathbb{R}, \boldsymbol{\xi}\in\mathbb{R}^N} J(g, b, \boldsymbol{\xi}, \boldsymbol{\alpha}, \boldsymbol{\mu}). \tag{3.11}$$

Then, Lagrangian duality states that the solution to (3.9) coincides with that of the dual problem

$$\max_{\boldsymbol{\alpha}\geq\mathbf{0}, \boldsymbol{\mu}\geq\mathbf{0}} J_D(\boldsymbol{\alpha}, \boldsymbol{\mu}). \tag{3.12}$$

Since $J(\boldsymbol{g}, b, \boldsymbol{\xi}, \boldsymbol{\alpha}, \boldsymbol{\mu})$ is convex with respect to the primal variables, the value of $J_D(\boldsymbol{\alpha}, \boldsymbol{\mu})$ in (3.11) can be determined via the necessary first-order optimality conditions (see (2.30)), i.e., by solving the equations

$$\frac{\partial J(\boldsymbol{g}, b, \boldsymbol{\xi}, \boldsymbol{\alpha}, \boldsymbol{\mu})}{\partial \boldsymbol{g}} = \boldsymbol{g} - \sum_{k=1}^{N}\alpha_k q_k \boldsymbol{x}_k = \mathbf{0} \tag{3.13}$$

$$\frac{\partial J(\boldsymbol{g}, b, \boldsymbol{\xi}, \boldsymbol{\alpha}, \boldsymbol{\mu})}{\partial b} = -\sum_{k=1}^{N}\alpha_k q_k = 0 \tag{3.14}$$

$$\frac{\partial J(\boldsymbol{g}, b, \boldsymbol{\xi}, \boldsymbol{\alpha}, \boldsymbol{\mu})}{\partial \xi_k} = C - \alpha_k - \mu_k = 0, \; k = 1, \ldots, N. \tag{3.15}$$

The first one directly yields

$$\boldsymbol{g} = \sum_{k=1}^{N}\alpha_k q_k \boldsymbol{x}_k, \tag{3.16}$$

while by (3.15), we have $\mu_k = C - \alpha_k$, $k = 1, \ldots, N$. Thus, we can rewrite the Lagrange dual function as

$$\begin{aligned} J_D(\boldsymbol{\alpha}, \boldsymbol{\mu}) = \frac{1}{2}\left\|\sum_{k=1}^{N}\alpha_k q_k \boldsymbol{x}_k\right\|_2^2 + C\sum_{k=1}^{N}\xi_k - \sum_{k=1}^{N}\sum_{i=1}^{N}\alpha_k\alpha_i q_k q_i \boldsymbol{x}_k^\top \boldsymbol{x}_i \\ - b\sum_{k=1}^{N}\alpha_k q_k + \sum_{k=1}^{N}\alpha_k - \sum_{k=1}^{N}\alpha_k\xi_k - \sum_{k=1}^{N}(C - \alpha_k)\xi_k, \end{aligned}$$

which, by using (3.14) to remove the term with b and

$$\left\|\sum_{k=1}^{N}\alpha_k q_k \boldsymbol{x}_k\right\|_2^2 = \left(\sum_{k=1}^{N}\alpha_k q_k \boldsymbol{x}_k\right)^\top \left(\sum_{i=1}^{N}\alpha_i q_i \boldsymbol{x}_i\right) = \sum_{k=1}^{N}\sum_{i=1}^{N}\alpha_k\alpha_i q_k q_i \boldsymbol{x}_k^\top \boldsymbol{x}_i,$$

simplifies into

$$J_D(\boldsymbol{\alpha}, \boldsymbol{\mu}) = -\frac{1}{2}\sum_{k=1}^{N}\sum_{i=1}^{N}\alpha_k\alpha_i q_k q_i \boldsymbol{x}_k^\top \boldsymbol{x}_i + \sum_{k=1}^{N}\alpha_k.$$

We see that this expression does not involve the dual variables μ_k anymore. However, the constraint on their positivity in (3.12) implies with (3.15) that $\alpha_k \leq C$, $k = 1, \dots, N$. Gathering all the constraints, an SVM can thus be trained by solving the dual problem (3.12) formulated as

$$\begin{aligned} \max_{\boldsymbol{\alpha}\in\mathbb{R}^N}\ & -\frac{1}{2}\sum_{k=1}^{N}\sum_{i=1}^{N}\alpha_k\alpha_i q_k q_i \boldsymbol{x}_k^\top \boldsymbol{x}_i + \sum_{k=1}^{N}\alpha_k \\ \text{s.t. } & \sum_{k=1}^{N}\alpha_k q_k = 0 \\ & 0 \leq \alpha_k \leq C, \quad k = 1, \dots, N. \end{aligned} \tag{3.17}$$

From the solution to this problem in terms of $\boldsymbol{\alpha}$, the classifier parameters $\boldsymbol{g}$ are given by (3.16). The parameter b is computed in a final step from the so-called complementary slackness conditions

$$\begin{aligned} & \alpha_k\left[q_k(\boldsymbol{g}^\top \boldsymbol{x}_k + b) - 1 + \xi_k\right] = 0, \qquad k = 1, \dots, N, \\ & \mu_k\xi_k = 0, \qquad k = 1, \dots, N, \end{aligned}$$

that hold at the optimum. These conditions express the fact that a nonzero Lagrange multiplier is associated with a saturated constraint (a constraint holding with equality instead of inequality). Thus, for k such that $0 < \alpha_k < C$, we have $\mu_k = C - \alpha_k > 0$, which implies $\xi_k = 0$ and $q_k(\boldsymbol{g}^\top \boldsymbol{x}_k + b) = 1$, so that (since $q_k^2 = 1$)

$$b = q_k - \boldsymbol{g}^\top \boldsymbol{x}_k.$$

The points $\boldsymbol{x}_k$ with index k such that $\alpha_k > 0$ are called *support vectors* and those for which $0 < \alpha_k < C$ are the ones that lie exactly on the border of the margin, i.e., the ones that encode relevant information ($\alpha_k \neq 0$) but do not violate the margin constraint ($\xi_k = 0$).

3.1.1.3 Linear Programming SVM

Linear classifiers can also be trained by solving linear programming problems. The linear programming SVM starts by assuming that the parameter vector is expressed as a linear combination of the data points, as in (3.16),

$$g = \sum_{i=1}^{N} \beta_i \boldsymbol{x}_i$$

and penalizes the ℓ_1-norm of the weight vector $\boldsymbol{\beta} = \begin{bmatrix}\beta_1 & \dots & \beta_N\end{bmatrix}^\top$ instead of the ℓ_2-norm of $\boldsymbol{g}$. Using the margin constraints of (3.9), this leads to the training problem

$$\min_{\boldsymbol{\beta}\in\mathbb{R}^N, b\in\mathbb{R}, \boldsymbol{\xi}\geq\boldsymbol{0}} \quad \|\boldsymbol{\beta}\|_1 + C\sum_{k=1}^{N} \xi_k \tag{3.18}$$
$$\text{s.t. } q_k \sum_{i=1}^{N} \beta_i \boldsymbol{x}_i^\top \boldsymbol{x}_k + q_k b \geq 1 - \xi_k, \quad k = 1, \dots, N,$$

which can easily be cast as a linear program (see the discussion around (2.32) in Sect. 2.1.3).

3.1.2 Multi-class Problems

A discrimination problem is said to be multi-class when the number of categories, $|\mathcal{Q}| = s$, is above two. In this case, various approaches can be considered to produce a multi-class decision function g as in (3.1). These come in two major flavors: decomposition methods and multi-class methods. On the one hand, a decomposition method combines multiple binary classifiers, with output in $\{-1, +1\}$ and typically trained separately, to predict the class label of a given input pattern. On the other hand, a multi-class method directly tackles the multi-class problem and estimates a classifier g at once.

3.1.2.1 One-Versus-All Decomposition

A popular decomposition method is the one-versus-all strategy, where a binary classifier is built for each category $j \in \mathcal{Q}$ in order to separate the patterns of this category from the others. Given a multi-class data set, $\{(\boldsymbol{x}_k, q_k)\}_{k=1}^N$ with $q_k \in \mathcal{Q}$, this is obtained by training, for every $j \in \mathcal{Q}$, the jth binary classifier on the data set $\{(\boldsymbol{x}_k, q_{jk})\}_{k=1}^N$ relabeled with

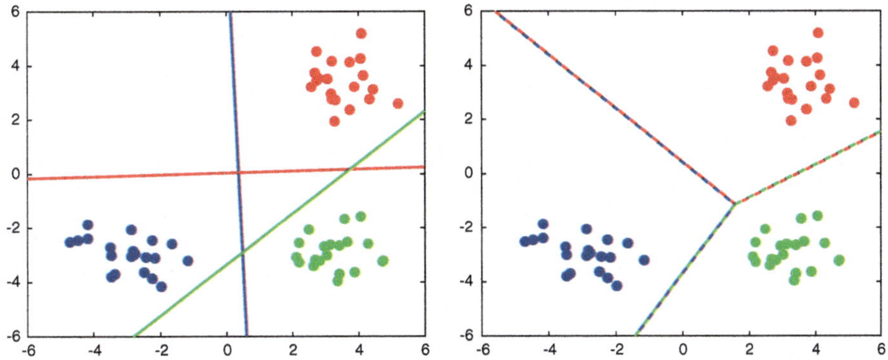

Fig. 3.3 Multi-class problem with three categories (blue, red, and green). *Left:* hyperplanes corresponding to the binary classifiers trained to separate one category from the two others (plain lines). *Right:* the pairwise boundaries between categories of the resulting multi-class one-vs-all classifier (dashed lines) computed as in (3.20). Each boundary is plotted as a half line since the remaining half is irrelevant: It lies in a region where the two other boundaries already determine the class

$$q_{jk} = 2\mathbb{1}_{q_k=j} - 1 \in \{-1, +1\}.$$

Binary classifiers based on the sign of a real-valued score function $g_j : \mathcal{X} \to \mathbb{R}$, such as the SVM of Sect. 3.1.1.2, are typically considered. Then, the final classification corresponds to the maximum score:

$$g(\boldsymbol{x}) = \operatorname*{argmax}_{j\in\mathcal{Q}} g_j(\boldsymbol{x}). \tag{3.19}$$

The corresponding pairwise boundaries are given, for all pairs of categories (j, l), by

$$\left\{\boldsymbol{x} \in \mathcal{X} \ : \ g_j(\boldsymbol{x}) - g_l(\boldsymbol{x}) = 0\right\}. \tag{3.20}$$

Indeed, for a point $\boldsymbol{x}$ to be classified in category j instead of l by g in (3.19), we need $g_j(\boldsymbol{x}) > g_l(\boldsymbol{x})$ and vice versa, with a decision boundary at $g_j(\boldsymbol{x}) = g_l(\boldsymbol{x})$.

Figure 3.3 shows the separating hyperplanes of the binary linear classifiers obtained from a data set with three categories, together with the corresponding pairwise boundaries, which, in this case, also take the form of hyperplanes.

3.1.2.2 Multi-class Support Vector Machine

The SVM described in Sect. 3.1.1.2 can be extended to directly handle multi-class problems. Such an extension typically starts with (3.19), but then learns all the functions g_j simultaneously. In this case, the constraints ensuring correct classification and margin exclusion read

$$\forall j \in \mathcal{Q} \setminus \{q_k\}, \quad g_{q_k}(\boldsymbol{x}_k) \geq g_j(\boldsymbol{x}_k) + 1,$$

which can be relaxed to

$$\forall j \in \mathcal{Q} \setminus \{q_k\}, \quad g_{q_k}(\boldsymbol{x}_k) - g_j(\boldsymbol{x}_k) \geq 1 - \xi_{jk}$$

in the nonseparable case by introducing $(s-1)N$ positive slack variables ξ_{jk}. Focusing on linear classifiers with $g_j(\boldsymbol{x}_k) = \boldsymbol{g}_j^\top \boldsymbol{x}_k + b_j$ and considering the regularization of all score functions g_j in the objective function then yield a training problem of the form

$$\begin{aligned}
&\min_{\{\boldsymbol{g}_j \in \mathbb{R}^d, b_j \in \mathbb{R}\}_{j=1}^s, \xi_{jk} \geq 0} \frac{1}{2} \sum_{j=1}^s \|\boldsymbol{g}_j\|_2^2 + C \sum_{k=1}^N \sum_{j \neq q_k} \xi_{jk} \\
&\text{s.t. } \forall k \in [N], \ \forall j \in \mathcal{Q} \setminus \{q_k\}, \quad (\boldsymbol{g}_{q_k} - \boldsymbol{g}_j)^\top \boldsymbol{x}_k + b_{q_k} - b_j \geq 1 - \xi_{jk}.
\end{aligned}$$

3.1.3 Nonlinear Classification (★)

Many pattern recognition applications call for nonlinear classifiers implementing more complex boundaries than hyperplanes between the different classes. The SVM algorithms described above belong to the family of so-called kernel machines. These learning machines can be easily extended to the nonlinear case via the kernel trick.

Recall the feature space interpretation of kernel functions presented in Sect. 2.4.2.3, and note that a nonlinear classification can be decomposed into two basic operations: first a mapping of the data points by a nonlinear function $\boldsymbol{\phi} : \mathcal{X} \mapsto \mathcal{X}_\phi$ and then a linear classification in $\mathcal{X}_\phi$ instead of $\mathcal{X}$. Thus, nonlinear SVMs are obtained by simply replacing all occurrences of data points $\boldsymbol{x}_k$ in the derivations of Sect. 3.1.1.2 by their images $\boldsymbol{\phi}(\boldsymbol{x}_k)$ in $\mathcal{X}_\phi$. The (kernel) trick is that all such occurrences are involved in inner products between data points, both in the training problem (3.17) and in predictions based on

$$g_0(\boldsymbol{x}) = \boldsymbol{g}^\top \boldsymbol{x} + b = \sum_{k=1}^N \alpha_k q_k \boldsymbol{x}_k^\top \boldsymbol{x} + b.$$

Therefore, the only operation required to implement a nonlinear SVM is the computation of inner products of the form $\langle \boldsymbol{\phi}(\boldsymbol{x}), \boldsymbol{\phi}(\boldsymbol{x}') \rangle_{\mathcal{X}_\phi}$, which we know from (2.76) can be conveniently done thanks to the introduction of a kernel function. In particular, there is no need to explicitly compute the images $\boldsymbol{\phi}(\boldsymbol{x})$, thus allowing for very high-dimensional feature spaces $\mathcal{X}_\phi$ (associated with more complex boundaries in $\mathcal{X}$) without hindering the computations.

Given a kernel function K as in Definition 2.1, the final form of a nonlinear (binary) SVM is thus

$$g(\boldsymbol{x}) = \text{sign}\left(\sum_{k=1}^{N} \alpha_k q_k K(\boldsymbol{x}_k, \boldsymbol{x}) + b\right)$$

with weights α_k determined by solving (3.17) with $\boldsymbol{x}_k^\top \boldsymbol{x}_i$ replaced by $K(\boldsymbol{x}_k, \boldsymbol{x}_i)$.

Following the presentation of regularized learning and reproducing kernel Hilbert spaces in Sect. 2.4.2, a nonlinear SVM can also be defined as the classifier $g(\boldsymbol{x}) = \text{sign}(h(\boldsymbol{x}) + b)$ based on the solution of the functional optimization problem

$$\min_{h \in \mathcal{H}, b \in \mathbb{R}} \sum_{k=1}^{N} \max\{0, 1 - q_k(h(\boldsymbol{x}_k) + b)\} + \lambda \|h\|_{\mathcal{H}}^2.$$

Here, $\mathcal{H}$ stands for the RKHS associated with K and $\max\{0, 1 - q_k(h(\boldsymbol{x}_k) + b)\}$ is the hinge loss function that penalizes negative values of $q_k(h(\boldsymbol{x}_k) + b)$ corresponding to misclassifications and those below 1 corresponding to margin violations, as in (3.10).

3.2 Clustering

The general aim of clustering is to divide a set of objects (typically represented as points in $\mathbb{R}^d$) into several groups on the basis of some measure of similarity between objects. Many instances of this problem can be found across the fields of science and engineering.

The clustering problem also appears as an inner step of some hybrid system identification procedures. In particular, Sect. 7.3 will present an approach which amounts to solving the clustering problem in some particular feature space. However, finding the exact solution of the clustering problem can be as difficult as solving the hybrid system identification problem.

There are two levels of difficulty when considering a clustering problem. First, one should define a criterion used to evaluate the quality of a solution. While this is almost straightforward for discrimination (or regression), here, this criterion intrinsically represents the optimal classification of the data which highly depends on the assumptions or goals that one sets for clustering. For instance, should we maximize the concentration of data points within each group of points, the distance between the groups, or both? Also, an important issue, which can however be inferred from the application, is how to define the similarity between two data points. The second level of difficulty concerns the optimization of the chosen criterion. Here, the standard criteria often call for nonconvex optimization and the use of suboptimal heuristics.

This section describes one of the most popular choices for the criterion and its optimization algorithm. For the case where the number of groups is fixed to s, the so-called k-means criterion measures the dispersion of points within each group. This dispersion is computed as the sum of distances to the centers of the groups, $\{\boldsymbol{c}_j\}_{j=1}^s$. With the Euclidean distance, the k-means method thus amounts to solving

$$\min_{\{\boldsymbol{c}_j\}_{j=1}^s \subset \mathbb{R}^d, \boldsymbol{q} \in [s]^N} \sum_{j=1}^{s} \sum_{\{k:\ q_k=j\}} \|\boldsymbol{x}_k - \boldsymbol{c}_j\|_2^2, \tag{3.21}$$

where $\boldsymbol{q} = \begin{bmatrix} q_1 & \dots & q_N \end{bmatrix}^T$ contains the estimated labels (representing the group memberships) of the N data points $\boldsymbol{x}_k$. The k-means algorithm, depicted in its most basic form in Algorithm 2, provides a fast but suboptimal solution to (3.21). It basically alternates between a classification step solving (3.21) with respect to $\boldsymbol{q}$ for fixed centers and a step updating the centers by solving (3.21) w.r.t. $\{\boldsymbol{c}_j\}_{j=1}^s$ for a fixed $\boldsymbol{q}$. Various refinements can be considered, for instance, to deal with the case where a group becomes empty.

Figure 3.4 illustrates how Algorithm 2 converges to the solution from a random initialization of the centers.

Algorithm 2 k-means

Require: the data set $\{\boldsymbol{x}_k\}_{k=1}^N \subset \mathbb{R}^d$, the number of groups s and an initial vector $\boldsymbol{c}^0 = \begin{bmatrix} \boldsymbol{c}_1^{0\top} & \dots & \boldsymbol{c}_s^{0\top} \end{bmatrix}^\top$.

Initialize $i \leftarrow 0$.

repeat

Classify the data points according to

$$q_k^i = \operatorname*{argmin}_{j \in [s]} \|\boldsymbol{x}_k - \boldsymbol{c}_j\|_2^2, \qquad k = 1, \dots, N. \tag{3.22}$$

for $j = 1$ to s **do**

Update the center of group j with

$$\boldsymbol{c}_j^{i+1} = \operatorname*{argmin}_{\boldsymbol{c}_j \in \mathbb{R}^d} \sum_{\{k:\ q_k^i=j\}} \|\boldsymbol{x}_k - \boldsymbol{c}_j\|_2^2 = \frac{1}{|\{k : q_k^i = j\}|} \sum_{\{k: q_k^i=j\}} \boldsymbol{x}_k. \tag{3.23}$$

end for

Increase the counter $i \leftarrow i + 1$.

until convergence, e.g., until

$$\|\boldsymbol{c}^i - \boldsymbol{c}^{i-1}\|_2 \le \epsilon,$$

or no more changes occur in the classification, i.e., $\boldsymbol{q}^i = \boldsymbol{q}^{i-1}$.

return $\hat{\boldsymbol{c}} = \boldsymbol{c}^i$ and the classification $\hat{\boldsymbol{q}} = \boldsymbol{q}^i$.

Notes

The presentation in this chapter is a short introduction to classification that follows the machine learning perspective. More information in this vein can be found in the textbooks [1–3], which also provide a number of additional classification methods. A more interactive presentation of this field can be found in the Webbook [4].

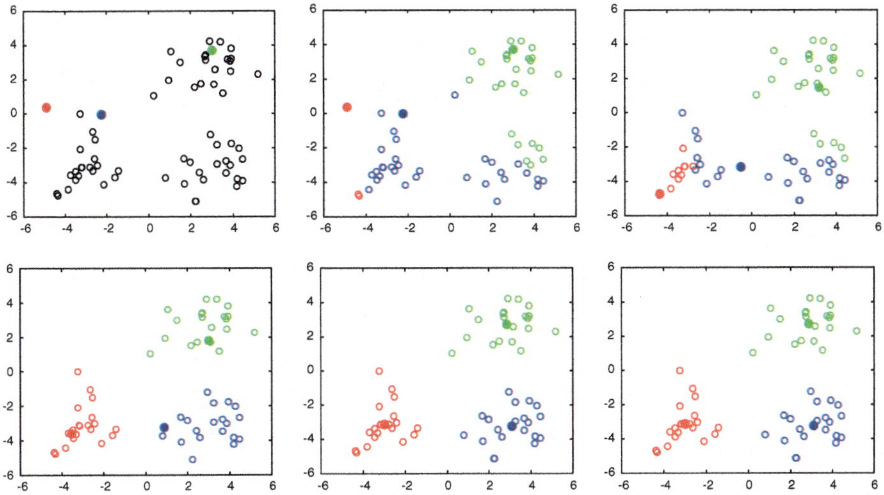

Fig. 3.4 K-means Algorithm 2 in action, from top left to bottom right. Top row: The algorithm starts with random centers (filled dots). Then, the points (black circles) are assigned to the group of their closest center, and the centers of these groups are updated. Bottom row: Subsequent iterations draw the centers toward their final location as the centers of the desired groups

Support vector machines were originally proposed in [5, 6] and came with a sound theory of generalization based on the work of Vapnik and Chervonenkis [7]. Linear programming versions were developed in parallel by [8, 9]. Lemma 3.1 (combinatorial result on the number of possible classifications) is commonly referred to as "Sauer's lemma" or "Sauer–Shelah lemma" but was independently derived in various contexts by several authors [10–12] (see Leon Bottou's discussion [13] on the difficult question of first authorship for this result).

For more details on the convex optimization problems encountered in this chapter and Lagrangian duality, we refer the reader to [14].

While there is (mostly) a single definition of the binary SVM, multiple variants of the multi-class SVM (M-SVM) can be found in the literature. The one presented here is taken from [15] and other major variants are the ones of [16–18], which were unified in a generic model by [19].

Clustering is also extensively discussed in [1]. The k-means algorithm described in Sect. 3.2 is attributed to [20].

Software

A number of efficient implementations of the SVM algorithms for classification are available. We here mention a few among the most well known. For binary classification and the decomposition approach based on binary classifiers, LIBSVM

[21] (http://www.csie.ntu.edu.tw/~cjlin/libsvm/) and SVM^{light} [22] (http://svmlight.joachims.org/) are good choices, for which MATLAB interfaces are also available. For the specific case of linear classification in moderate dimension, LIBLINEAR [23] (https://www.csie.ntu.edu.tw/~cjlin/liblinear/) can be much more efficient than these two by working directly on the primal form (3.10) of the problem. For the direct approach to the multi-class problem with M-SVMs, the MSVMpack software [24] implements all the variants of M-SVMs in a single package. This is a very efficient parallel implementation with a MATLAB interface available at https://members.loria.fr/FLauer/files/MSVMpack/. As another MATLAB alternative, the SSVM toolbox implements the M-SVM variant of [17], but has a limitation on the number of data due to large memory requirements for the entire kernel matrix.

K-means is a famous algorithm with many implementations available. It is, for instance, a standard command (`kmeans`) in the Matlab Statistics and Machine Learning Toolbox.

References

1. Duda, R.O., Hart, P.E., Stork, D.G.: Pattern Classification, 2nd edn. Wiley, New York (2000)
2. Hastie, T., Tibshirani, R., Friedman, J.: The Elements of Statistical Learning: Data Mining, Inference, and Prediction, 2nd edn. Springer, New York (2009)
3. Shalev-Shwartz, S., Ben-David, S.: Understanding Machine Learning: From Theory to Algorithms. Cambridge University Press, Cambridge (2014)
4. Lauer, F.: An interactive journey into machine learning (2017). http://mlweb.loria.fr/book/
5. Boser, B.E., Guyon, I.M., Vapnik, V.N.: A training algorithm for optimal margin classifiers. In: Proceedings of the fifth Annual Workshop on Computational Learning Theory (COLT), Pittsburgh, PA, USA, pp. 144–152 (1992)
6. Cortes, C., Vapnik, V.: Support-vector networks. Mach. Learn. **20**(3), 273–297 (1995)
7. Vapnik, V.N.: Statistical Learning Theory. Wiley, New York (1998)
8. Bennet, K.P., Mangasarian, O.L.: Robust linear programming discrimination of two linearly inseparable sets. Optim. Methods Softw. **1**(1), 23–34 (1992)
9. Bennet, K.P., Mangasarian, O.L.: Multicategory discrimination via linear programming. Optim. Methods Softw. **3**(1–3), 27–39 (1994)
10. Vapnik, V.N., Chervonenkis, A.: On the uniform convergence of relative frequencies of events to their probabilities. Theory Probab. Appl. **16**(2), 264–280 (1971)
11. Sauer, N.: On the density of families of sets. J. Comb. Theory Ser. A **13**(1), 145–147 (1972)
12. Shelah, S.: A computational problem: stability and order of models and theory of infinitary languages. Pac. J. Math. **41**(1), 247–261 (1972)
13. Bottou, L.: On the Vapnik–Chevonenkis–Sauer lemma. Online discussion (2012)
14. Boyd, S., Vandenberghe, L.: Convex Optimization. Cambridge University Press, Cambridge (2004)
15. Weston, J., Watkins, C.: Multi-class support vector machines. Technical Report CSD-TR-98-04, Royal Holloway, University of London (1998)
16. Crammer, K., Singer, Y.: On the algorithmic implementation of multiclass kernel-based vector machines. J. Mach. Learn. Res. **2**, 265–292 (2001)
17. Lee, Y., Lin, Y., Wahba, G.: Multicategory support vector machines: theory and application to the classification of microarray data and satellite radiance data. J. Am. Stat. Assoc. **99**(465), 67–81 (2004)
18. Guermeur, Y., Monfrini, E.: A quadratic loss multi-class SVM for which a radius-margin bound applies. Informatica **22**(1), 73–96 (2011)

19. Guermeur, Y.: A generic model of multi-class support vector machine. Int. J. Intell. Inf. Database Syst. **6**(6), 555–577 (2012)
20. MacQueen, J.: Some methods for classification and analysis of multivariate observations. In: Proceedings of the 5th Berkeley Symposium on Mathematical Statistics and Probability, Berkeley, CA, USA, pp. 281–297 (1967)
21. Chang, C.C., Chih-Jen, L.: LIBSVM: a library for support vector machines. ACM Trans. Intell. Syst. Technol. **2**(3), 27:1–27:27 (2001). http://www.csie.ntu.edu.tw/~cjlin/libsvm/
22. Joachims, T.: SVM^{light} (1998). http://svmlight.joachims.org/
23. Fan, R.E., Chang, K.W., Hsieh, C.J., Wang, X.R., Lin, C.J.: LIBLINEAR: a library for large linear classification. J. Mach. Learn. Res. **9**, 1871–1874 (2008). https://www.csie.ntu.edu.tw/~cjlin/liblinear/
24. Lauer, F., Guermeur, Y.: MSVMpack: a multi-class support vector machine package. J. Mach. Learn. Res. **12**, 2293–2296 (2011)

Chapter 4
Hybrid System Identification

Hybrid system identification is the subfield of system identification (see Chap. 2) dedicated to hybrid dynamical systems, or hybrid systems for short. Hybrid systems, as outlined in Chap. 1, are dynamical systems that can abruptly switch from one behavior or operating mode to another. This chapter formally introduces hybrid systems and their models most encountered in automatic control, and in particular for system identification. These models can take many different forms. On the one hand, they come with all the variants already seen for classical systems in Chap. 2 (input–output vs. state-space, linear vs. nonlinear, etc.). And on the other hand, they can have different switching mechanisms.

As a consequence, hybrid system identification includes the study of a large variety of problems. Indeed, in addition to its dedication to a specific model, a precise hybrid system identification problem also entails a particular choice of viewpoint, a choice that is connected to the inherent trade-off between the number of modes and the error of the model. This chapter makes this trade-off explicit before discussing the various problem formulations and viewpoints. Thus, for both switched and piecewise-defined models, two cases are detailed: for a fixed number of submodels and for a given upper bound on the error. The chapter also highlights the relationships between hybrid system identification and other fields, such as subspace clustering, in which similar trade-offs can be found.

Chapter Outline

The models of hybrid systems considered in this book are detailed in Sect. 4.1. Then, Sect. 4.2 focuses on their identification with different viewpoints and criteria to optimize. Particularly, from the trade-off, illustrated in Sect. 4.2.2, between the number of submodels and the bound on the error, the two cases are presented in Sect. 4.2.3 and Sect. 4.2.4, respectively. The chapter ends with a few comments on related problems from other fields, nonlinear system identification and subspace clustering, in Sect. 4.3.

F. Lauer and G. Bloch, *Hybrid System Identification*,
Lecture Notes in Control and Information Sciences 478,
https://doi.org/10.1007/978-3-030-00193-3_4

4.1 Hybrid System Models

As indicated in Chap. 1, we only focus on two hybrid system classes: piecewise-defined systems for which the discrete state is a function of the continuous state and arbitrarily switched systems for which the discrete state is independent of the continuous state. In a nutshell, piecewise-defined hybrid systems are nonlinear systems, the behavior of which changes according to the operating point, i.e., the inclusion of the continuous state in one of several regions. Such systems can be more easily handled when represented by several models than by a unique one, particularly for control. It is worth noting that piecewise affine (PWA) maps have universal approximation properties: Any smooth function can be arbitrarily well approximated by a PWA map. Rather, in arbitrarily switched hybrid systems, an external signal, given, for example, by an upper-level supervisor, drives the change from a model to another one.

In the following, hybrid models are detailed in discrete time, which is the most common and convenient form for the identification.

4.1.1 *State-Space Versus Input–Output Models*

Hybrid system models in discrete-time k can be first presented either in state-space representation or in input–output form. In both classes, they are given in a general (possibly nonlinear) noisy form, with *additive* noise.

4.1.1.1 State-Space Form

Following (1.5), a discrete-time hybrid model in state-space (SS) form is described by

$$\begin{cases} \mathbf{x}_{k+1} = \mathbf{f}_{q_k}(\mathbf{x}_k, \boldsymbol{u}_k) + \boldsymbol{w}_k \\ \mathbf{y}_k = \mathbf{g}_{q_k}(\mathbf{x}_k, \boldsymbol{u}_k) + \boldsymbol{v}_k, \end{cases} \tag{4.1}$$

where $\mathbf{x}_k \in \mathbb{R}^{n_x}$, $\boldsymbol{u}_k \in \mathbb{R}^{n_u}$, and $\mathbf{y}_k \in \mathbb{R}^{n_y}$ are, respectively, the continuous state, input and output vectors, and where $\boldsymbol{w}_k \in \mathbb{R}^{n_x}$ and $\boldsymbol{v}_k \in \mathbb{R}^{n_y}$ are noise vectors.

At time k, the system is in the particular mode j, i.e., $q_k = j$, and the "submodel" (or "mode") $(\mathbf{f}_j, \mathbf{g}_j)$ is active. There is a finite number s of possible modes: $q_k \in \{1, \ldots, s\}$ and $(\mathbf{f}_{q_k}, \mathbf{g}_{q_k}) \in \{(\mathbf{f}_1, \mathbf{g}_1), \ldots, (\mathbf{f}_s, \mathbf{g}_s)\}$, with $\mathbf{f}_j : \mathbb{R}^{n_x+n_u} \to \mathbb{R}^{n_x}$ and $\mathbf{g}_j : \mathbb{R}^{n_x+n_u} \to \mathbb{R}^{n_y}$.

4.1.1.2 Input–Output Form

Now from (1.6), the discrete-time models of hybrid systems can be written in the input–output (I/O) form as

$$\mathbf{y}_k = \boldsymbol{f}_{q_k}(\boldsymbol{x}_k) + \boldsymbol{\varepsilon}_k,$$

where the output $\mathbf{y}_k \in \mathbb{R}^{n_y}$ depends on the continuous regression vector $\boldsymbol{x}_k \in \mathbb{R}^d$ and the discrete state q_k, and where $\boldsymbol{\varepsilon}_k \in \mathbb{R}^{n_y}$ is a noise vector.

Again, at time k, the system is in the particular mode j, i.e., $q_k = j$, and the submodel $\boldsymbol{f}_j$ is active. There is a finite number s of possible modes or submodels, $q_k \in \{1, \dots, s\}$, and $\boldsymbol{f}_{q_k} \in \{\boldsymbol{f}_1, \dots, \boldsymbol{f}_s\}$, with $\boldsymbol{f}_j : \mathbb{R}^d \to \mathbb{R}^{n_y}$.

In the single output case, which is the focus of the sequel, this model simply reduces to

$$y_k = f_{q_k}(\boldsymbol{x}_k) + \varepsilon_k, \tag{4.2}$$

where $y_k \in \mathbb{R}$, $\varepsilon_k \in \mathbb{R}$, and $f_{q_k} : \mathbb{R}^d \to \mathbb{R}$, with again $q_k \in \{1, \dots, s\}$.

4.1.2 Linear Versus Nonlinear Submodels

Hybrid systems can also be distinguished according to the nature of the continuous dynamics.

4.1.2.1 Linear Submodels

Input–Output Form

As discussed in Sect. 2.1.1, the class of SISO "linear" models in I/O form regroups several truly linear models and some that are only pseudo-linear. Only two cases are now detailed for hybrid models.

The general form with autoregressive with exogenous input (ARX) submodels is derived from (2.9) as follows:

$$y_k = \boldsymbol{x}_k^\top \boldsymbol{\theta}_{q_k} + \varepsilon_k, \tag{4.3}$$

where $y_k \in \mathbb{R}$ is the output, $\boldsymbol{x}_k \in \mathbb{R}^d$ is the regression vector, $q_k \in \{1, \dots, s\}$ is the discrete state, with s the number of submodels, $\boldsymbol{\theta}_j \in \mathbb{R}^d$, $j = 1, \dots, s$, are the vectors of parameters defining each submodel, and $\varepsilon_k \in \mathbb{R}$ is a noise term.

As in (2.15), the regressor $\boldsymbol{x}_k \in \mathbb{R}^d$, $d = n_a + n_b$, with the model orders n_a and n_b, and the pure delay n_d, is given by:

$$\boldsymbol{x}_k = \begin{bmatrix} -y_{k-1} & \dots & -y_{k-n_a} & u_{k-n_\mathrm{d}} & \dots & u_{k-n_\mathrm{d}-n_b+1} \end{bmatrix}^\top,$$

where the u_{k-i}'s are the delayed inputs. The pure delay will be often fixed to $n_\mathrm{d} = 1$, leading, as in (2.11), to a regression vector defined as

$$\boldsymbol{x}_k = \left[-y_{k-1} \ \dots \ -y_{k-n_a} \ u_{k-1} \ \dots \ u_{k-n_b}\right]^\top. \tag{4.4}$$

The parameter vectors $\boldsymbol{\theta}_j$, $j = 1, \dots, s$, are derived from (2.12), giving:

$$\boldsymbol{\theta}_j = \left[\theta_{j1} \ \dots \ \theta_{jn_a} \ \theta_{j(n_a+1)} \ \dots \ \theta_{j(n_a+n_b)}\right] = \left[a_{j1} \ \dots \ a_{jn_a} \ b_{j1} \ \dots \ b_{jn_b}\right]. \tag{4.5}$$

Then, the associated predictor is simply given by

$$\hat{y}_k = \boldsymbol{x}_k^\top \boldsymbol{\theta}_{q_k}.$$

For I/O models, this ARX form is almost always considered in hybrid system identification. Note nevertheless that setting the model structure "hyperparameter" n_d to an appropriate value, in addition to the model orders n_a and n_b, remains an issue, as usual.

Remark 4.1 In the case of I/O *affine* systems, the regression vector $\boldsymbol{x}_k$ is simply extended to $\tilde{\boldsymbol{x}}_k = \left[\boldsymbol{x}_k^\top \ 1\right]^\top$. Except when mentioned otherwise, we only consider linear (ARX) submodels and assume that the extension to affine submodels is straightforward.

In the MIMO case, model (4.3) becomes

$$\mathbf{y}_k = \boldsymbol{\Theta}_{q_k} \boldsymbol{x}_k + \boldsymbol{\varepsilon}_k,$$

where $\mathbf{y}_k \in \mathbb{R}^{n_y}$ is the output vector, $\boldsymbol{\varepsilon}_k \in \mathbb{R}^{n_y}$ an additive disturbance vector, the regression vector $\boldsymbol{x}_k$, of dimension $d = n_y n_a + n_u n_b$, is formed from the past output ($\boldsymbol{y}_{k-i}$) and input ($\boldsymbol{u}_{k-i}$) vectors as

$$\boldsymbol{x}_k = \left[-\mathbf{y}_{k-1}^\top \ \dots \ -\mathbf{y}_{k-n_a}^\top \ \boldsymbol{u}_{k-n_\mathrm{d}}^\top \ \dots \ \boldsymbol{u}_{k-n_\mathrm{d}-n_b+1}^\top\right]^\top,$$

and where $\boldsymbol{\Theta}_{q_k} \in \mathbb{R}^{n_y \times d}$ is the parameter matrix for submodel q_k.

Note that considering a regression vector in that form implies an identical delay order n_a for the n_y outputs, and similar delay order n_b and pure delay n_d for the n_u inputs, which is a common, but often unrealistic, assumption.

With output error (OE) submodels, the hybrid model is derived from (2.19) in the following form:

$$\begin{cases} z_k = \boldsymbol{r}_k^\top \boldsymbol{\theta}_{q_k} \\ y_k = z_k + \varepsilon_k, \end{cases} \tag{4.6}$$

where z_k is the noise-free unmeasured output, y_k the measured output, ε_k the noise term, and, as for the ARX form, $q_k \in \{1, \dots, s\}$ is the discrete state, with s the number of submodels, $\boldsymbol{\theta}_j \in \mathbb{R}^d$, $j = 1, \dots, s$, are the parameter vectors.

If the pure delay n_d equals 1, using the order of (4.4), the regressor $\boldsymbol{r}_k \in \mathbb{R}^d$, $d = n_a + n_b$, with the model orders n_a and n_b, can be written as

$$\boldsymbol{r}_k = \left[-z_{k-1} \ \dots \ -z_{k-n_a} \ u_{k-1} \ \dots \ u_{k-n_b}\right]^\top, \tag{4.7}$$

allowing the parameter vectors $\boldsymbol{\theta}_j$, $j = 1, \dots, s$, to be written with the same notations as (4.5). The associated predictor

$$\hat{y}_k = \boldsymbol{x}_k^\top \boldsymbol{\theta}_{q_k}$$

is obtained by replacing the unknown quantities in $\boldsymbol{r}_k$ by their estimates:

$$\boldsymbol{x}_k = \left[-\hat{y}_{k-1} \ \dots \ -\hat{y}_{k-n_a} \ u_{k-1} \ \dots \ u_{k-n_b}\right]^\top. \tag{4.8}$$

Note nevertheless that hybrid OE model identification has been little investigated until now. A recursive identification approach is presented in Sect. 8.1.1.1.

State-Space Form

In state-space form, the model of affine hybrid systems is derived from (4.1), as

$$\begin{cases} \mathbf{x}_{k+1} = \boldsymbol{A}_{q_k}\mathbf{x}_k + \boldsymbol{B}_{q_k}\boldsymbol{u}_k + \tilde{\boldsymbol{f}}_{q_k} + \boldsymbol{w}_k \\ \mathbf{y}_k = \boldsymbol{C}_{q_k}\mathbf{x}_k + \boldsymbol{D}_{q_k}\boldsymbol{u}_k + \tilde{\boldsymbol{g}}_{q_k} + \boldsymbol{v}_k, \end{cases} \tag{4.9}$$

where $\mathbf{x}_k \in \mathbb{R}^{n_x}$, $\boldsymbol{u}_k \in \mathbb{R}^{n_u}$, and $\mathbf{y}_k \in \mathbb{R}^{n_y}$ are, respectively, the continuous state, input and output vectors of the system, and $\boldsymbol{w}_k \in \mathbb{R}^{n_x}$ and $\boldsymbol{v}_k \in \mathbb{R}^{n_y}$ are noise/error terms. The matrices $\boldsymbol{A}_{q_k}$, $\boldsymbol{B}_{q_k}$, $\boldsymbol{C}_{q_k}$, $\boldsymbol{D}_{q_k}$ and vectors $\tilde{\boldsymbol{f}}_{q_k}$ and $\tilde{\boldsymbol{g}}_{q_k}$, $q_k = 1, \dots, s$, with appropriate dimensions, describe each subsystem. For linear hybrid systems, the terms $\tilde{\boldsymbol{f}}_{q_k}$ and $\tilde{\boldsymbol{g}}_{q_k}$ are omitted.

4.1.2.2 Nonlinear Submodels

For nonlinear hybrid systems, subsystems should be modeled by nonlinear functions. More precisely, these nonlinear submodels must be smooth functions, i.e., of class C^∞ (infinitely differentiable). This restriction implies that possible jumps in the continuous behavior of the system are taken into account from the hybrid nature of the model by switches of submodels and not in the submodels themselves.

Thus, a SISO nonlinear hybrid system is expressed in I/O form by a set of s smooth functions $\{f_j\}_{j=1}^s$ as

$$y_k = f_{q_k}(\boldsymbol{x}_k) + \varepsilon_k, \tag{4.10}$$

where y_k, $\boldsymbol{x}_k$ and ε_k, and q_k, are defined as in (4.3).

4.1.3 Piecewise Smooth Versus Arbitrarily Switched Systems

Two cases can finally be distinguished according to the dependence of the discrete state q_k on the continuous regression vector $\boldsymbol{x}_k$ (I/O form), or the continuous state $\mathbf{x}_k$ and input $\boldsymbol{u}_k$ (SS form). These cases are described here in I/O form.

4.1.3.1 Arbitrarily Switched Systems

For arbitrarily switched systems, the discrete state is independent of the continuous state and can be switched arbitrarily among a finite set of modes, i.e., $q_k \in \{1, \ldots, s\}$.

In (I/O) ARX form, this leads to switched nonlinear ARX (SNARX) models or switched affine (or linear) ARX (SARX) models, according to the (non)linearity of the submodels. Switched output error (SOE) models can also be encountered.

4.1.3.2 Piecewise Smooth (PWS) Systems

In a piecewise smooth (PWS) system, the discrete state q_k depends on the continuous regression vector $\boldsymbol{x}_k$, i.e.,

$$q_k = \begin{cases} 1 & \text{if } \boldsymbol{x}_k \in \mathcal{X}_1, \\ \vdots & \\ s & \text{if } \boldsymbol{x}_k \in \mathcal{X}_s, \end{cases} \tag{4.11}$$

where $\mathcal{X}_j$, $j \in \{1, \ldots, s\}$, are regions that form a partition of the regression space $\mathcal{X}$.

In each region $\mathcal{X}_j$, the submodel f_j must be a smooth function. If this is not the case, then we split the nonsmooth submodel into two or more smooth submodels to recover a PWS representation. If this cannot be done with a reasonable number of submodels, this means that the system is mostly nonsmooth and that there is not much hope for its identification from a finite data sample (see footnote 8 on p. 44).

A popular subclass of piecewise smooth (PWS) systems is the piecewise affine (PWA) system class, in which the subsystems f_j are affine functions of $\boldsymbol{x}_k$, or, with (linear) ARX subsystems, to the piecewise ARX (PWARX) system class, frequently abbreviated in piecewise affine (PWA).

Convex Polyhedra

The regions $\mathcal{X}_j$ are most often chosen as convex polyhedra. In this case, the sets of the complete polyhedral partition in the regression space $\mathcal{X}$ are defined as

$$\mathcal{X}_j = \left\{ \boldsymbol{x} \in \mathcal{X} \,:\, \boldsymbol{H}_j \begin{bmatrix} \boldsymbol{x} \\ 1 \end{bmatrix} \leq \mathbf{0} \right\}, \; j = 1, \ldots, s, \; \text{s.t.} \begin{cases} \cup_{j=1}^{s} \mathcal{X}_j = \mathcal{X}, \\ \mathcal{X}_i \cap \mathcal{X}_j = \emptyset, \forall i \neq j, \end{cases} \tag{4.12}$$

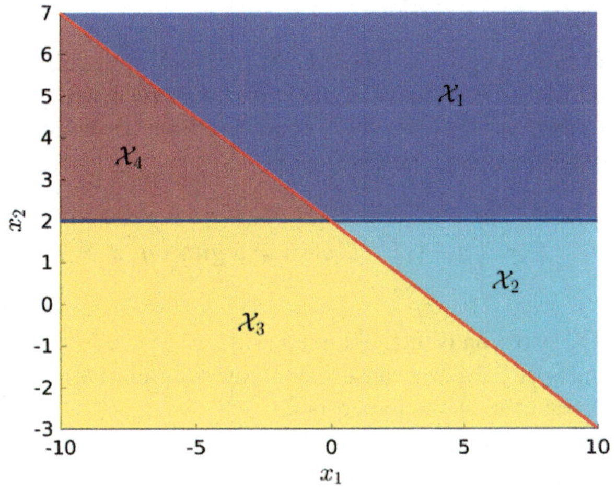

Fig. 4.1 Partition of the regression space into four convex polyhedra in dimension 2 for Examples 4.1–4.2

where each row of a matrix $\boldsymbol{H}_j$ defines a separating hyperplane between $\mathcal{X}_j$ and another region (therefore, the maximum number of rows in a matrix $\boldsymbol{H}_j$ is $s-1$). Strict inequalities can replace inequalities for some regions in order to avoid that the boundaries belong to two adjacent regions.

Example 4.1 (*Convex polyhedra*) As an illustrative example, we consider a regression vector $\boldsymbol{x} = \begin{bmatrix} x_1 & x_2 \end{bmatrix}^\top$ in a regression space divided into $s = 4$ regions $\mathcal{X}_1$, $\mathcal{X}_2$, $\mathcal{X}_3$, and $\mathcal{X}_4$, represented in Fig. 4.1 with x-axis x_1. The two separating lines are defined by $a_1x_1 + a_2x_2 + a_3 = 0$, with $a_1 = 0, a_2 = 1, a_3 = -2$, i.e., $x_2 - 2 = 0$ (blue horizontal line), and $a'_1x_1 + a'_2x_2 + a'_3 = 0$, with $a'_1 = 1, a'_2 = 2, a'_3 = -4$, i.e., $x_1 + 2x_2 - 4 = 0$ (red line). The matrices of the inequalities (4.12), $\boldsymbol{H}_j \begin{bmatrix} \boldsymbol{x}^\top & 1 \end{bmatrix}^\top \leq 0$, are then

$$\boldsymbol{H}_1 = \begin{bmatrix} -a_1 & -a_2 & -a_3 \\ -a'_1 & -a'_2 & -a'_3 \end{bmatrix} = \begin{bmatrix} 0 & -1 & 2 \\ -1 & -2 & 4 \end{bmatrix}, \boldsymbol{H}_2 = \begin{bmatrix} a_1 & a_2 & a_3 \\ -a'_1 & -a'_2 & -a'_3 \end{bmatrix} = \begin{bmatrix} 0 & 1 & -2 \\ -1 & -2 & 4 \end{bmatrix},$$

$$\boldsymbol{H}_3 = \begin{bmatrix} a_1 & a_2 & a_3 \\ a'_1 & a'_2 & a'_3 \end{bmatrix} = \begin{bmatrix} 0 & 1 & -2 \\ 1 & 2 & -4 \end{bmatrix}, \text{ and } \boldsymbol{H}_4 = \begin{bmatrix} -a_1 & -a_2 & -a_3 \\ a'_1 & a'_2 & a'_3 \end{bmatrix} = \begin{bmatrix} 0 & -1 & 2 \\ 1 & 2 & -4 \end{bmatrix}.$$

Note that in this example, each region shares a boundary with only two others, so that these matrices have only two rows. If the regions $\mathcal{X}_3$ and $\mathcal{X}_4$ were only one region, the region $\mathcal{X}_3$, then the matrices $\boldsymbol{H}_1$ and $\boldsymbol{H}_2$ would be unchanged and $\boldsymbol{H}_3 = \begin{bmatrix} a'_1 & a'_2 & a'_3 \end{bmatrix} = \begin{bmatrix} 1 & 2 & -4 \end{bmatrix}$.

Linear Classifiers

In most identification methods for PWA systems, as those described in Chap. 7, the partition is estimated at the end of the procedure using standard pattern recognition tools. For $\mathcal{X} \subseteq \mathbb{R}^d$, these tools (introduced in Chap. 3) typically implement linear classifiers from the family

$$\mathcal{G} = \left\{ g \in [s]^{\mathcal{X}} \ : \ g(\boldsymbol{x}) = \operatorname*{argmax}_{j \in [s]} \boldsymbol{g}_j^\top \boldsymbol{x} + b_j, \ \boldsymbol{g}_j \in \mathbb{R}^d, \ b_j \in \mathbb{R} \right\}. \tag{4.13}$$

The partition is then obtained as $\mathcal{X}_j = \{\boldsymbol{x} \in \mathcal{X} : g(\boldsymbol{x}) = j\}$, $j = 1, \dots, s$. By recalling Sect. 3.1.2, we can also compute the pairwise region boundaries from the classifier parameters as the hyperplanes

$$H_{ji} = \{\boldsymbol{x} \ : \ (\boldsymbol{g}_j - \boldsymbol{g}_i)^\top \boldsymbol{x} + b_j - b_i = 0\}, \quad 1 \le j < i \le s. \tag{4.14}$$

Example 4.2 (*Linear classifiers—convex polyhedra continued*) Consider the convex polyhedral regions of Example 4.1 plotted in Fig. 4.1. We will show how to implement this partition with a linear classifier from (4.13). The definition of the convex polyhedra via linear inequalities as in (4.12) directly gives the equations of the pairwise separating hyperplanes (here, merely lines), in which the coefficients can be identified with the difference of parameters involved in (4.14). For instance, for the blue line which separates the regions $\mathcal{X}_1$ and $\mathcal{X}_2$, this yields

$$\boldsymbol{g}_1 - \boldsymbol{g}_2 = \begin{bmatrix} a_1 \\ a_2 \end{bmatrix} = \begin{bmatrix} 0 \\ 1 \end{bmatrix}, \quad b_1 - b_2 = a_3 = -2.$$

Writing down these equations for all pairs of adjacent regions yields the linear system

$$\begin{bmatrix}
1 & 0 & -1 & 0 & 0 & 0 & 0 & 0 & 0 & 0 & 0 & 0 \\
0 & 1 & 0 & -1 & 0 & 0 & 0 & 0 & 0 & 0 & 0 & 0 \\
0 & 0 & 0 & 0 & 0 & 0 & 0 & 0 & 1 & -1 & 0 & 0 \\
1 & 0 & 0 & 0 & 0 & 0 & -1 & 0 & 0 & 0 & 0 & 0 \\
0 & 1 & 0 & 0 & 0 & 0 & 0 & -1 & 0 & 0 & 0 & 0 \\
0 & 0 & 0 & 0 & 0 & 0 & 0 & 0 & 1 & 0 & 0 & -1 \\
0 & 0 & 1 & 0 & -1 & 0 & 0 & 0 & 0 & 0 & 0 & 0 \\
0 & 0 & 0 & 1 & 0 & -1 & 0 & 0 & 0 & 0 & 0 & 0 \\
0 & 0 & 0 & 0 & 0 & 0 & 0 & 0 & 0 & 1 & -1 & 0 \\
0 & 0 & 0 & 0 & 1 & 0 & -1 & 0 & 0 & 0 & 0 & 0 \\
0 & 0 & 0 & 0 & 0 & 1 & 0 & -1 & 0 & 0 & 0 & 0 \\
0 & 0 & 0 & 0 & 0 & 0 & 0 & 0 & 0 & 0 & 1 & -1 \\
1 & 0 & 1 & 0 & 1 & 0 & 1 & 0 & 0 & 0 & 0 & 0 \\
0 & 1 & 0 & 1 & 0 & 1 & 0 & 1 & 0 & 0 & 0 & 0 \\
0 & 0 & 0 & 0 & 0 & 0 & 0 & 0 & 1 & 1 & 1 & 1
\end{bmatrix}
\begin{bmatrix} \boldsymbol{g}_1 \\ \boldsymbol{g}_2 \\ \boldsymbol{g}_3 \\ \boldsymbol{g}_4 \\ b_1 \\ b_2 \\ b_3 \\ b_4 \end{bmatrix}
=
\begin{bmatrix} 0 \\ 1 \\ -2 \\ 1 \\ 2 \\ -4 \\ 1 \\ 2 \\ -4 \\ 0 \\ -1 \\ 2 \\ 0 \\ 0 \\ 0 \end{bmatrix}, \tag{4.15}$$

in which the three last rows correspond to the additional constraints

$$\sum_{j=1}^{s} \boldsymbol{g}_j = \mathbf{0}, \quad \sum_{j=1}^{s} b_j = 0,$$

that can always be enforced without influencing the classification by $g(\boldsymbol{x})$. Without these, the system would be underdetermined since constraining the differences of parameters only defines them up to the addition of a constant. Solving the system of equations (4.15) yields the parameters $\boldsymbol{g}_1 = \begin{bmatrix}0.5 & 1.5\end{bmatrix}^\top$, $\boldsymbol{g}_2 = \begin{bmatrix}0.5 & 0.5\end{bmatrix}^\top$, $\boldsymbol{g}_3 = \begin{bmatrix}-0.5 & -1.5\end{bmatrix}^\top$, $\boldsymbol{g}_4 = \begin{bmatrix}-0.5 & -0.5\end{bmatrix}^\top$, $b_1 = -3$, $b_2 = -1$, $b_3 = 3$, and $b_4 = 1$ for the linear classifier $g(\boldsymbol{x})$ whose heat map is shown in Fig. 4.1.

From NARX to PWARX

As previously said, any smooth function can be arbitrarily well approximated by a PWA map. We now show a way to obtain a PWARX description of a SISO nonlinear ARX (NARX) model by linearization around a number of operating points. Consider the NARX model $y_k = f(\boldsymbol{x}_k) + \varepsilon_k$. Assuming that f is one time differentiable, the first-order Taylor series expansion of f is given by

$$f(\boldsymbol{x}) = f(\boldsymbol{c}_j) + (\boldsymbol{x} - \boldsymbol{c}_j)^\top \nabla f(\boldsymbol{c}_j) + e(\boldsymbol{x}, \boldsymbol{c}_j),$$

where $\nabla f(\boldsymbol{c}_j)$ is the gradient of f computed at $\boldsymbol{c}_j$. When $\boldsymbol{x}$ lies in a ball of radius ϵ_j around $\boldsymbol{c}_j$, $B_{\epsilon_j}(\boldsymbol{c}_j) = \{\boldsymbol{x} \in \mathcal{X} : \|\boldsymbol{x} - \boldsymbol{c}_j\| \leq \epsilon_j\}$, for ϵ_j sufficiently small, the remainder term $e(\boldsymbol{x}, \boldsymbol{c}_j)$ is negligible and f can be approximated by

$$\hat{f}(\boldsymbol{x}) = \begin{bmatrix}\boldsymbol{x}^\top & 1\end{bmatrix} \boldsymbol{\theta}_j \quad \text{if } \boldsymbol{x} \in B_{\epsilon_j}(\boldsymbol{c}_j),$$

with

$$\boldsymbol{\theta}_j = \begin{bmatrix} \nabla f(\boldsymbol{c}_j) \\ f(\boldsymbol{c}_j) - \boldsymbol{c}_j^\top \nabla f(\boldsymbol{c}_j) \end{bmatrix}.$$

Then, the regression domain $\mathcal{X}$ can be decomposed into a set of a large enough number s of regions $B_{\epsilon_j}(\boldsymbol{c}_j)$, $j = 1, \ldots, s$, on which affine approximations hold. Nevertheless, when defined in this manner, the regions can overlap. Thus, the s balls are reshaped into nonoverlapping cells centered at the $\boldsymbol{c}_j$'s to form a so-called Voronoï partition of the regression space $\mathcal{X}$:

$$\mathcal{X}_j = \left\{ \boldsymbol{x} \in \mathcal{X} \ : \ j = \operatorname*{argmin}_{i \in [s]} \|\boldsymbol{x} - \boldsymbol{c}_i\|_2 \right\}, \quad j = 1, \ldots, s. \tag{4.16}$$

These regions correspond to convex polyhedra of the form (4.12):

$$\mathcal{X}_j = \left\{ \boldsymbol{x} \in \mathcal{X} \ : \ \begin{bmatrix}\tilde{\boldsymbol{H}}_j & \boldsymbol{b}_j\end{bmatrix} \begin{bmatrix}\boldsymbol{x} \\ 1\end{bmatrix} \leq \mathbf{0} \right\}, \tag{4.17}$$

where

$$\tilde{\boldsymbol{H}}_j = 2 \begin{bmatrix} \boldsymbol{c}_1^\top - \boldsymbol{c}_j^\top \\ \vdots \\ \boldsymbol{c}_{j-1}^\top - \boldsymbol{c}_j^\top \\ \boldsymbol{c}_{j+1}^\top - \boldsymbol{c}_j^\top \\ \vdots \\ \boldsymbol{c}_s^\top - \boldsymbol{c}_j^\top \end{bmatrix}, \quad \boldsymbol{b}_j = \begin{bmatrix} \boldsymbol{c}_j^\top \boldsymbol{c}_j - \boldsymbol{c}_1^\top \boldsymbol{c}_1 \\ \vdots \\ \boldsymbol{c}_j^\top \boldsymbol{c}_j - \boldsymbol{c}_{j-1}^\top \boldsymbol{c}_{j-1} \\ \boldsymbol{c}_j^\top \boldsymbol{c}_j - \boldsymbol{c}_{j+1}^\top \boldsymbol{c}_{j+1} \\ \vdots \\ \boldsymbol{c}_j^\top \boldsymbol{c}_j - \boldsymbol{c}_s^\top \boldsymbol{c}_s \end{bmatrix}.$$

Indeed, each row in the inequality of (4.17) thus guarantees that

$$-2\boldsymbol{c}_j^\top \boldsymbol{x}_j + \boldsymbol{c}_j^\top \boldsymbol{c}_j - (-2\boldsymbol{c}_i^\top \boldsymbol{x} + \boldsymbol{c}_i^\top \boldsymbol{c}_i) = \|\boldsymbol{x} - \boldsymbol{c}_j\|_2^2 - \|\boldsymbol{x} - \boldsymbol{c}_i\|_2^2 \leq 0$$

for some $i \neq j$ such that (4.17) coincides with (4.16). Note however, that, since (4.16) involves fewer parameters than the general form (4.12), it is restrictive and cannot implement all convex polyhedral partitions into s modes.

Other Partitioning Schemes

Other types of regression domain partitioning have been proposed, namely polytopic, simplicial, hyperrectangular. Besides, the regions can be more complex, with nonlinear separating curves. Such a nonlinear partitioning can be obtained by considering nonlinear classifiers $g(\boldsymbol{x}) = \operatorname{argmax}_{j\in[s]} g_j(\boldsymbol{x})$ as in (4.13) but with nonlinear functions g_j (see Sect. 3.1.3).

4.2 Identification Problems

In the case where the mode q_k is observed and known, the problem of hybrid system identification degenerates, for a given time interval in a particular mode, to a standard identification problem, as discussed in Chap. 2. Then, for s modes, this problem reduces to s "simple," nonhybrid, identification problems, even if there are complications for state-space hybrid systems, as presented in Sect. 8.2. In this chapter and in the rest of the book, we focus on the challenging case where the mode q_k is unknown and must be estimated from the input–output data. Note nevertheless that the idea of identifying a hybrid system by first determining the mode from the data in order to recover a set of more simple problems can be kept in mind.

4.2.1 Hybrid System Identification with Unknown Mode

When the mode is unknown, the choice of model classes for hybrid systems is more complicated than in the classical cases of linear and nonlinear systems. The model can be a combination of various submodel types. Moreover, the switching between subsystems causes the data set to contain data points from multiple modes, thus preventing the direct identification of each subsystem separately. Therefore, if the switching sequence is unknown, the identification of hybrid systems becomes a difficult problem which comprises two intertwined tasks: the classification one of grouping the data points into subsets corresponding to the modes and the regression one of estimating the submodels.

In the following, we restrict the discussion to SISO models in ARX form. In this case, the identification of a switched system amounts to the following problem. Given a data set of input–output pairs (u_k, y_k), generated by a switched system, estimate the model orders n_a and n_b, the number of submodels s, the submodels $\{f_j \in \mathcal{F}_j\}_{j=1}^s$, and the discrete state q_k for each input–output pair.

Commonly, with some prior knowledge about the system, the model orders n_a and n_b are fixed. Then, the preceding problem simplifies to a switching regression one with a fixed regressor structure and a data set $\mathcal{D} = \{(\boldsymbol{x}_k, y_k)\}_{k=1}^N$ as follows.

Problem 4.1 Given a data set $\mathcal{D} = \{(\boldsymbol{x}_k, y_k)\}_{k=1}^N$, generated by a switched system, estimate the number of submodels s, the submodels $\{f_j \in \mathcal{F}_j\}_{j=1}^s$, and the switching sequence $\boldsymbol{q} = \{q_k\}_{k=1}^N \in [s]^N$.

When further focusing on switched linear systems with models of the form (4.10), the regression task boils down to the estimation of the submodel parameter vectors $\{\boldsymbol{\theta}_j\}_{j=1}^s$ instead of arbitrary nonlinear submodels f_j.

Finally, for piecewise affine/smooth systems, we have to consider the additional task of estimating the regions $\{\mathcal{X}_j\}_{j=1}^s$ in (4.11).

4.2.2 *The Trade-Off Between the Number of Modes and the Error*

Problem 4.1 as posed above is ill-posed: It is not precise enough and may have many very different but valid solutions. Indeed, this problem entails a trade-off between the number of submodels and the error with which the global model fits the data.

For instance, a trivial solution consists in taking as many submodels as data points, i.e., $s = N$, thus resulting in a perfect fit of all data points. This trivial solution is obviously useless as it is most probably far from the true system and not accurate in prediction.

At the other end of the spectrum of solutions, we can choose $s = 1$ and easily estimate a single submodel with standard methods from Chap. 2. However, here again this solution is not very helpful as it does not resemble a hybrid system and probably has a large error.

Without more information on the sought solution, or, equivalently, a more precise statement of the problem, it is very difficult to discriminate between satisfactory and trivial solutions and select the best solution among a whole range of intermediate ones. This amounts to a model selection problem, for which standard techniques rely on an additional and independent data set, called a validation set, as described in Sect. 2.5. For instance, one can apply an algorithm estimating a fixed number s of submodels for all values of s in a given range and measure the error of the resulting models on the validation set. At the end, the model with smallest validation error is retained.

The procedure just described already entails one fundamental idea used by many methods: In order to formulate an effective algorithm, we first had to fix the number of modes. Then, the problem could be set as the minimization of the error and became well posed. Yet, one can think of the converse approach in which the error is fixed and the number of submodels is minimized by the algorithm. In this case, the model selection procedure depicted above remains the same except that a set of values for the error has to be scanned (which might not be as easy since this implies a choice of discretization).

These two approaches to recover well-posedness are now detailed in the following sections.

4.2.3 Fixing the Number of Submodels

As seen above, fixing the number of submodels is the most straightforward manner to recover well-posedness. Another advantage of this approach is that it naturally gives rise to a finite set of discrete values for the hyperparameter s to be tuned in the model selection phase.

4.2.3.1 Arbitrarily Switched Systems

Given Problem 4.1, this approach yields the (more) well-posed problem[1] below, where ℓ is a loss function as defined in Sect. 2.1.2.

Problem 4.2 Given a data set $\mathcal{D} = \{(\boldsymbol{x}_k, y_k)\}_{k=1}^N$ and a number of submodels s, estimate the submodels $\{f_j\}_{j=1}^s$, $f_j \in \mathcal{F}_j$, and the switching sequence $\boldsymbol{q} = \{q_k\}_{k=1}^N \in [s]^N$ by minimizing the error:

$$\min_{\{f_j \in \mathcal{F}_j\}_{j=1}^s, \boldsymbol{q} \in [s]^N} \frac{1}{N} \sum_{k=1}^N \ell(y_k - f_{q_k}(\boldsymbol{x}_k)). \tag{4.18}$$

Figure 4.2 illustrates this problem in the (static) switching linear regression case with two modes.

[1] Note that the well-posedness of Problem 4.2 also depends on the choice of function classes $\mathcal{F}_j$. In particular, if the structure of the f_j's is not fixed, the problem remains ill-posed (see the discussion on the nonparametric setting of Sect. 2.4.2).

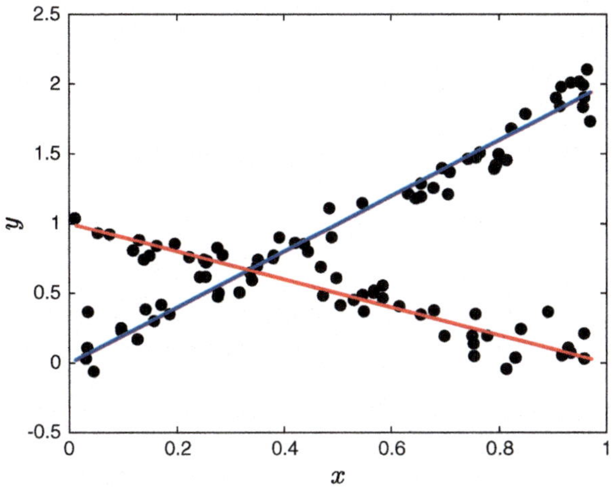

Fig. 4.2 A one-dimensional switching linear regression problem with two modes. The aim is to recover, from the data points (black dots) the two target functions (blue and red lines) without knowing which of the two was used to generate each data point

Though possibly well posed, the optimization problem (4.18) remains nontrivial as it involves both continuous and integer variables, and the latter appear as function indexes. A more standard formulation of (4.18) from an optimization viewpoint is as a Mixed-Integer NonLinear Program (MINLP):

$$\begin{aligned} \min_{\{f_j \in \mathcal{F}_j\}_{j=1}^s, \{\beta_{jk}\}} \quad & \frac{1}{N} \sum_{k=1}^{N} \sum_{j=1}^{s} \beta_{jk}\, \ell(y_k - f_j(\boldsymbol{x}_k)) \\ \text{s.t.} \quad & \beta_{jk} \in \{0, 1\}, \ j = 1, \dots, s, \ k = 1, \dots, N, \\ & \sum_{j=1}^{s} \beta_{jk} = 1, \ k = 1, \dots, N, \end{aligned} \tag{4.19}$$

where β_{jk} is a binary variable encoding the assignment of the kth point to the jth submodel.

It is also possible to reformulate these problems without integer variables. To see this, note that (4.18) can be solved explicitly with respect to $\boldsymbol{q}$ for fixed $\{f_j\}_{j=1}^s$ by assigning each point to the model with minimum error as

$$q_k \in \operatorname*{argmax}_{j \in [s]} \ell(y_k - f_j(\boldsymbol{x}_k)), \quad k = 1, \dots, N. \tag{4.20}$$

Embedding this into the problem formulation then yields the so-called Minimum-of-Errors (ME) formulation,

$$\min_{\{f_j \in \mathcal{F}_j\}_{j=1}^s} \ \frac{1}{N} \sum_{k=1}^{N} \min_{j \in [s]} \ell(y_k - f_j(\boldsymbol{x}_k)), \tag{4.21}$$

which only involves the submodels f_j as variables.

4.2.3.2 Piecewise Smooth Systems

As mentioned in Sect. 4.2.1, the identification of piecewise affine/smooth models also involves the estimation of the regions $\mathcal{X}_j$ of activity for each mode. We can, without loss of generality, assume that the partition of $\mathcal{X}$ into $\mathcal{X}_j$'s is given by a partitioning function $g : \mathcal{X} \to [s]$ such that

$$g(\boldsymbol{x}) = j \Leftrightarrow \boldsymbol{x} \in \mathcal{X}_j.$$

Adopting the machine learning terminology, we call such a function g a *classifier*, as it classifies the points $\boldsymbol{x}$ into the different modes (see Chap. 3).

In order to recover a well-posed problem, one has to put constraints on the forms of the regions $\mathcal{X}_j$. Otherwise, even with a fixed number of regions, unrealistic partitions of $\mathcal{X}$ can be obtained. In our classification-based setup, this amounts to restricting g to belong to some subset $\mathcal{G}$ of the class of functions from $\mathcal{X}$ into $[s]$.

We can now state the well-posed[2] piecewise smooth regression problem with a fixed number of modes.

Problem 4.3 Given a data set $\mathcal{D} = \{(\boldsymbol{x}_k, y_k)\}_{k=1}^N$ and a number of submodels s, estimate the submodels $\{f_j\}_{j=1}^s$ and the partitioning function g by minimizing the error:

$$\min_{\{f_j \in \mathcal{F}_j\}_{j=1}^s, g \in \mathcal{G}} \frac{1}{N} \sum_{k=1}^N \ell(y_k - f_{g(\boldsymbol{x}_k)}(\boldsymbol{x}_k)). \tag{4.22}$$

Figure 4.3 illustrates this problem in the (static) PWA regression case with two affine modes.

Also note that problem (4.22) can be rewritten using the indicator function $\mathbb{1}$ ($\mathbb{1}_E = 1$ if E holds and 0 otherwise) as

$$\min_{\{f_j \in \mathcal{F}_j\}_{j=1}^s, g \in \mathcal{G}} \frac{1}{N} \sum_{k=1}^N \sum_{j=1}^s \mathbb{1}_{g(\boldsymbol{x}_k)=j}\, \ell(y_k - f_j(\boldsymbol{x}_k)). \tag{4.23}$$

4.2.4 Fixing a Bound on the Error

As seen above, an alternative to fixing the number of submodels is to fix the error (or accuracy) with which we would like the model to fit the data and minimize the number of submodels.

---

[2] As for the arbitrarily switched case, the well-posedness of the problem still depends on the parametric vs. nonparametric nature of the $\mathcal{F}_j$'s, and here also of $\mathcal{G}$.

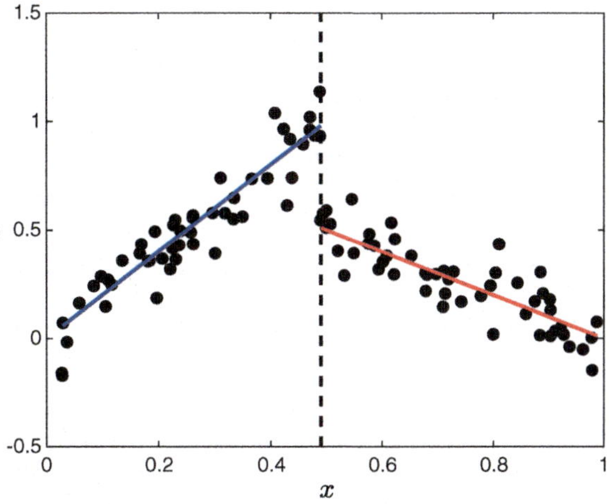

Fig. 4.3 A one-dimensional PWA regression problem with two modes. The aim is to recover the two target functions (blue and red lines) and the classifier partitioning the x-axis (at the dashed line) from the data points (black dots)

4.2.4.1 Arbitrarily Switched Systems

For switched systems, the so-called bounded-error approach amounts to solving the following problem.

Problem 4.4 Given a data set $\mathcal{D} = \{(\boldsymbol{x}_k, y_k)\}_{k=1}^N$ and a bound ϵ on the mean loss, estimate the minimal number of submodels $\{f_j\}_{j=1}^s$ and the switching sequence needed to achieve that bound:

$$\begin{aligned}
&\min_{s\in\mathbb{N}, \{f_j \in \mathcal{F}\}, \boldsymbol{q}\in\mathbb{N}^N} \quad s \\
&\text{s.t. } \boldsymbol{q} \in [s]^N \\
&\frac{1}{N}\sum_{k=1}^N \ell(y_k - f_{q_k}(\boldsymbol{x}_k)) \leq \epsilon.
\end{aligned}$$

Note that we could also constrain the maximum loss instead of the mean and that we explicitly made all the submodels f_j belong to the same model class $\mathcal{F}$ in order to limit the impact of the ordering of the modes.

This problem could be solved via a sequence of problems with a fixed number of modes, in which s is increased until the mean loss satisfies the bound. However, the bounded-error viewpoint also gave rise to dedicated methods. These often work in a greedy manner to gain in computational efficiency. Instead of solving the complete problem with s submodels for all s, the submodels are estimated one by one while maintaining the previous ones unchanged and removing the data points well approximated from the data at each iteration. Such a greedy approach is suboptimal in nature, but often offers a good compromise between accuracy and numerical efficiency in practice.

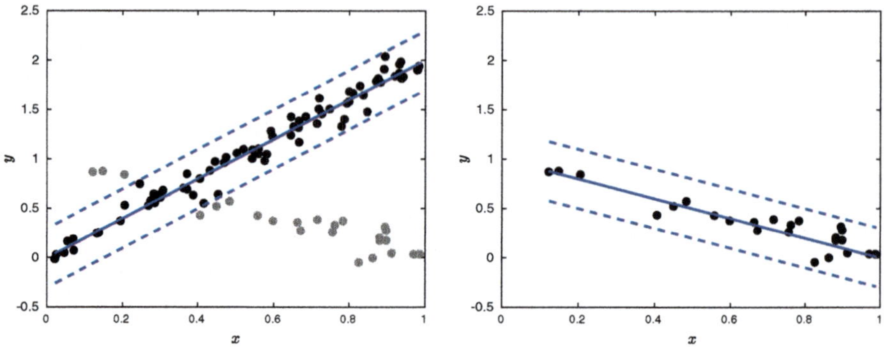

Fig. 4.4 A one-dimensional switching regression problem with two modes tackled by a greedy bounded-error approach. The first iteration (left) yields the first submodel (plain line) by considering all data points outside a tube of width ϵ around it as outliers (gray dots). Then, the inliers (black dots) are removed before applying the second iteration (right) which finds the second submodel (plain line) from the remaining points

The crux of the bounded-error approach is the robust estimation of the submodels. Indeed, one cannot estimate the first submodel from the entire data set with standard tools, since this would not yield a relevant submodel for the subsequent iterations but merely a "mean model" for all the modes. Instead, one uses robust estimation and the following rationale. Given a data set containing points generated by different modes, the points from the dominant mode can be considered as inliers and those from the other modes as outliers. Here, the dominant mode is the one from which the majority of the points comes. With that viewpoint, robust estimation applied to the entire data set should return an accurate model of the dominant mode. Once this is obtained, the points sufficiently well approximated by this model are removed from the data set and a similar operation takes place on the remaining data to estimate the next submodel.

Robust estimation is a complex task in itself, independently of the hybrid system context. Here, we limit the discussion to methods based on the minimization of a (nonconvex) saturated loss function as defined in Table 2.1, i.e., an $\ell_{p,\epsilon}$-loss function with $p \in \{0, 1, 2\}$.

Given such a loss, the submodels are estimated by iteratively solving

$$\min_{f_j \in \mathcal{F}} \sum_{k \in I_j} \ell_{p,\epsilon}(y_k - f_j(\boldsymbol{x}_k)), \quad j = 1, 2, \ldots, \tag{4.24}$$

where $I_1 = [N]$ and

$$I_j = \{k \in I_{j-1} : |y_k - f_{j-1}(\boldsymbol{x}_k)| > \epsilon\}, \quad j = 2, 3, \ldots \tag{4.25}$$

Figure 4.4 illustrates this greedy bounded-error approach in the (static) switching regression case with two modes.

4.2.4.2 Piecewise Smooth Systems

For PWS systems, the bounded-error approach above amounts to solving the following problem.

Problem 4.5 Given a data set $\mathcal{D} = \{(\boldsymbol{x}_k, y_k)\}_{k=1}^N$ and a bound ϵ on the mean loss, estimate the minimal number of submodels $\{f_j\}_{j=1}^s$ and the partitioning function g needed to achieve that bound:

$$\begin{aligned}\min_{s\in\mathbb{N},\{f_j\in\mathcal{F}\},g} \quad & s\\ \text{s.t. } & g \in \mathcal{G}(s)\\ & \frac{1}{N}\sum_{k=1}^N \ell(y_k - f_{g(\boldsymbol{x}_k)}(\boldsymbol{x}_k)) \leq \epsilon,\end{aligned}$$

where each $\mathcal{G}(s) \subset [s]^{\mathcal{X}}$ is a set of s-category classifiers with output in $[s]$.

The greedy approach depicted above for switched systems is not so simple to apply to PWS systems. Indeed, the inliers should be detected by taking two factors into account: the ability of the model f_j to accurately estimate y_k and the shape of the boundary between the regression vectors $\boldsymbol{x}_k$ in the set of inliers and the others. For instance, if the aim is to obtain a partition into convex polyhedra, then the convex hull of the inliers in $\mathcal{X}$ should not contain regression vectors from outliers.

Instead, the opposite approach can be considered: Starting from a set of $s = N$ submodels f_j that can perfectly fit the data, the number of submodels is decreased as much as possible while maintaining the error below the given threshold. Methods described in Sect. 7.3 do not directly implement this idea but can be understood with this point of view in mind.

4.2.5 *Hybrid Model Assessment*

The criteria described in Sect. 2.5 for assessing the models can be extended to hybrid systems in a more or less straightforward manner.

4.2.5.1 In Simulated Case Studies

In simulations, when the true parameters, $\{\boldsymbol{\theta}_j\}_{j=1}^s$, of the linear hybrid model are known, the estimates, $\{\hat{\boldsymbol{\theta}}_j\}_{j=1}^s$, can be evaluated by averaging the NPE criterion (2.80) over the s submodels:

$$MNPE = \frac{1}{s}\sum_{j=1}^{s} NPE_j = \frac{1}{s}\sum_{j=1}^{s}\frac{\|\hat{\boldsymbol{\theta}}_j - \boldsymbol{\theta}_j\|_2}{\|\boldsymbol{\theta}_j\|_2}. \tag{4.26}$$

Furthermore, when the modes $\{q_k\}_{k=1}^N$ are known, another important criterion in hybrid system identification is the classification error rate (CE):

$$CE = \frac{1}{N}\sum_{k=1}^{N} \mathbb{1}_{\hat{q}_k \neq q_k}, \tag{4.27}$$

where

$$\mathbb{1}_{\hat{q}_k \neq q_k} = \begin{cases} 1 & \text{if } \hat{q}_k \neq q_k \\ 0 & \text{if } \hat{q}_k = q_k. \end{cases} \tag{4.28}$$

This error rate measures the quality of the estimation of the switching sequence $\{q_k\}_{k=1}^N$ and thus allows us to analyze the ability of the identification methods to distinguish between the modes.

Symmetries and Reordering of the Modes

When computing the criteria above, one must pay attention to the ordering of the modes. Indeed, without specific constraints and if the function classes $\mathcal{F}_j$ are similar, the estimated modes can typically be arbitrarily ordered. For instance, the problems (4.19) and (4.21) involve symmetries and have multiple global solutions that differ only in the ordering of the modes.

Therefore, the criteria $MNPE$ (4.26) and CE (4.27) should be computed as their minimum value over all permutations σ of $[s]$:

$$MNPE' = \min_{\sigma:[s]\rightarrowtail\!\!\!\!\twoheadrightarrow[s]} \frac{1}{s}\sum_{j=1}^{s}\frac{\|\hat{\boldsymbol{\theta}}_{\sigma(j)} - \boldsymbol{\theta}_j\|_2}{\|\boldsymbol{\theta}_j\|_2} \tag{4.29}$$

and

$$CE' = \min_{\sigma:[s]\rightarrowtail\!\!\!\!\twoheadrightarrow[s]} \frac{1}{N}\sum_{k=1}^{N} \mathbb{1}_{\sigma(\hat{q}_k)\neq q_k}. \tag{4.30}$$

4.2.5.2 In Practice

In practice, without additional information beside the data itself, we can use the MSE (2.82) and/or FIT (2.83) criteria computed on an independent test set. For piecewise-defined systems, this is straightforward, since the model prediction is a deterministic function of the regression vector:

$$\hat{y}_k = f_{g(\boldsymbol{x}_k)}(\boldsymbol{x}_k),$$

where f_j is the jth submodel and g the classifier determining the active mode.

But for arbitrarily switched systems, the model prediction $\hat{y}_k$ is undetermined without information on the mode q_k. To remedy to this situation, we could apply a two-step procedure: First estimate the mode $\hat{q}_k$ as in (4.20), and then compute the prediction $\hat{y}_k = f_{\hat{q}_k}(\boldsymbol{x}_k)$ for this mode. This is equivalent to compute the following mean squared error based on a switching loss function implementing the selection of the active mode:

$$MSE_{\text{switched}} = \frac{1}{N} \sum_{k=1}^{N} \min_{j \in [s]} \left(y_k - f_j(\boldsymbol{x}_k) \right)^2 . \tag{4.31}$$

4.3 Other Related Problems (★)

Hybrid system identification can also be looked at from different perspectives and cast into problems typically studied by different communities. We here highlight two major connections with other fields.

4.3.1 Nonlinear System Identification

It is straightforward to observe that a PWA or PWS system is a particular form of a nonlinear system; i.e., it can be expressed as $y_k = f(\boldsymbol{x}_k)$ with a nonlinear function f having a PWA/S structure. Therefore, the numerous methods for nonlinear system identification could be applied. However, doing so one does not necessarily obtain an easy-to-handle model such as a PWA one for which many control techniques are available.

In addition, identification techniques for general nonlinear and PWA/S systems are very different for two reasons. First, a PWA/S system has a strong structure which can be used as prior knowledge in order to improve the accuracy of the identified model. Second, generic methods for identifying nonlinear models as depicted in Sect. 2.4.2 are not suitable for PWA/S systems.

Indeed, most methods estimating nonlinear models rely on a very basic assumption, namely that the target function is smooth. For PWA/S functions that can include discontinuities, this assumption is violated. While there might still be smooth approximations to PWA/S functions, finding these is not only a nontrivial task, but also one for which the standard approaches are not designed. For instance, regularization techniques commonly used in nonlinear regression measure the function smoothness at a global level, which penalizes even very smooth solutions with occasional large jumps such as PWA/S functions. Figure 4.5 illustrates this issue in the (static) PWA regression case with two modes.

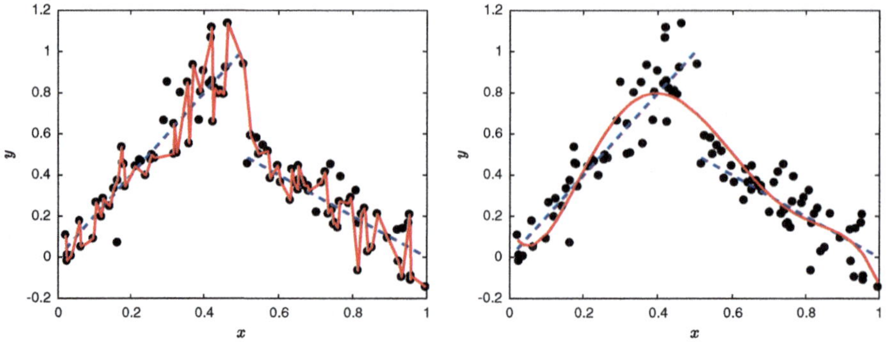

Fig. 4.5 A one-dimensional PWA regression problem with two modes tackled by a generic nonlinear regression method. Either the nonlinear model accurately estimates the jump between the two modes and also learns the noise (left) or it remains robust to the noise but fails to accurately identify the jump (right)

The method described in Sect. 7.2 attempts to circumvent these issues by forging a different regularization scheme dedicated to PWA/S functions. However, this approach remains practical only in low dimension.

4.3.2 Subspace Clustering

Clustering as described in Sect. 3.2 is very close to the switched system identification problem in the sense that it also relies on the solution of two intrinsically mixed subproblems: the classification of the points into groups and the estimation of a model of each group (the center of the group $\boldsymbol{c}_j$ in clustering, the submodel f_j in hybrid system identification). This relationship is explicitly used by the approach of Sect. 6.1.3 where an algorithm in the spirit of k-means (Algorithm 2) is formulated for switched system identification.

Subspace clustering is a particular form of clustering (with many applications in computer vision) that is even more closely related to hybrid system identification. In subspace clustering, the aim is to divide a set of points, $\{\boldsymbol{x}_k\}_{k=1}^N \subset \mathbb{R}^d$, into a number of groups such that each group of points can be well approximated by a subspace. Observing that such subspaces act as linear models of the data highlights the connection with switched system identification and switching regression.

Formally, a p-dimensional subspace in $\mathbb{R}^d$ can be represented by a basis whose p vectors $\boldsymbol{b}_i$, $i = 1, \dots, p$, are gathered in a matrix $\boldsymbol{B} = \begin{bmatrix}\boldsymbol{b}_1 & \dots & \boldsymbol{b}_p\end{bmatrix} \in \mathbb{R}^{d\times p}$, or, alternatively, a matrix $\boldsymbol{N} = \begin{bmatrix}\boldsymbol{n}_1 & \dots & \boldsymbol{n}_{d-p}\end{bmatrix} \in \mathbb{R}^{d\times(d-p)}$ containing normal vectors that form a basis for the orthogonal complement of the subspace (see Appendix B.2.3). Then, points on the subspace belong to the span of $\boldsymbol{B}$ and the null space of $\boldsymbol{N}^\top$. With this formalism, the aim of subspace clustering is to estimate s matrices $\{\boldsymbol{B}_j\}_{j=1}^s$ such that all data points $\{\boldsymbol{x}_k\}_{k=1}^N$ belong to the union of the corresponding s subspaces.

The major difference with switching regression is that in subspace clustering there is no output variable y_k. However, if we restrict the subspaces to hyperplanes, i.e., subspaces of maximal dimension $p = d - 1$, the normal matrix $\boldsymbol{N}$ reduces to a single vector $\boldsymbol{n}$, which gives rise to the linear model

$$\boldsymbol{n}^\top \tilde{\boldsymbol{x}}_k = 0$$

for points $\tilde{\boldsymbol{x}}_k$ on the subspace. We see that this is equivalent to the linear regression model $y_k = \boldsymbol{x}_k^\top \boldsymbol{\theta}$ if we let $\tilde{\boldsymbol{x}}_k = \begin{bmatrix} y_k \ \boldsymbol{x}_k^\top \end{bmatrix}^\top$ and $\boldsymbol{n} = \begin{bmatrix} -1 \ \boldsymbol{\theta}^\top \end{bmatrix}^\top$. Therefore, a switching regression problem in dimension d can be cast as a subspace clustering one in dimension $d + 1$. Yet, this is not always the case. Differences start to appear as soon as we consider noisy data.

The most standard model of noise for regression is an additive centered noise ε_k on the output, i.e.,

$$y_k^{noisy} = y_k + \varepsilon_k,$$

while the standard assumption in subspace clustering is for the noise to be orthogonal to the subspace, i.e., $\tilde{\boldsymbol{x}}_k^{noisy} = \tilde{\boldsymbol{x}}_k + \boldsymbol{N}\boldsymbol{\varepsilon}_k$, where $\boldsymbol{\varepsilon}_k$ is a random vector. For hyperplanes, this can be written as

$$\tilde{\boldsymbol{x}}_k^{noisy} = \tilde{\boldsymbol{x}}_k + \boldsymbol{n}\varepsilon_k,$$

which impacts all components of $\tilde{\boldsymbol{x}}_k^{noisy}$. Figure 4.6 illustrates the difference between these two assumptions: The data are generated by a one-dimensional linear system $y_k = \theta x_k$, and a model is estimated either by linear regression or subspace estimation, which minimize different error terms.

In this example, both estimations yield approximately the same model. However, when comparing switching regression and subspace clustering, different classifications can result from the choice of a different distance for computing the error. Figure 4.7 illustrates this issue. This difference is even more pronounced when increasing the dimension, in which case the regression error, $|y_k - \boldsymbol{x}_k^\top \boldsymbol{\theta}|$, has less weight in the orthogonal distance that also depends on the norm of the parameter vector. Indeed, the orthogonal distance to the hyperplane corresponding to the model $y_k = \boldsymbol{x}_k^\top \boldsymbol{\theta}$ is

$$\frac{|\boldsymbol{n}^\top \tilde{\boldsymbol{x}}_k|}{\|\boldsymbol{n}\|_2} = \frac{|y_k - \boldsymbol{x}_k^\top \boldsymbol{\theta}|}{\sqrt{1 + \|\boldsymbol{\theta}\|_2^2}}.$$

Therefore, the larger the dimension is, the larger the norm $\|\boldsymbol{\theta}\|_2$ can be and the difference between orthogonal distances computed with respect to two hyperplanes becomes more influenced by the difference in this norm than by the regression error.

Another drawback of subspace clustering is that to obtain a subspace representation, one must estimate one additional parameter, the weight n_1 assigned to the y

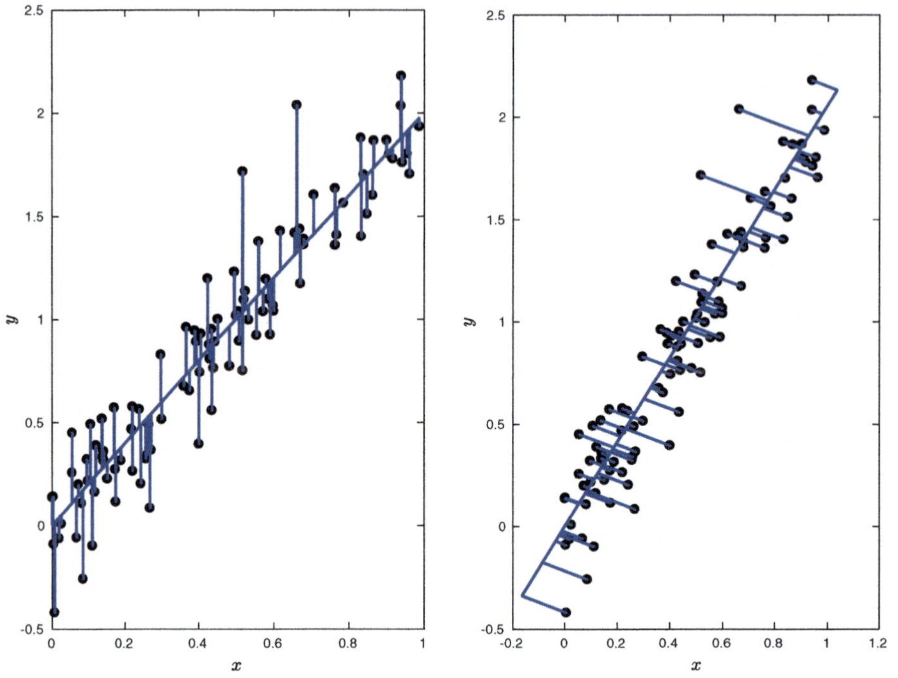

Fig. 4.6 Illustration of the difference between regression and subspace estimation. Regression considers errors between the model and data points along the y-axis (left), whereas subspace estimation considers the orthogonal distance between data points and the model (right)

Fig. 4.7 Illustration of the difference between switching regression and subspace clustering. For switching regression, the data point should be assigned to the first model (blue line) due to a smaller distance along the y-axis. For subspace clustering, the point would be assigned to the other model (red line) with a smaller orthogonal distance, leading to a very large regression error

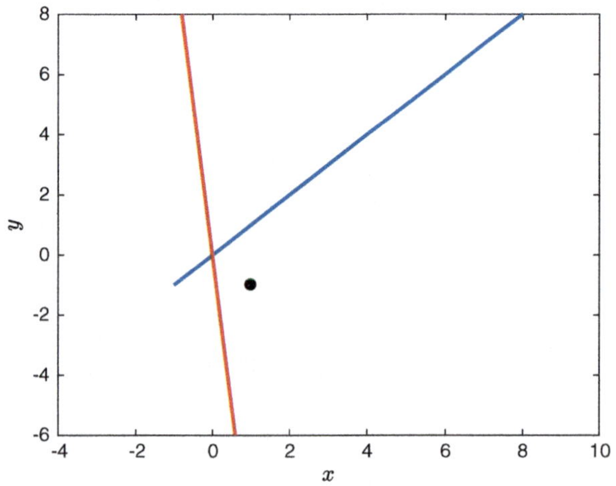

variable. In addition, this can lead to numerical issues when computing $\boldsymbol{\theta} = -\boldsymbol{n}_2/n_1$ from $\boldsymbol{n} = \begin{bmatrix} n_1 \\ \boldsymbol{n}_2 \end{bmatrix}$ for values of n_1 close to zero.

Therefore, we will not further discuss the numerous subspace clustering methods from the literature that could also apply to some extent to switching regression but rather focus on methods dedicated to the regression viewpoint.

Yet, the connection between switching regression and subspace clustering has been at the core of the seminal work on algebraic methods, which were first introduced for the noiseless case, as described in Sect. 6.1.1, and which continue to inspire research in both fields.

Notes

PWA systems are introduced in [1] and arbitrarily switched systems in [2]. The universal approximation capability of PWA models was proved in [3, 4]. The approximation of a NARX model by a PWARX one as described in this chapter is taken from [5]. The various types of partitions of the regression domain for PWA systems have been evaluated in terms of computational complexity for digital implementation in [6]. System identification with nonlinear partitions was considered in [7].

Switching regression was introduced by [8], and early algorithms include the one of [9] and the expectation-maximization methods of [10] for the two-mode case, further generalized by [11, 12].

Regression trees [13], and subsequent improvements [14, 15], are well-known early examples of piecewise regression models, together with the mixtures of experts [16], which however usually consider smooth switchings.

In the machine learning community, the mixture-of-experts framework and expectation-maximization algorithm were extended to model hybrid dynamical systems with hidden Markov models in [17]. A number of works based on such a probabilistic view can be found in [18, 19] or in [20, 21], where so-called jump Markov linear systems are considered.

More recently, most of the work in this field was produced by the control community for hybrid system identification. A first review of the methods operating in that context appeared in 1982 in [22], while the revival of the field was described in the tutorial of [2] with a review of some of the new major techniques at the time that are also experimentally compared in [23]. An updated survey followed in 2012 in [24].

The bounded-error approach in which the submodels are estimated one by one until a predefined error is achieved was first considered in this context in [25] and then followed in [5, 26] with different methods to estimate the jth submodel at each step.

The Minimum-of-Errors formulation in (4.21) was considered in [27], where its direct optimization with the multilevel coordinate search solver of [28] was proposed (see Sect. 6.1.2).

Several works aiming at directly modeling PWA systems with nonlinear models can be found in [29], and particularly multilayer neural networks with PWA activation functions in [30, 31].

The connection between subspace clustering and hybrid system identification was first discussed in [32] and resulted in the so-called algebraic methods. For more details on subspace clustering and related methods, see [33, 34].

References

1. Sontag, E.D.: Nonlinear regulation: the piecewise linear approach. IEEE Trans. Autom. Control. **26**(2), 346–358 (1981)
2. Paoletti, S., Juloski, A.Lj., Ferrari-Trecate, G., Vidal, R.: Identification of hybrid systems: a tutorial. Eur. J. Control. **13**(2–3), 242–260 (2007)
3. Lin, J.N., Unbehauen, R.: Canonical piecewise-linear approximations. IEEE Trans. Circuits Syst. I: Fundam. Theory Appl. **39**(8), 697–699 (1992)
4. Breiman, L.: Hinging hyperplanes for regression, classification, and function approximation. IEEE Trans. Inf. Theory **39**(3), 999–1013 (1993)
5. Bako, L.: Identification of switched linear systems via sparse optimization. Automatica **47**(4), 668–677 (2011)
6. Comaschi, F., Genuit, B.A.G., Oliveri, A., Heemels, W.P.M.H., Storace, M.: FPGA implementations of piecewise affine functions based on multi-resolution hyperrectangular partitions. IEEE Trans. Circuits Syst.- I Regul. Pap. **59**(12), 2920–2933 (2012)
7. Lauer, F., Bloch, G.: Switched and piecewise nonlinear hybrid system identification. In: Proceedings of the 11th International Workshop on Hybrid Systems: Computation and Control (HSCC), St. Louis, MO, USA. Lecture Notes in Computer Science, vol. 4981, pp. 330–343 (2008)
8. Quandt, R.E.: The estimation of the parameters of a linear regression system obeying two separate regimes. J. Am. Stat. Assoc. **53**(284), 873–880 (1958)
9. Späth, H.: Clusterwise linear regression. Computing **22**(4), 367–373 (1979)
10. Hosmer, D.W.: Maximum likelihood estimates of the parameters of a mixture of two regression lines. Commun. Stat. **3**(10), 995–1006 (1974)
11. DeSarbo, W.S., Cron, W.L.: A maximum likelihood methodology for clusterwise linear regression. J. Classif. **5**(2), 249–282 (1988)
12. Gaffney, S., Smyth, P.: Trajectory clustering with mixtures of regression models. In: Proceedings of the fifth ACM SIGKDD International Conference on Knowledge discovery and data mining, San Diego, CA, USA, pp. 63–72 (1999)
13. Breiman, L., Friedman, J.H., Olshen, R.A., Stone, C.J.: Classification and Regression Trees. Chapman & Hall/CRC, Boca Raton (1984)
14. Friedman, J.H.: Multivariate adaptive regression splines. Ann. Stat. **19**(1), 1–67 (1991)
15. Rao, A.V., Miller, D.J., Rose, K., Gersho, A.: A deterministic annealing approach for parsimonious design of piecewise regression models. IEEE Trans. Pattern Anal. Mach. Intell. **21**(2), 159–173 (1999)
16. Jacobs, R.A., Jordan, M.I., Nowlan, S.J., Hinton, G.E.: Adaptive mixtures of local experts. Neural Comput. **3**(1), 79–87 (1991)
17. Ghahramani, Z., Hinton, G.E.: Variational learning for switching state-space models. Neural Comput. **12**(4), 831–864 (2000)
18. Billio, M., Monfort, A., Robert, C.P.: Bayesian estimation of switching ARMA models. J. Econ. **93**(2), 229–255 (1999)
19. Blake, A., North, B., Isard, M.: Learning multi-class dynamics. In: Advances in Neural Information Processing Systems 11, pp. 389–395. MIT Press, Cambridge (1999)

20. Doucet, A., Logothetis, A., Krishnamurthy, V.: Stochastic sampling algorithms for state estimation of jump Markov linear systems. IEEE Trans. Autom. Control. **45**(2), 188–202 (2000)
21. Cinquemani, E., Porreca, R., Ferrari-Trecate, G., Lygeros, J.: A general framework for the identification of jump Markov linear systems. In: Proceedings of the 46th IEEE Conference on Decision and Control (CDC), New Orleans, LA, USA, pp. 5737–5742 (2007)
22. Tugnait, J.K.: Detection and estimation for abruptly changing systems. Automatica **18**(5), 607–615 (1982)
23. Juloski, A.Lj., Heemels, W.P.M.H., Ferrari-Trecate, G., Vidal, R., Paoletti, S., Niessen, J.H.G.: Comparison of four procedures for the identification of hybrid systems. In: Proceedings of the 8th International Workshop on Hybrid Systems: Computation and Control (HSCC), Zürich, Switzerland. Lecture Notes in Computer Science, vol. 3414, pp. 354–369 (2005)
24. Garulli, A., Paoletti, S., Vicino, A.: A survey on switched and piecewise affine system identification. In: Proceedings of the 16th IFAC Symposium on System Identification (SYSID), Brussels, Belgium, IFAC Proceedings Volumes, vol. 45(16), pp. 344–355 (2012)
25. Bemporad, A., Garulli, A., Paoletti, S., Vicino, A.: A bounded-error approach to piecewise affine system identification. IEEE Trans. Autom. Control. **50**(10), 1567–1580 (2005)
26. Ozay, N., Sznaier, M., Lagoa, C.M., Camps, O.I.: A sparsification approach to set membership identification of switched affine systems. IEEE Trans. Autom. Control. **57**(3), 634–648 (2012)
27. Lauer, F., Bloch, G., Vidal, R.: A continuous optimization framework for hybrid system identification. Automatica **47**(3), 608–613 (2011)
28. Huyer, W., Neumaier, A.: Global optimization by multilevel coordinate search. J. Glob. Optim. **14**(4), 331–355 (1999)
29. Nelles, O.: Nonlinear System Identification: From Classical Approaches to Neural Networks and Fuzzy Models. Springer, Berlin (2001)
30. Batruni, R.: A multilayer neural network with piecewise-linear structure and back-propagation learning. IEEE Trans. Neural Netw. **2**(3), 395–403 (1991)
31. Gad, E.F., Atiya, A.F., Shaheen, S., El-Dessouki, A.: A new algorithm for learning in piecewise-linear neural networks. Neural Netw. **13**(4–5), 485–505 (2000)
32. Vidal, R., Soatto, S., Ma, Y., Sastry, S.: An algebraic geometric approach to the identification of a class of linear hybrid systems. In: Proceedings of the 42nd IEEE Conference on Decision and Control (CDC), Maui, HI, USA, pp. 167–172 (2003)
33. Vidal, R.: Subspace clustering. IEEE Signal Process. Mag. **28**(2), 52–68 (2011)
34. Vidal, R., Ma, Y., Sastry, S.: Generalized Principal Component Analysis. Springer, Berlin (2016)

Chapter 5
Exact Methods for Hybrid System Identification

In this chapter, we investigate the possibility of actually solving the problems we set for ourselves in Chap. 4. In particular, we consider the following three representative problems: Problems 4.2 and 4.3 introduced in Sect. 4.2.3 for the estimation of arbitrarily switched and piecewise smooth systems with a fixed number of modes and Problem 4.4 introduced in Sect. 4.2.4 for the bounded-error estimation of switched systems. More precisely, for the latter, we concentrate on the robust regression Problem (4.24) corresponding to a single iteration of the greedy approach.

Here, we focus on regression and optimization issues while leaving aside dynamical system ones. The aim is to provide some understanding of the core problems encountered in hybrid system identification with a computational perspective and discuss the opportunity of computing exact solutions.

As this chapter puts the emphasis on the difficulty of the problems, we also restrict the discussion to the allegedly most simple case, where the submodels are linear. Therefore, the whole chapter is dedicated to the estimation of a collection of linear models

$$f_j(\boldsymbol{x}) = \boldsymbol{x}^\top \boldsymbol{\theta}_j$$

from N data pairs made of regression vectors $\boldsymbol{x}_k$ of dimension d and scalar target outputs y_k.

Before attacking hybrid system identification, one must understand the importance of the combinatorial issues involved. Indeed, most of the classical issues for system identification are also present in hybrid system identification, but they usually constitute the "easy to deal with" part of the problem. Here, the necessity to estimate the mode or to associate the data to the submodels, which is a combinatorial problem, is what makes hybrid system identification so complex. We show in this chapter that this complexity indeed corresponds to that of the most complex problems in computer science. However, this negative result can be used to gain insights into the role of the fundamental parameters of the problem and we will also show in particular that the number of data is not as limiting as their dimensionality: There are algorithms that solve the problems exactly in time polynomial in the number of data, but exponential in the dimension. These algorithms rely on explicit

F. Lauer and G. Bloch, *Hybrid System Identification*,
Lecture Notes in Control and Information Sciences 478,
https://doi.org/10.1007/978-3-030-00193-3_5

connections between the problems of interest and linear classification (as introduced in Sect. 3.1.1), which allow us to solve the combinatorial problems by complete enumeration in an "efficient" manner. Yet, "efficient" here mostly means "theoretically reasonable" and these algorithms can only be applied in practice in a limited set of situations.

By embedding the combinatorial component into an optimization program defined over continuous variables only, we can also tackle the problems via standard global optimization approaches. Of particular interest are the branch-and-bound methods that can certify the global optimality of the yielded solution. This chapter describes how these methods can be used for hybrid system identification. Compared to the combinatorial algorithms discussed above, their advantage is rather practical: The worst-case computing time remains exponential in the dimension, but the average computing time can be much more reasonable in practice.

Chapter Outline

In every section of this chapter, the results are divided into three subsections, one for each of the core problems we discuss: switching linear regression, piecewise affine regression and bounded-error linear estimation.

The chapter starts in Sect. 5.1 with preliminary observations on how to solve these three problems exactly. Then, Sect. 5.2 formally analyzes the computational complexity of the problems. This section relies heavily on theoretical computer science concepts rather than on control or system identification theory. As such, it can be skipped upon first reading, while keeping in mind its summary: most problems of interest in hybrid system identification are $\mathcal{NP}$-hard, meaning that, *in general*, we cannot hope to compute an exact solution in reasonable time.

As detailed in Sect. 5.3, by fixing some parameters such as the dimension of the data, the problems become "tractable," meaning that an exact solution can be computed in time polynomial in the number of data. However, in most cases, the degrees of the polynomial complexities are too large to consider the corresponding algorithms as practical solutions. In Sect. 5.4, we investigate branch-and-bound methods for global optimization that can leverage this issue with better efficiency on average.

Finally, a few numerical results in Sect. 5.5 make it clear that exact solutions shall remain reachable only in some specific conditions and, in particular, when the data dimensionality is not too high. This explains why so many practical methods with more or less accuracy and computational efficiency have been proposed, such as the heuristics/approximations developed in the next chapters.

5.1 Straightforward Solutions

The following subsections detail how one could, *in principle*, solve the three main problems studied in this book: switching linear regression, bounded-error linear estimation, and PWA regression. We also highlight the computational limitations of such naive solutions that fail in most practical situations.

5.1.1 Switching Regression with Fixed s

We first concentrate on the switching regression problem in which one must estimate a fixed number of submodels. The alternative of estimating the number of submodels with a bounded error will be discussed in Sect. 5.1.2. We further focus on the approach in which the parameters are estimated by minimizing the error over the data. In this case, the problem can be setup as follows:

Problem 5.1 (*Switching linear regression*) Given a data set $\{(\boldsymbol{x}_k, y_k)\}_{k=1}^N \subset \mathbb{R}^d \times \mathbb{R}$ and a positive integer s, find a global solution to

$$\min_{\{\boldsymbol{\theta}_j \in \mathbb{R}^d\}_{j=1}^s, \boldsymbol{q} \in [s]^N} \frac{1}{N} \sum_{k=1}^N \ell(y_k - \boldsymbol{x}_k^\top \boldsymbol{\theta}_{q_k}). \tag{5.1}$$

First, note that the interesting values for the number of modes s in Problem 5.1 are in the interval $[2, N/d]$. Obviously, if $s = 1$, Problem 5.1 reduces to a standard linear regression problem. But if $s > N/d$, then Problem 5.1 also has a trivial solution. Indeed, in this case, the N data points can be arbitrarily classified into s groups of less than d points. For each group, a linear model in dimension d can perfectly fit those points, which leads to a mean error over all points of zero and thus the global minimum of the cost function in Problem 5.1.

In general (and in particular when $s \in [2, N/d]$), Problem 5.1 can be solved explicitly with respect to $\boldsymbol{q}$ for fixed $\{\boldsymbol{\theta}_j\}_{j=1}^s$ by assigning each point to the model with minimum error as

$$q_k \in \operatorname*{argmin}_{j \in [s]} \ell(y_k - \boldsymbol{x}_k^\top \boldsymbol{\theta}_j), \quad k = 1, \dots, N. \tag{5.2}$$

Conversely, Problem 5.1 can be solved with respect to the $\boldsymbol{\theta}_j$'s for fixed $\boldsymbol{q}$ as s independent linear regression subproblems

$$\min_{\boldsymbol{\theta}_j \in \mathbb{R}^d} \sum_{k \in \{i : q_i = j\}} \ell(y_k - \boldsymbol{x}_k^\top \boldsymbol{\theta}_j), \quad j = 1, \dots, s. \tag{5.3}$$

Thus, two global optimization approaches can be readily formulated. The first one tests all possible classifications $\boldsymbol{q}$ and solves the problem with respect to the $\boldsymbol{\theta}_j$'s for each of them. But, this leads to $s \times s^N$ linear regression subproblems (5.3) and quickly becomes intractable when N increases. The second approach applies a continuous global optimization strategy to directly estimate $\{\boldsymbol{\theta}_j\}_{j=1}^s$ under the optimal classification rule (5.2), which is equivalent to solving

$$\min_{\{\boldsymbol{\theta}_j \in \mathbb{R}^d\}_{j=1}^s} \frac{1}{N} \sum_{k=1}^N \min_{j \in [s]} \ell(y_k - \boldsymbol{x}_k^\top \boldsymbol{\theta}_j) \tag{5.4}$$

and then recovering the mode estimates with (5.2). However, global optimality is difficult to guarantee. For instance, the complexity remains exponential in the number of variables sd, for a grid search to obtain a solution with an error that is only guaranteed to be close to the global optimum.

These observations illustrate the difficulty of the problem, which will be more formally quantified in Sect. 5.2. The first strategy depicted above will then be further investigated in Sect. 5.3 to obtain via classification arguments an algorithm that runs in polynomial time with respect to the number of data for fixed s and d. The second strategy will be detailed in Sect. 5.4 with a branch-and-bound approach offering a practical method with optimality guarantees for low-dimensional problems.

Remark 5.1 We refer to an exact method for solving Problem 5.1 as an exact switching regression method. However, note that exactly solving the optimization Problem 5.1 does not yield the optimal solution to the regression problem of estimating the parameter vectors. Many factors come into play, including for instance the persistence of excitation, the identifiability, the choice of the loss function, the noise model.

However, being able to solve Problem 5.1 is already a big step toward hybrid system identification and can often be generalized to regularized versions of the problem. For instance, adding an ℓ_2-regularization term, $\sum_{j=1}^{s} \|\boldsymbol{\theta}_j\|_2^2$, to the objective function does not change much the polynomial-time algorithm of Sect. 5.3.2 (using regularized least squares). The branch-and-bound approach of Sect. 5.4.1 could also be extended with a regularized objective.

5.1.2 Bounded-Error Estimation

The bounded-error approach to switching regression of Sect. 4.2.4.1 amounts to solving Problem 4.4. As previously discussed, this is a method of choice when one prefers to set a tolerance on the error and leave the number of modes free to be estimated. However, this formulation is rarely handled directly. Instead, the greedy approach depicted in Sect. 4.2.4.1, in which the submodels are estimated one by one, is most often considered. It amounts to iteratively solve problems (4.24), which, for linear submodels, are of the following form.[1]

Problem 5.2 (*Bounded-error linear estimation*) Given a data set $\{(\boldsymbol{x}_k, y_k)\}_{k=1}^N \subset \mathbb{R}^d \times \mathbb{R}$ and $\epsilon \geq 0$, find a global solution to

$$\min_{\boldsymbol{\theta}\in\mathbb{R}^d} \sum_{k=1}^{N} \ell_{p,\epsilon}(y_k - \boldsymbol{x}_k^\top \boldsymbol{\theta}). \tag{5.5}$$

[1]Recall that in a complete procedure for hybrid system identification, at each iteration the data set is reduced to leave only the points that are not well approximated by already estimated submodels, so that N changes in Problem 5.2.

By letting

$$\mathcal{I}_1(\boldsymbol{\theta}) = \{k \in [N] : |y_k - \boldsymbol{x}_k^\top \boldsymbol{\theta}| \le \epsilon\}, \tag{5.6}$$

Problem (5.5) can be equivalently written as

$$\min_{\boldsymbol{\theta}\in\mathbb{R}^d} \epsilon^p(N - |\mathcal{I}_1(\boldsymbol{\theta})|) + \sum_{k\in\mathcal{I}_1(\boldsymbol{\theta})} \ell_p(y_k - \boldsymbol{x}_k^\top \boldsymbol{\theta}) \tag{5.7}$$

for $p \in \{1, 2\}$ and

$$\min_{\boldsymbol{\theta}\in\mathbb{R}^d} \epsilon^p(N - |\mathcal{I}_1(\boldsymbol{\theta})|) \tag{5.8}$$

for $p = 0$.

The equivalence between (5.5) and (5.7) or (5.8) comes from the definition of $\ell_{p,\epsilon}$ as a saturated loss function (see Table 2.1, Sect. 2.1.2). In particular, the set $\mathcal{I}_1(\boldsymbol{\theta})$ gathers the indexes of data points that are well approximated by the linear model of parameter $\boldsymbol{\theta}$. Thus, $N - |\mathcal{I}_1(\boldsymbol{\theta})|$ coincides with the number of data points for which the loss function $\ell_{p,\epsilon}$ saturates while the loss at all points with index in $\mathcal{I}_1(\boldsymbol{\theta})$ can be computed with a standard (non-saturated) ℓ_p loss.

These equivalent formulations emphasize the connection between saturated loss minimization and the maximization of the number of points approximated with a bounded error. Indeed, these points are here marked with index in $\mathcal{I}_1(\boldsymbol{\theta})$ and maximizing their number is equivalent to minimizing the number of points with index not in $\mathcal{I}_1(\boldsymbol{\theta})$, i.e., $N - |\mathcal{I}_1(\boldsymbol{\theta})|$.

This also draws a connection with the classification problem of separating between points that are approximated with a bounded error by an optimal model $\boldsymbol{\theta}^*$ and those that are not. In particular, given the solution to this classification problem, i.e., $\mathcal{I}_1(\boldsymbol{\theta}^*)$ for some optimal $\boldsymbol{\theta}^*$, a global solution $\hat{\boldsymbol{\theta}}$ (possibly different from $\boldsymbol{\theta}^*$) can be recovered by solving (5.7) or (5.8) under the constraint $\mathcal{I}_1(\boldsymbol{\theta}) = \mathcal{I}_1(\boldsymbol{\theta}^*)$. Then, for $p = 0$, the cost in (5.8) is a mere constant and it suffices to find a $\boldsymbol{\theta}$ such that

$$\max_{k\in\mathcal{I}_1(\boldsymbol{\theta}^*)} |y_k - \boldsymbol{x}_k^\top \boldsymbol{\theta}| < \epsilon$$

to satisfy the constraint. Conversely, for $p \in \{1, 2\}$, the cost in (5.7) simplifies to a constant plus a sum of errors over a fixed set of points. Hence, given $\mathcal{I}_1(\boldsymbol{\theta}^*)$, these problems amount to standard regression problems with a non-saturated loss and $\hat{\boldsymbol{\theta}}$ can be computed as

$$\hat{\boldsymbol{\theta}} \in \operatorname*{argmin}_{\boldsymbol{\theta}\in\mathbb{R}^d} \begin{cases} \max\limits_{k\in\mathcal{I}_1(\boldsymbol{\theta}^*)} |y_k - \boldsymbol{x}_k^\top \boldsymbol{\theta}|, & \text{if } p = 0 \\ \sum\limits_{k\in\mathcal{I}_1(\boldsymbol{\theta}^*)} \ell_p(y_k - \boldsymbol{x}_k^\top \boldsymbol{\theta}), & \text{otherwise.} \end{cases} \tag{5.9}$$

As for switching regression, a naive algorithm can thus be devised by considering all classifications of the data into two groups, those with index in $\mathcal{I}_1$ and the others. Then, one minimizes the ℓ_p loss over all points with index in $\mathcal{I}_1$ and computes the cost function value as in (5.7) or (5.8). Finally, the solution with lowest cost is retained. However, here again the complexity of this approach is proportional to the number of classifications and in $\mathcal{O}(2^N)$, thus much too large for practical purposes. Reducing this number to a polynomial function of N will be the topic of Sect. 5.3.3.

5.1.3 Piecewise Affine Regression with Fixed s

With affine submodels,

$$f_j(\boldsymbol{x}) = \tilde{\boldsymbol{x}}^\top \boldsymbol{\theta}_j,$$

where $\tilde{\boldsymbol{x}} = \begin{bmatrix}\boldsymbol{x}^\top & 1\end{bmatrix}^\top$, and linear classifiers, the PWS regression Problem 4.3 of Sect. 4.2.3 can be rewritten as follows.

Problem 5.3 (*PWA regression*) Given a data set $\{(\boldsymbol{x}_k, y_k)\}_{k=1}^N \subset \mathbb{R}^d \times \mathbb{R}$ and a number of modes, s, find a global solution to

$$\min_{\boldsymbol{\theta}\in\mathbb{R}^{sd}, g\in\mathcal{G}} \frac{1}{N} \sum_{k=1}^{N} \sum_{j\in[s]} \mathbb{1}_{g(\boldsymbol{x}_k)=j} \ell(y_k - \boldsymbol{x}_k^\top \boldsymbol{\theta}_j) \tag{5.10}$$

with $\boldsymbol{\theta} = \begin{bmatrix}\boldsymbol{\theta}_1^\top & \dots & \boldsymbol{\theta}_s^\top\end{bmatrix}^\top$ and the set $\mathcal{G}$ of linear classifiers as in (4.13).

This problem could also be tackled (at least in principle) with a straightforward approach: testing the s^N classifications of the data points into s modes and solving s independent linear regression subproblems for each of them. The only required modification when compared to the arbitrarily switching case of Sect. 5.1.1 is to check for each classification that it is indeed consistent with the set of classifiers $\mathcal{G}$ before going through with the regression. For the set of linear classifiers as in (4.13), this amounts to a linear separability problem that can be posed as a feasibility problem solved with linear programming techniques. For instance, in the binary case with $s = 2$ discussed in Sect. 3.1.1, it is common to consider the set of modes $\mathcal{Q} = \{-1, +1\}$ instead of $\{1, 2\}$. Then, a classification $\boldsymbol{q} \in \{-1, +1\}^N$ can be tested by determining whether the linear system of inequalities

$$q_k(\boldsymbol{g}^\top \boldsymbol{x}_k + b) > 0, \quad k = 1, \dots, N$$

has a feasible solution in $(\boldsymbol{g}, b) \in \mathbb{R}^d \times \mathbb{R}$ (see also (3.6)).

5.2 Hardness Results (⋆)

The aim of this section is to formally analyze the algorithmic complexity of the problems of interest. We start with basic definitions from the field of computational complexity, before analyzing each one of the three problems discussed above in subsequent subsections.

5.2.1 Basics in Computational Complexity

We here only introduce concepts from computational complexity at a rather high level in order to understand the meaning and value of the results below.

The complexity of algorithms or problems can only be analyzed with respect to a given model of computation, which defines how the data is encoded in memory and what are the basic operations on these data that can be found in a program.

5.2.1.1 Encoding Scheme

We restrict the discussion to the encoding of numbers and shall furthermore limit ourselves to rational numbers. The reason is that rational numbers can be represented by two integers, for which classical encoding schemes and computational models are typically devised. Models of computation over the reals also exist, but these lead to different classes of complexity.

The standard encoding scheme for integers is the binary encoding, or bit model, with which an integer i is encoded as a sequence of $n = \lceil \log i \rceil$ zeros and ones, called bits. Another classical choice not considered here is the unary encoding, in which an integer i is encoded as a sequence of $n = i$ symbols from a single-letter alphabet.

5.2.1.2 Model of Computation

A model of computation is an abstraction of a computer that should be both sufficiently simple to allow for its analysis and sufficiently realistic to give meaning to the results of the analysis. The most common model of computation is the Turing machine. It consists of a linear tape, which is an infinite sequence of cells playing the role of the memory, and a head that has a finite number of possible internal states. Each cell contains a symbol from a finite alphabet (say $\{0, 1\}$) or a "blank". At the beginning of a program, the input is written on finitely many cells of the tape (the others are left blank), the head is in a particular "start" state, and it scans the first cell. Then, at each time step, a transition occurs: (i) the internal state changes according to the predefined program as a function of the read symbol in the scanned cell and the internal state, (ii) a new symbol is written in that cell, and (iii) the head moves one cell left or right. The computation terminates when the head enters the special "halt" state and the tape content is considered as the output of the program.

Though rather basic, this machine is capable of performing arithmetic operations on n-bit integers in time which is polynomial in n. Hence, it can simulate any reasonable model of computation with only polynomial slowdown and is sufficient for the purpose of time complexity analysis.

The Turing machine just described is a *deterministic* one. In its nondeterministic counterpart, the transitions only occur with a certain probability, which results in possibly different outputs for a given input.

5.2.1.3 Time Complexity

The **time complexity of an algorithm** is a function $T(n)$ of its input size n whose value corresponds to the maximal number of steps occurring in its computation over all inputs of size n. Thus, it is a *worst-case* measure of the algorithm running time. In particular, under the nondeterministic model of computation, the maximal number of steps over both all inputs of size n and all possible paths of computations on these inputs is considered. Under the bit model described above, the size of an input u given as a list of integers u_i is understood as the number of bits used to encode that input in binary format and satisfies $n = \mathcal{O}(\log \max_i u_i)$.

Time complexities are often expressed using asymptotic notations as in

$$T(n) = \mathcal{O}(f(n)),$$

which means that there exist positive numbers n_0 and c such that $T(n) \leq cf(n)$ for all $n \geq n_0$. Then, an algorithm is said to have a linear time complexity if $T(n) = \mathcal{O}(n)$, and more generally, to have a polynomial-time complexity (or simply to be a *polynomial algorithm*) if $T(n) = \mathcal{O}(p(n))$ for some polynomial p.

The **time complexity of a problem** is defined as the smallest time complexity of an algorithm that solves any instance of that problem.[2]

5.2.1.4 Classes of Problems

We now define the common classes of problems and terminology.

- A **decision problem** is one for which the solution (or answer) can be either "yes" or "no".
- $\mathcal{P}$ is the class of deterministic polynomial-time decision problems, i.e., the set of decision problems whose time complexity on a deterministic Turing machine is no more than polynomial in the input size.

[2]One can similarly define the space complexity of algorithms and problems to analyze the memory requirements rather than the running time. Throughout the book, the term "complexity" refers to the time complexity.

- $\mathcal{NP}$ is the class of nondeterministic polynomial-time decision problems, i.e., the set of decision problems whose time complexity on a nondeterministic Turing machine is no more than polynomial in the input size.
- A problem (not necessarily a decision one) is **$\mathcal{NP}$–hard** if it is at least as hard as any problem in $\mathcal{NP}$.
- A decision problem is **$\mathcal{NP}$–complete** if it is both in $\mathcal{NP}$ and $\mathcal{NP}$-hard.

The class $\mathcal{NP}$ can be understood as the set of problems for which a candidate solution can be certified in polynomial time. For instance, deciding whether there is a vector $\boldsymbol{x} \in \{0, 1\}^d$ satisfying a given linear system of equations $\boldsymbol{Ax} = \boldsymbol{b}$, $\boldsymbol{A} \in \mathbb{Z}^{p\times d}$, is a difficult problem, but which can be solved by testing all the 2^d possible values for $\boldsymbol{x}$. Therefore, a nondeterministic machine can start by drawing a random $\boldsymbol{x}$ and then verify whether $\boldsymbol{Ax} = \boldsymbol{b}$. Since there is at least one path of random computation (i.e., one random draw of $\boldsymbol{x}$) that yields the correct answer, this machine solves the problem. And since all computations take only the time needed to check $\boldsymbol{Ax} = \boldsymbol{b}$, which is linear in p and d, the problem has a polynomial complexity over a nondeterministic machine and belongs to $\mathcal{NP}$.

The class $\mathcal{NP}$ naturally includes $\mathcal{P}$, since the deterministic machine solving a problem from $\mathcal{P}$ is also a nondeterministic one with a single path of computation. The converse is not so trivial and it is in fact conjectured that $\mathcal{P} \neq \mathcal{NP}$.

An **optimization problem**, $\min_{\boldsymbol{\theta}\in\Theta} J(\boldsymbol{\theta})$, is usually proved to be $\mathcal{NP}$-hard by showing that its decision form,

$$\text{Given } \epsilon, \text{ is there some } \boldsymbol{\theta} \in \Theta, \text{ such that } J(\boldsymbol{\theta}) \leq \epsilon?$$

is $\mathcal{NP}$-hard. Indeed, solving the optimization problem also yields the answer to the decision problem and thus cannot be easier.

Remark 5.2 Following the discussion on the encoding scheme above, problem formulations below and throughout Sect. 5.2 are considered over rational data, i.e., with all occurrences of $\mathbb{R}$ replaced by $\mathbb{Q}$.

5.2.2 *Hardness of Switching Regression*

The following result characterizes the difficulty of switching regression as formulated in Problem 5.1 (over rational numbers).

Theorem 5.1 *With a loss function ℓ such that*

$$\begin{cases} \ell(0) = 0, \\ \forall e \in \mathbb{Q},\ \ell(-e) = \ell(e), \\ \forall (e, e') \in \mathbb{Q}^2,\ \ell(e) < \ell(e') \Leftrightarrow |e| < |e'|, \end{cases} \tag{5.11}$$

Problem 5.1 is $\mathcal{NP}$-hard.

Note that the form of the loss function in this theorem is not very restrictive. In particular, it includes the standard squared and absolute losses.

$\mathcal{NP}$-hardness is usually proved by reducing a known $\mathcal{NP}$-hard problem to the problem of interest in such a way that the $\mathcal{NP}$-hard problem can be solved as an instance of the other. If such a reduction exists and can be constructed in polynomial time, then this means that the problem of interest is at least as hard as an $\mathcal{NP}$-hard one and thus also $\mathcal{NP}$-hard. Here, we consider reductions from the partition problem, known to be $\mathcal{NP}$-hard.

Problem 5.4 (*Partition*) Given a multiset (a set with possibly multiple instances of its elements) of d positive integers, $\mathcal{S} = \{n_k\}_{k=1}^d$, decide whether there is a multisubset $\mathcal{S}_1 \subset \mathcal{S}$ such that

$$\sum_{n_k \in \mathcal{S}_1} n_k = \sum_{n_k \in \mathcal{S} \setminus \mathcal{S}_1} n_k,$$

or, equivalently, such that

$$\sum_{n_k \in \mathcal{S}_1} n_k = \frac{1}{2} \sum_{n_k \in \mathcal{S}} n_k.$$

The original proof of Theorem 5.1 involves a reduction from the partition problem to a noiseless instance of the decision form of switching regression.

Problem 5.5 (*Decision form of switching regression*) Given a data set $\{(\boldsymbol{x}_k, y_k)\}_{k=1}^N \in (\mathbb{Q}^d \times \mathbb{Q})^N$, an integer $s \in [2, N/d]$ and a threshold $\epsilon \geq 0$, decide whether there is a set of vectors $\{\boldsymbol{\theta}_j\}_{j=1}^s \subset \mathbb{Q}^d$ and a labeling $\boldsymbol{q} \in [s]^N$ such that

$$\frac{1}{N} \sum_{k=1}^N \ell(y_k - \boldsymbol{x}_k^\top \boldsymbol{\theta}_{q_k}) \leq \epsilon. \tag{5.12}$$

However, the proof can be adapted to the restriction of the problem excluding noiseless instances, which is a stronger result (Theorem 5.2 implies Theorem 5.1).

Theorem 5.2 *With a loss function ℓ as in* (5.11), *Problem 5.1 is $\mathcal{NP}$-hard, even when excluding noiseless instances for which the global minimum is zero.*

The reduction used in the original proof for the noiseless case relies on a specific construction of the data set for switching regression from the data of the partition problem, illustrated in Fig. 5.1. For each value n_k, this data set is made with two points $(\boldsymbol{x}_k, y_k)$ and $(\boldsymbol{x}_{k+d}, y_{k+d})$ differing only in the y value such that each point in the pair must be fitted by a different linear regression model. Then, an additional point $(\boldsymbol{x}_{2d+1}, y_{2d+1})$ is added such that if a linear model goes through it, the sum of n_k values for which this model fits one of the two previously described points equals half of the total sum of the n_k's, hence yielding a valid partition.

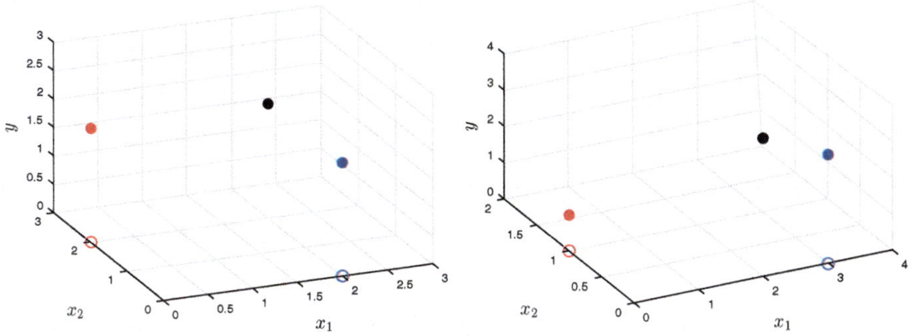

Fig. 5.1 Construction of the reduction (5.13) to a switching regression problem of the (toy) Partition Problem 5.4: with $\mathcal{S} = \{n_1 = 2, n_2 = 2\}$ (left) or $\mathcal{S} = \{n_1 = 3, n_2 = 1\}$ (right). The blue points are built from n_1 and the red ones from n_2; they are plotted with filled disks for $k \leq d$ and empty circles for $k > d$. The black point is $(\boldsymbol{x}_5, y_5)$. *Left:* a linear model can fit the blue point (•) at (2, 0, 0), the red one (○) at (0, 2, 2) and the black one (•) at (2, 2, 2), while another linear model fits the remaining two points (○ and •) plus the black one. This yields a valid partition by taking in $\mathcal{S}_1$ the values n_k for which $(\boldsymbol{x}_k, y_k)$ with $k \leq d$ is fitted by one of the two models. Note that there is no other way to perfectly fit all the points with two linear models passing through the origin. *Right:* the partition problem has no solution and these data points cannot be all fitted by a pair of linear models

We now give the formal proof of Theorem 5.2 for the noisy case, which is based on a similar construction with noise added to the y values.

Proof Since given a candidate solution $\left(\{\boldsymbol{\theta}_j\}_{j=1}^s, \boldsymbol{q}\right)$ the condition (5.12) can be verified in polynomial time, Problem 5.5 is in $\mathcal{NP}$. Then, the proof of its $\mathcal{NP}$-completeness proceeds by showing that the Partition Problem 5.4 has an affirmative answer if and only if a particular instance of Problem 5.5 has an affirmative answer.

Given an instance of Problem 5.4, build an instance of Problem 5.5 with $s = 2$, $N = 2d + 1$, $\epsilon > 0$ (to be defined below) and a noisy data set such that

$$(\boldsymbol{x}_k, y_k) = \begin{cases} (n_k \boldsymbol{i}_k,\ n_k + \varepsilon_k), & \text{if } 1 \leq k \leq d \\ (n_{k-d} \boldsymbol{i}_{k-d},\ \varepsilon_k), & \text{if } d < k \leq 2d \\ \left(\boldsymbol{n} = \sum_{i=1}^d n_i \boldsymbol{i}_i\ ,\ \frac{1}{2}\sum_{i=1}^d n_i + \varepsilon_k\right), & \text{if } k = 2d + 1, \end{cases} \tag{5.13}$$

where $\boldsymbol{i}_k$ is the kth unit vector of the canonical basis for $\mathbb{Q}^d$ (i.e., a vector of zeros with a single one at the kth entry) and $\varepsilon_k \in [-\sigma, \sigma]$ is a bounded noise term with $\sigma > 0$ chosen such that $\ell(\sigma) = \epsilon$. If Problem 5.4 has an affirmative answer, let $\mathcal{I}_1$ be the set of indexes of the elements of $\mathcal{S}$ in $\mathcal{S}_1$ and $\mathcal{I}_2$ the set of indexes of the elements of $\mathcal{S}$ not in $\mathcal{S}_1$. Then, we can set $\boldsymbol{\theta}_1 = \sum_{k \in \mathcal{I}_1} \boldsymbol{i}_k$ and $\boldsymbol{\theta}_2 = \sum_{k \in \mathcal{I}_2} \boldsymbol{i}_k$, which gives

$$\boldsymbol{x}_k^\top \boldsymbol{\theta}_1 = \begin{cases} n_k = y_k - \varepsilon_k, & \text{if } k \leq d \text{ and } k \in \mathcal{I}_1 \\ 0, & \text{if } k \leq d \text{ and } k \in \mathcal{I}_2 \\ n_{k-d}, & \text{if } k > d \text{ and } k - d \in \mathcal{I}_1 \\ 0 = y_k - \varepsilon_k, & \text{if } k > d \text{ and } k - d \in \mathcal{I}_2 \\ \sum_{i \in \mathcal{I}_1} n_i = \frac{1}{2} \sum_{i=1}^{d} n_i = y_k - \varepsilon_k, & \text{if } k = 2d + 1 \end{cases}$$

and

$$\boldsymbol{x}_k^\top \boldsymbol{\theta}_2 = \begin{cases} 0, & \text{if } k \leq d \text{ and } k \in \mathcal{I}_1 \\ n_k = y_k - \varepsilon_k, & \text{if } k \leq d \text{ and } k \in \mathcal{I}_2 \\ 0 = y_k - \varepsilon_k, & \text{if } k > d \text{ and } k - d \in \mathcal{I}_1 \\ n_{k-d}, & \text{if } k > d \text{ and } k - d \in \mathcal{I}_2 \\ \sum_{i \in \mathcal{I}_2} n_i = \frac{1}{2} \sum_{i=1}^{d} n_i = y_k - \varepsilon_k, & \text{if } k = 2d + 1. \end{cases}$$

Therefore, for all points, either $\boldsymbol{x}_k^\top \boldsymbol{\theta}_1 = y_k - \varepsilon_k$ or $\boldsymbol{x}_k^\top \boldsymbol{\theta}_2 = y_k - \varepsilon_k$, and $\min_{j \in \{1,2\}} \ell(y_k - \boldsymbol{x}_k^\top \boldsymbol{\theta}_j) \leq \ell(\sigma) = \epsilon$. Thus, (5.12) holds with $\boldsymbol{q}$ set as in (5.2), yielding an affirmative answer for Problem 5.5.

Now, assume that Problem 5.5 has an affirmative answer for some $\epsilon > 0$. Then, we can ensure that $\min_{j \in \{1,2\}} \ell(y_k - \boldsymbol{x}_k^\top \boldsymbol{\theta}_j) \leq \epsilon N$, $k = 1, \dots, N$. Since ℓ is strictly increasing with the absolute value of its argument, this implies that $\min_{j \in \{1,2\}} |y_k - \boldsymbol{x}_k^\top \boldsymbol{\theta}_j| \leq \ell^{-1}(\epsilon N) = \beta$, where ℓ^{-1} denotes the inverse of ℓ over the positive reals. Thus, we obtain for all k that

$$\boldsymbol{x}_k^\top \boldsymbol{\theta}_1 \in [y_k - \beta, y_k + \beta] \quad \text{or} \quad \boldsymbol{x}_k^\top \boldsymbol{\theta}_2 \in [y_k - \beta, y_k + \beta]. \tag{5.14}$$

By construction, for $k \leq d$ such that $\boldsymbol{x}_k^\top \boldsymbol{\theta}_1 \in [y_k - \beta, y_k + \beta]$, we have $\boldsymbol{x}_k^\top \boldsymbol{\theta}_1 = n_k \theta_{1k}$ and thus

$$\theta_{1k} \in \left[\frac{y_k}{n_k} - \frac{\beta}{n_k}, \frac{y_k}{n_k} + \frac{\beta}{n_k}\right] = \left[1 + \frac{\varepsilon_k}{n_k} - \frac{\beta}{n_k}, 1 + \frac{\varepsilon_k}{n_k} + \frac{\beta}{n_k}\right].$$

On the one hand, by supposing that

$$\beta < \frac{n_k}{2} - \sigma, \tag{5.15}$$

this yields $\theta_{1k} \in \left(\frac{1}{2}, \frac{3}{2}\right)$. On the other hand, this also implies that $\boldsymbol{x}_{k+d}^\top \boldsymbol{\theta}_1 = n_k \theta_{1k}$ is in the interval $[n_k + \varepsilon_k - \beta, n_k + \varepsilon_k + \beta]$ whose intersection with $[y_{k+d} - \beta, y_{k+d} + \beta] = [\varepsilon_{k+d} - \beta, \varepsilon_{k+d} + \beta]$ is empty if $\beta < n_k/2 + (\varepsilon_k - \varepsilon_{k+d})/2$, which is the case whenever (5.15) holds. In this case, under (5.15), (5.14) implies $\boldsymbol{x}_{k+d}^\top \boldsymbol{\theta}_2 = n_k \theta_{2k} \in [\varepsilon_{k+d} - \beta, \varepsilon_{k+d} + \beta]$ and

$$\theta_{2k} \in \left[\frac{\varepsilon_{k+d} - \beta}{n_k}, \frac{\varepsilon_{k+d} + \beta}{n_k}\right] \subseteq \left[\frac{-(\sigma + \beta)}{n_k}, \frac{\sigma + \beta}{n_k}\right] \subset \left(-\frac{1}{2}, \frac{1}{2}\right).$$

Similarly, for $k \leq d$ such that $\boldsymbol{x}_k^\top \boldsymbol{\theta}_2 \in [y_k - \beta, y_k + \beta]$, we can show that $\theta_{2k} \in [1 + \varepsilon_k/n_k - \beta/n_k, 1 + \varepsilon_k/n_k + \beta/n_k]$ and $\theta_{1k} \in [(\varepsilon_{k+d} - \beta)/n_k, (\varepsilon_{k+d} + \beta)/n_k] \subset \left(-\frac{1}{2}, \frac{1}{2}\right)$. This means that we can detect which part of (5.14) holds by checking whether $\theta_{1k} \geq 1/2$.

For $k = 2d + 1$, we obtain at least one of the two inclusions

$$\boldsymbol{x}_{2d+1}^\top \boldsymbol{\theta}_1 \in \left[\frac{1}{2}\sum_{i=1}^{d} n_i - \beta,\ \frac{1}{2}\sum_{i=1}^{d} n_i + \beta\right] = I \tag{5.16}$$

$$\boldsymbol{x}_{2d+1}^\top \boldsymbol{\theta}_2 \in I. \tag{5.17}$$

Assume that (5.16) holds (a similar reasoning applies to (5.17)) and notice that if $\beta < 1/8d$ and $\sigma < 1/8d$, we have

$$I \subset \left(\frac{-1}{4d} + \frac{1}{2}\sum_{i=1}^{d} n_i,\ \frac{1}{4d} + \frac{1}{2}\sum_{i=1}^{d} n_i\right)$$

while the dot product $\boldsymbol{x}_{2d+1}^\top \boldsymbol{\theta}_1$ lives in $[u_1, u_2]$ with

$$\begin{aligned} u_1 &= \sum_{i \in \{k \leq d:\ \theta_{1k} \geq 1/2\}} (n_i + \varepsilon_i - \beta) + \sum_{i \in \{k \leq d:\ \theta_{1k} < 1/2\}} \left(\varepsilon_i - \frac{\beta}{n_i}\right) \\ &> \frac{-1}{4} + \sum_{i \in \{k \leq d:\ \theta_{1k} \geq 1/2\}} n_i \end{aligned}$$

and

$$\begin{aligned} u_2 &= \sum_{i \in \{k \leq d:\ \theta_{1k} \geq 1/2\}} (n_i + \varepsilon_i + \beta) + \sum_{i \in \{k \leq d:\ \theta_{1k} < 1/2\}} \left(\varepsilon_i + \frac{\beta}{n_i}\right) \\ &< \frac{1}{4} + \sum_{i \in \{k \leq d:\ \theta_{1k} \geq 1/2\}} n_i. \end{aligned}$$

For the inclusion (5.16) to hold, we need $[u_1, u_2] \cap I \neq \emptyset$ and thus

$$\left| \sum_{i \in \{k \leq d:\ \theta_{1k} \geq 1/2\}} n_i - \frac{1}{2}\sum_{i=1}^{d} n_i \right| < \frac{1}{4} + \frac{1}{4d} \leq \frac{1}{2}.$$

Since the left sum is an integer and the right one is a multiple of 1/2, their distance cannot be less than 1/2 unless they are equal. Thus, we obtain a valid partition for Problem 5.4 by taking $\mathcal{S}_1 = \{n_k : \theta_{1k} \geq 1/2,\ k \leq d\}$. The assumption $\beta < 1/8d$ amounts to $\ell^{-1}(\epsilon N) < 1/8d$ and thus easily holds by setting ϵ such that, for instance, $0 < \epsilon < 1/(8d)^p(2d+1))$ for the ℓ_p-losses $\ell(e) = |e|^p$ (recall that $N = 2d + 1$).

The assumption (5.15) trivially holds for all $\sigma < 1/8d$ due to $n_k \geq 1$ and $d \geq 1$. Finally, since ℓ^{-1} is an increasing function, the assumption $\sigma < 1/8d$ is satisfied as $\sigma = \ell^{-1}(\epsilon) \leq \ell^{-1}(\epsilon N) = \beta < 1/8d$.

5.2.3 Hardness of PWA Regression

The PWA regression Problem 5.3 can be shown to be $\mathcal{NP}$-hard by following the reasoning we applied for switching regression in Sect. 5.2.2. Thus, we construct a reduction from the Partition Problem 5.4 to a PWA regression one such that a valid partition in the first can be found if and only if a perfect fit of the data can be obtained with a PWA model. This will involve a PWA model with $s = 2$ modes, and thus, we will use the standard set of labels $\mathcal{Q} = \{-1, 1\}$ for binary classification instead of $[s]$ in Problem 5.3.

More precisely, given an instance of Problem 5.4, we let $N = 2d + 3$, $s = 2$, $\epsilon = 0$ and build a data set with

$$(\boldsymbol{x}_k, y_k) = \begin{cases} (n_k \boldsymbol{i}_k, n_k), & \text{if } 1 \leq k \leq d \\ (-n_{k-d}\boldsymbol{i}_{k-d}, n_{k-d}), & \text{if } d < k \leq 2d \\ (\boldsymbol{n}, 0), & \text{if } k = 2d+1 \\ (-\boldsymbol{n}, 0), & \text{if } k = 2d+2 \\ (\boldsymbol{0}, 0), & \text{if } k = 2d+3, \end{cases} \tag{5.18}$$

where $\boldsymbol{i}_k$ is the kth unit vector of the canonical basis for $\mathbb{Q}^d$ and $\boldsymbol{n} = \sum_{i=1}^d n_i \boldsymbol{i}_i$. Figure 5.2 illustrates the construction. For each value n_k, the data set contains two points $(\boldsymbol{x}_k, y_k)$ and $(\boldsymbol{x}_{k+d}, y_{k+d})$ differing only in the $\boldsymbol{x}$ value such that each point in the pair must be fitted by a different linear model in a solution fitting all data points. The precise values of $\boldsymbol{x}_k$ and the additional points with $k > 2d$ are chosen such that, if a pair of linear models fits all data points, then the sum of n_k's for which one model fits $(\boldsymbol{x}_k, y_k)$ with $k \leq d$ equals the sum of the remaining values, which provides a valid partition.

Conversely, if Problem 5.4 has an affirmative answer, then using a linear classifier $g(\boldsymbol{x}) = \operatorname{sign}(\boldsymbol{g}^\top \boldsymbol{x} + b)$ as in (3.5), we can set

$$\boldsymbol{\theta}_1 = \sum_{k \in \mathcal{I}_1} \tilde{\boldsymbol{e}}_k - \sum_{k \in \mathcal{I}_{-1}} \tilde{\boldsymbol{e}}_k, \quad \boldsymbol{\theta}_{-1} = -\boldsymbol{\theta}_1,$$

$$\boldsymbol{g} = \sum_{k \in \mathcal{I}_1} \boldsymbol{i}_k - \sum_{k \in \mathcal{I}_{-1}} \boldsymbol{i}_k, \quad b = 0,$$

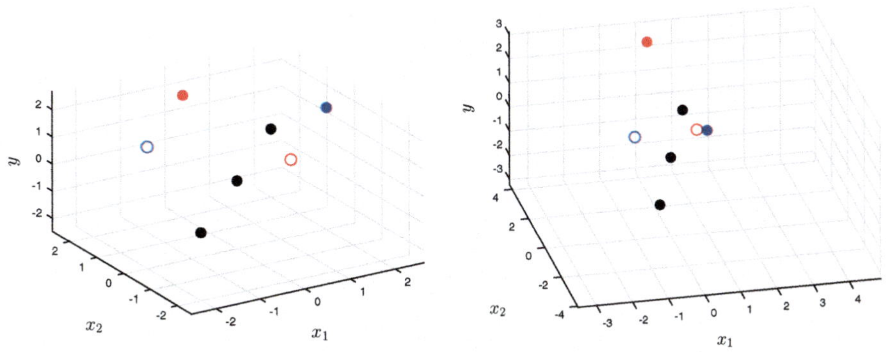

Fig. 5.2 PWA regression data set built as in (5.18) for the reduction of the (toy) partition problem: with $\mathcal{S} = \{n_1 = 2, n_2 = 2\}$ (left) and $\mathcal{S} = \{n_1 = 1, n_2 = 3\}$ (right). The blue points are built from n_1 and the red ones from n_2; they are plotted with filled disks for $k \leq d$ and empty cricles for $k > d$. The black points are the three last ones in the data set. *Left:* a linear model can fit the filled blue and empty red points on the right, while another one fits the empty blue and filled red points on the left and the black points are fitted by both models. This yields a solution to the partition problem by taking in $\mathcal{S}_1$ the values n_k for which one of these models fits $(\boldsymbol{x}_k, y_k)$ with $k \leq d$ (filled disks). Note that there is no other way to fit all the points with two linear models. *Right:* the partition problem has no solution and no pair of linear models can fit all these data points

where $\tilde{\boldsymbol{e}}_k = \left[\boldsymbol{i}_k^\top \; 0\right]^\top$, $\mathcal{I}_1$ is the set of indexes of the elements of $\mathcal{S}$ in $\mathcal{S}_1$ and $\mathcal{I}_{-1} = \{1, \dots, d\} \setminus \mathcal{I}_1$. This gives

$$\tilde{\boldsymbol{x}}_k^\top \boldsymbol{\theta}_1 = \begin{cases} n_k = y_k, & \text{if } k \leq d \text{ and } k \in \mathcal{I}_1 \\ -n_k, & \text{if } k \leq d \text{ and } k \in \mathcal{I}_{-1} \\ n_{k-d} = y_k, & \text{if } k > d \text{ and } k - d \in \mathcal{I}_{-1} \\ -n_{k-d}, & \text{if } k > d \text{ and } k - d \in \mathcal{I}_1 \\ \sum_{i \in \mathcal{I}_1} n_i - \sum_{i \in \mathcal{I}_{-1}} n_i = 0 = y_k, & \text{if } k = 2d + 1 \\ \sum_{i \in \mathcal{I}_{-1}} n_i - \sum_{i \in \mathcal{I}_1} n_i = 0 = y_k, & \text{if } k = 2d + 2 \\ 0 = y_k, & \text{if } k = 2d + 3 \end{cases}$$

and we can similarly show that

$$\tilde{\boldsymbol{x}}_k^\top \boldsymbol{\theta}_{-1} = y_k, \quad \text{if } k \in \mathcal{I}_{-1} \text{ or } k - d \in \mathcal{I}_1 \text{ or } k > 2d,$$

while $\boldsymbol{g}^\top \boldsymbol{x}_k$ is positive if $k \in \mathcal{I}_1$ or $k - d \in \mathcal{I}_{-1}$ and negative if $k \in \mathcal{I}_{-1}$ or $k - d \in \mathcal{I}_1$. Therefore, for all points, $\tilde{\boldsymbol{x}}_k^\top \boldsymbol{\theta}_{g(\boldsymbol{x}_k)} = y_k$, $k = 1, \dots, 2d + 3$, and the cost function of Problem 5.3 is zero, yielding an affirmative answer for its decision form.

5.2.4 *Hardness of Bounded-Error Estimation*

For bounded-error estimation, Problem 4.4 with a constraint on the maximum absolute loss instead of the mean amounts to partitioning the inequalities

$$|y_k - \boldsymbol{x}_k^\top \boldsymbol{\theta}| \leq \epsilon$$

into a minimum number of feasible subsystems. This so-called MIN PFS problem can be shown to be $\mathcal{NP}$-hard using also a reduction from the Partition Problem 5.4. Therefore, most practical methods follow the greedy iterative scheme depicted in Sects. 4.2.4 and 5.1.2. On the computational side, it is also a viable alternative when s grows large: Since the submodels are iteratively estimated one by one, the algorithmic complexity does not grow more than linearly with s.

However, if we aim at an exact solution, the worst-case complexity remains high in general. Indeed, at each iteration, Problem 5.2 with the $\ell_{0,\epsilon}$-loss (see Table 2.1), must be solved to estimate one of the submodels, but it can also be shown to be $\mathcal{NP}$-hard. Overall, the greedy approach replaced the $\mathcal{NP}$-hard Problem 4.4 by a sequence of $\mathcal{NP}$-hard Problems 5.2. Yet, the later is more amenable to practical solutions, as will be seen in Chapter 6, or even global optimization, as detailed in Sect. 5.4.2.

5.3 Polynomial-Time Algorithms for Fixed Dimensions

The $\mathcal{NP}$-hardness of the problems of interest would seem to imply that we cannot hope to compute an exact solution. However, we will now show that reality is not so bad. In particular, the following question remains: What parameter (among s, d and N) really influences the hardness of the problems? And, subsequently, what parameter could be fixed in order to make the problems tractable?

It seems that the number of modes s plays a particular role here. On the one hand, the hardness results were obtained with reductions using the smallest value $s = 2$, implying that fixing it to a small value will not help in overcoming the complexity. On the other hand, a larger s will typically incur an even larger complexity.

By a careful look at the proofs of $\mathcal{NP}$-hardness, one can see that the critical parameter is the dimension d. To emphasize this more clearly, we will now derive exact algorithms with polynomial complexity in the last parameter, i.e., the number of data N. The existence of these algorithms proves that with a fixed dimension, the problems can be solved in polynomial time and are not $\mathcal{NP}$-hard anymore (unless $\mathcal{P} = \mathcal{NP}$). The converse is not true: By fixing N, we do not obtain polynomial-time algorithms in d. And we shall not hope for this, since such algorithms used on the reductions of the partition problem described in Sect. 5.2 would imply $\mathcal{P} = \mathcal{NP}$, i.e., would contradict one of the most famous conjecture in computer science.

We start with the PWA regression problem, where the idea behind the polynomial-time algorithm is easier to motivate, before generalizing it to the arbitrarily switching and bounded-error settings.

Remark 5.3 In the following, we return in notation to real numbers. The complexity results thus obtained can be understood in terms of the number of floating-point operations (flops), which is the standard measure of complexity in numerical analysis. Conversely, the whole section could be written with rational numbers to stick to the definitions of the preceding section.

5.3.1 PWA Regression with Fixed s and d

In the PWA regression Problem 5.3, one must simultaneously estimate a classifier g and a submodel f_j for each mode. While these functions typically include continuous parameters, their optimization can be reduced to a combinatorial search over a finite (and polynomial) number of cases.

To see this, we need to focus on the classifier g. Indeed, given g, the solution to the problem can be easily obtained, as already detailed in Sect. 5.1.3. In addition, whatever the values of its parameters, g only influences the optimization problem through its value at the data points $\boldsymbol{x}_1, \dots, \boldsymbol{x}_N$. As a result, we need not search for its parameter values but merely for its output values, which belong to the finite set $\mathcal{Q} = [s]$.

Yet, the number of possible classifications, s^N, remains exponential in N and too high to allow for a direct enumeration. It might seem that the crux of the approach is thus to reduce this number to a polynomial function of N. However, this is not utterly true: We also need to be able to enumerate all these classifications. The distinction between these two goals will become clear by specifying the shape of g. Let us assume that g belongs to the family $\mathcal{G}$ of binary linear classifiers, as defined in (3.5). Then, Lemma 3.1, well-known in machine learning for almost 50 years, states that the number of linear classifications is bounded by a polynomial function of N (see (3.7)). However, the proof of this result is not constructive and does not yield an enumeration algorithm. For this, we need another equivalence result that was proved much more recently and that basically relates the classification obtained by any $g \in \mathcal{G}$ to one obtained by a separating hyperplane passing through a subset $\mathcal{S}_g$ of d points from $\mathcal{S} = \{\boldsymbol{x}_1, \dots, \boldsymbol{x}_N\} \subset \mathbb{R}^d$. Based on this equivalence, enumerating all possible classifications can be done by enumerating the $\binom{N}{d} = \mathcal{O}(N^d)$ subsets of d points of $\mathcal{S}$.

We leave some technical details aside and simply state the final result, relying on Algorithm 3 for $s = 2$ modes, in which we let $\tilde{\boldsymbol{x}}_k = \begin{bmatrix}\boldsymbol{x}_k^\top & 1\end{bmatrix}^\top$. The case $s > 2$ with $\mathcal{G}$ as in (4.13) is dealt with by searching for the $s(s-1)/2$ pairwise binary classifiers (see Sect. 3.1.2 and (4.14)). Algorithm 3 performs two intertwined loops. As suggested above, the first one (loop A) loops over all subsets $\mathcal{S}_g$ of d points to build the separating hyperplanes. However, as the points in $\mathcal{S}_g$ lie exactly on the hyperplane, their classification is undetermined (points in $\mathcal{S}_g$ are not in $\mathcal{S}_1$ nor in $\mathcal{S}_2$) and a second loop (loop B) ensures that all classifications of these points (into $\mathcal{S}_g^1$ and $\mathcal{S}_g^2$) are tested.

Theorem 5.3 *For any fixed number of modes s and dimension d, if the points* $\{\boldsymbol{x}_k\}_{k=1}^N$ *are in general position, i.e., no hyperplane of* $\mathbb{R}^d$ *contains more than d points, and if (5.19) can be solved in* $\mathcal{O}(N^c)$ *time with a constant* $c \geq 1$ *independent of N, then the time complexity of Problem 5.3 is no more than polynomial in the number of data N and in the order of*

$$\mathcal{O}\left(N^{c+ds(s-1)/2}\right).$$

Algorithm 3 Polynomial-time PWA regression for $s = 2$

Require: A data set $\{(\boldsymbol{x}_k, y_k)\}_{k=1}^N \subset (\mathbb{R}^d \times \mathbb{R})^N$.
  Initialize $\mathcal{S} \leftarrow \{\boldsymbol{x}_k\}_{k=1}^N$ and $J^* \leftarrow +\infty$.
  for all subsets of d points, i.e., for all $\mathcal{S}_g \subset \mathcal{S}$ such that $|\mathcal{S}_g| = d$ **do** ▷ loop A
    Compute the parameters $(\boldsymbol{g}_{\mathcal{S}_g}, b_{\mathcal{S}_g})$ of a hyperplane passing through the points in $\mathcal{S}_g$.[3]
    Classify the data points:

$$\mathcal{S}_1 = \{\boldsymbol{x}_k \in \mathcal{S} \ : \ \boldsymbol{g}_{\mathcal{S}_g}^\top \boldsymbol{x}_k + b_{\mathcal{S}_g} > 0\},$$

$$\mathcal{S}_2 = \{\boldsymbol{x}_k \in \mathcal{S} \ : \ \boldsymbol{g}_{\mathcal{S}_g}^\top \boldsymbol{x}_k + b_{\mathcal{S}_g} < 0\}.$$

    for all classification of $\mathcal{S}_g$ into $\mathcal{S}_g^1$ and $\mathcal{S}_g^2$ **do** ▷ loop B
      Solve the single-mode regression subproblems:

$$\boldsymbol{\theta}_j \in \underset{\boldsymbol{\theta} \in \mathbb{R}^{d+1}}{\operatorname{argmin}} \sum_{\boldsymbol{x}_k \in \mathcal{S}_j \cup \mathcal{S}_g^j} \ell(y_k - \tilde{\boldsymbol{x}}_k^\top \boldsymbol{\theta}), \quad j = 1, 2. \tag{5.19}$$

      Compute the cost function value $J = \frac{1}{N} \sum_{j=1}^{2} \sum_{\boldsymbol{x}_k \in \mathcal{S}_j \cup \mathcal{S}_g^j} \ell(y_k - \tilde{\boldsymbol{x}}_k^\top \boldsymbol{\theta}_j)$,

      and update the best solution as $(J^*, \boldsymbol{\theta}^*, \boldsymbol{g}^*, b^*) \leftarrow (J, \left[\boldsymbol{\theta}_1^\top \ \boldsymbol{\theta}_2^\top\right]^\top, \boldsymbol{g}_{\mathcal{S}_g}, b_{\mathcal{S}_g})$ if $J < J^*$.
    end for
  end for
  return $\boldsymbol{\theta}^*, \boldsymbol{g}^*, b^*$.

For standard (and convex) loss functions, Theorem 5.3 guarantees that the exact solution can be computed in polynomial time with respect to N. For instance, with the squared loss, $\ell(e) = e^2$, (5.19) are least squares problems that can be solved in $\mathcal{O}(d^2 N)$ operations and $c = 1$ in Theorem 5.3. In Algorithm 3, the inner loop over the binary classifications of the d points of $\mathcal{S}_g$ involves 2^d iterations. It is required

---

[3]The normal $\boldsymbol{g}_{\mathcal{S}_g}$ of the hyperplane $\{\boldsymbol{x} : \boldsymbol{g}_{\mathcal{S}_g}^\top \boldsymbol{x} + b_{\mathcal{S}_g} = 0\}$ passing through the d points $\{\boldsymbol{x}_{k_i}\}_{i=1}^d$ of $\mathbb{R}^d$ can be computed in $\mathcal{O}(d^3)$ as a unit vector in the null space of $\left[\boldsymbol{x}_{k_2} - \boldsymbol{x}_{k_1} \ \dots \ \boldsymbol{x}_{k_d} - \boldsymbol{x}_{k_1}\right]^\top$, while the offset is given by $b_{\mathcal{S}_g} = -\boldsymbol{g}_{\mathcal{S}_g}^\top \boldsymbol{x}_{k_i}$ for any of the $\boldsymbol{x}_{k_i}$'s.

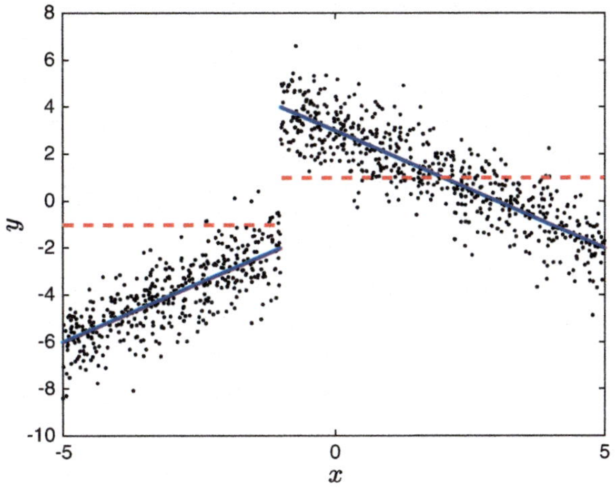

Fig. 5.3 One-dimensional example of binary PWA regression solved exactly by Algorithm 3: $f_{g(x)}(x)$ (plain line) and $g(x)$ (dashed line)

since the equivalence between the classifiers from $\mathcal{G}$ and the separating hyperplanes passing through the points in $\mathcal{S}_g$ holds only up to the classification of these points.

Example 5.1 Figure 5.3 illustrates the idea behind Algorithm 3 on a one-dimensional problem with $N = 1000$ data points satisfying $x_k \in [-5, 5]$. Even for this low-dimensional data, the direct optimization of the switching regression form involves a *four*-dimensional continuous search space for $\boldsymbol{\theta}$. By invoking the PWA assumption, the direct optimization of g is easier, but still involves an infinite number of possible values for its parameters. Meanwhile, the analysis of the linear classification problem tells us that there are only $4N = 4000$ possible classifications (half of which are opposite of the others) and thus that Algorithm 3 is guaranteed to find the global optimum in only 1000 iterations[4] with 2000 evaluations of the cost function (which took about 0.2 seconds on a standard laptop). On the other hand, using the mixed-integer programming approach presented in Sect. 5.4.3 requires solving a problem with 1000 binary variables, which is already intractable (in addition to the fact that this approach was not made to deal with discontinuous PWA maps).

While a number of other techniques could be used to obtain the result in the one-dimensional case similar to a time-series segmentation problem, Algorithm 3 generalizes the idea to higher-dimensional settings where an ordering of the points cannot easily be derived.

[4]Note that for the particular case of $d = 1$, the inner loop over the labelings of $\mathcal{S}_g$ could be avoided to reduce the number of cost function evaluations from $2N$ to N. Assume that the x_k's are sorted and fix the label of $x_k \in \mathcal{S}_g$ to $q_k = +1$ in the kth outer iteration and the ones of the points $x_i < x_k$ to $q_i = -1$. Then, we simply obtain the other labeling with $q_k = -1$ in the next iteration.

5.3.2 Switching Regression with Fixed s and d

We will now see how the results of the preceding section can be generalized to the switching regression setting of Sect. 5.1.1. Since these results are based on linear classification arguments, this might not seem straightforward. Indeed, recall that in PWA regression, the modes are typically assumed to be linearly separable in the regression space $\mathbb{R}^d$, whereas this is not the case in switching regression, where the mode sequence $\{q_k\}_{k=1}^N$ is arbitrary. However, the groups of data pairs $(\boldsymbol{x}_k, y_k)$ associated to different linear models can be "linearly separated" in some sense.

More precisely, the classification rule (5.2) used in switching regression implicitly entails a combination of two linear classifiers: one applying to the points $z_k = \left[\boldsymbol{x}_k^\top\ y_k\right]^\top$ in $\mathbb{R}^{d+1}$ and another one applying to the regression vectors $\boldsymbol{x}_k$ in $\mathbb{R}^d$. The equivalence between (5.2) and these linear classifiers will hold for all points that can be classified without ambiguity, i.e., those with index not in

$$\mathcal{E} = \left\{k \in [N] : \exists (j, i) \in [s]^2,\ j \neq i,\ |y_k - \boldsymbol{x}_k^\top \boldsymbol{\theta}_j| = |y_k - \boldsymbol{x}_k^\top \boldsymbol{\theta}_i|\right\}. \tag{5.20}$$

The cardinality of this set can be bounded as follows.

Lemma 5.1 *Let $\mathcal{E}$ be defined as in (5.20). If the points $\{\boldsymbol{x}_k\}_{k=1}^N$ are in general position, i.e., if no hyperplane of $\mathbb{R}^d$ contains more than d of these points, and if the points $\{z_k = \left[\boldsymbol{x}_k^\top\ y_k\right]^\top\}_{k=1}^N$ are also in general position in $\mathbb{R}^{d+1}$, then $|\mathcal{E}| \leq (2d+1)s(s-1)/2$.*

Proposition 5.1 *Given a set of parameter vectors $\{\boldsymbol{\theta}_j\}_{j=1}^s \subset \mathbb{R}^d$, let $\mathcal{E}$ be defined as in (5.20). Then, for all $k \notin \mathcal{E}$, the classification rule (5.2) with a loss function ℓ such that for all $(e, e') \in \mathbb{R}^2$, $\ell(e) < \ell(e') \Leftrightarrow |e| < |e'|$, is equivalent to the classification rule*

$$q_k = \operatorname*{argmax}_{j \in [s]} \sum_{i=1}^{j-1} \mathbb{1}_{c_{ij}(\boldsymbol{x}_k, y_k) = -1} + \sum_{i=j+1}^{s} \mathbb{1}_{c_{ji}(\boldsymbol{x}_k, y_k) = +1} \tag{5.21}$$

implementing a majority vote over a set of $s(s-1)/2$ pairwise classifiers $\{c_{ji}\}_{1 \leq j < i \leq s}$ of $\mathbb{R}^{d+1}$, where each c_{ji} is a product of binary linear classifiers defined as

$$\forall (\boldsymbol{x}, y) \in \mathbb{R}^d \times \mathbb{R},\ c_{ji}(\boldsymbol{x}, y) = g_{ji}(z) h_{ji}(\boldsymbol{x}),\quad 1 \leq j < i \leq s,$$

with linear classifiers respectively operating in $\mathbb{R}^{d+1}$ and $\mathbb{R}^d$ as

$$\forall z \in \mathbb{R}^{d+1},\ g_{ji}(z) = \operatorname{sign}\left(\left[-\overline{\boldsymbol{\theta}}_{ji}^\top\ 1\right] z\right),\quad 1 \leq j < i \leq s,$$

$$\forall \boldsymbol{x} \in \mathbb{R}^d,\ h_{ji}(\boldsymbol{x}) = \operatorname{sign}(\tilde{\boldsymbol{\theta}}_{ji}^\top \boldsymbol{x}),\quad 1 \leq j < i \leq s,$$

where $\overline{\boldsymbol{\theta}}_{ji} = (\boldsymbol{\theta}_j + \boldsymbol{\theta}_i)/2$ and $\tilde{\boldsymbol{\theta}}_{ji} = \boldsymbol{\theta}_j - \boldsymbol{\theta}_i$.

Proposition 5.1, illustrated by Fig. 5.4, provides a clear connection between switching regression and linear classification. Thanks to it, all the consistent classifi-

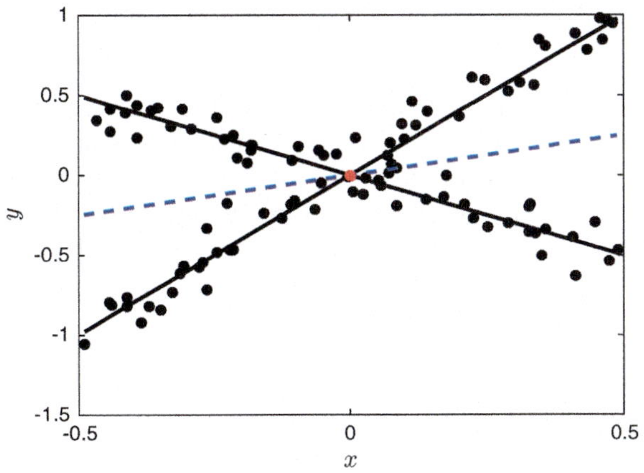

Fig. 5.4 Illustration of Proposition 5.1 for $s = 2$ and $d = 1$: The classification rule (5.2) assigning data points (•) to the linear models (plain lines) is equivalent to a combination of two linear classifiers: g_{12} applying to the z_k's in the (x, y) plane and h_{12} applying to the x_k's. The first one determines whether the point lies "above" or "below" the mean model (dashed line), while the second one indicates which of the two models is "above" or "below." Geometrically, g_{12} corresponds to the hyperplane in the (x, y) plane given by the graph of the mean model. The separating surface of h_{12} in the x axis is the projection onto that axis of the intersection of the two models, here, a single point (•) at the origin. The points with index in $\mathcal{E}$ and excluded from the proposition are those lying exactly on the dashed line

cations of N points for the switching regression problem can be formed by combining the linear classifications of the z_k's and those of the $\boldsymbol{x}_k$'s. According to the discussion of Sect. 5.3.1, for any pair (j, i), there are $\mathcal{O}(N^d)$ possible classifications by a g_{ji} and $\mathcal{O}(N^{d+1})$ possible classifications by a h_{ji}, which make $\mathcal{O}(N^{2d+1})$ possible classifications by a c_{ji}. By Proposition 5.1, there cannot be more classifications $\boldsymbol{q} \in [s]^N$ consistent with (5.2) than possible combinations of the outputs of $s(s-1)/2$ classifiers c_{ji}. Thus, the number of classifications to explore for switching regression is bounded by

$$\mathcal{O}\left(N^{(2d+1)s(s-1)/2}\right) = \mathcal{O}\left(N^{2ds(s-1)}\right).$$

Since Proposition 5.1 also provides a way to explicitly compute these classifications, we can directly apply the strategy depicted in Sect. 5.1.1 and solve (5.3) for all of them to find an exact solution in polynomial time. This reasoning can indeed be made rigorous to yield the following.

Theorem 5.4 *For any fixed number of modes s and dimension d, under the assumptions of Lemma 5.1 and Proposition 5.1, if* (5.3) *can be solved in $\mathcal{O}(N^c)$ time for a constant $c \geq 1$ independent of N, then the time complexity of Problem 5.1 is no more than polynomial in the number of data N and in the order of*

$$\mathcal{O}\left(N^{c+2ds(s-1)}\right).$$

5.3.3 Bounded-Error Estimation with Fixed d

Recall the formulation of the bounded-error estimation Problem 5.2 and its equivalent form given in (5.7). In order to derive a polynomial-time algorithm for this problem, we first need to connect it to linear classification. This can be done by mapping the data to a feature space in which the distinction between points with index in $\mathcal{I}_1(\boldsymbol{\theta})$ (5.6) and the others is given by a linear classifier.

Specifically, given a regression data set $\{(\boldsymbol{x}_k, y_k)\}_{k=1}^N$, we construct the classification data set

$$\mathcal{Z} = \{z_k\}_{k=1}^{2N}, \quad \text{with} \begin{cases} z_k &= \begin{bmatrix} y_k - \epsilon & -\boldsymbol{x}_k^\top \end{bmatrix}^\top \\ z_{k+N} &= \begin{bmatrix} -y_k - \epsilon & \boldsymbol{x}_k^\top \end{bmatrix}^\top \end{cases}, \quad k = 1, \dots, N. \tag{5.22}$$

Then, the connection is formally obtained as follows.

Lemma 5.2 *Given a parameter vector $\boldsymbol{\theta} \in \mathbb{R}^d$, the set $\mathcal{I}_1(\boldsymbol{\theta})$ defined in* (5.6) *is given by*

$$\mathcal{I}_1(\boldsymbol{\theta}) = \{k \in [N] : g(z_k) = g(z_{k+N}) = -1\}, \tag{5.23}$$

where $g(z_k) = \text{sign}(\boldsymbol{g}_\theta^\top z_k)$, $\boldsymbol{g}_\theta = \begin{bmatrix} 1 & \boldsymbol{\theta}^\top \end{bmatrix}^\top$ and the z_k's are as in (5.22).

Proof For any $k \in [N]$, we have $g(z_k) = \text{sign}(\boldsymbol{g}_\theta^\top z_k) = \text{sign}(y_k - \epsilon - \boldsymbol{x}_k^\top \boldsymbol{\theta})$, while $g(z_{k+N}) = \text{sign}(\boldsymbol{g}_\theta^\top z_{k+N}) = \text{sign}(-y_k - \epsilon + \boldsymbol{x}_k^\top \boldsymbol{\theta})$. Thus,

$$|y_k - \boldsymbol{x}_k^\top \boldsymbol{\theta}| < \epsilon \Leftrightarrow \begin{cases} y_k - \epsilon - \boldsymbol{x}_k^\top \boldsymbol{\theta} < 0 \\ -y_k - \epsilon + \boldsymbol{x}_k^\top \boldsymbol{\theta} < 0 \end{cases}$$
$$\Leftrightarrow \begin{cases} g(z_k) = -1 \\ g(z_{k+N}) = -1 \end{cases}$$

and recalling the definition of $\mathcal{I}_1(\boldsymbol{\theta})$ in (5.6) completes the proof.

Lemma 5.2 is illustrated in Fig. 5.5. Equipped with this, we can devise a regression algorithm based on the enumeration of all linear classifications of $\mathcal{Z}$ in (5.22). By following the discussion of Sect. 5.3.1, these classifications can be produced by hyperplanes passing through the origin and d points of $\mathcal{Z}$. Therefore, we construct Algorithm 4 to scan all subsets of d points in a loop (loop A). By taking into account all ambiguities that are due to points lying exactly on the hyperplane in a second loop (loop B), we thus guarantee that all classifications are indeed tested and that the set $\mathcal{I}_1^c$ evaluates to the optimal $\mathcal{I}_1(\boldsymbol{\theta}^*)$ at some iteration.

The exactness and complexity of Algorithm 4 are characterized by Theorem 5.5 below. Note that due to how the set $\mathcal{Z}$ is built, we cannot assume that $\mathcal{Z}$ is in general

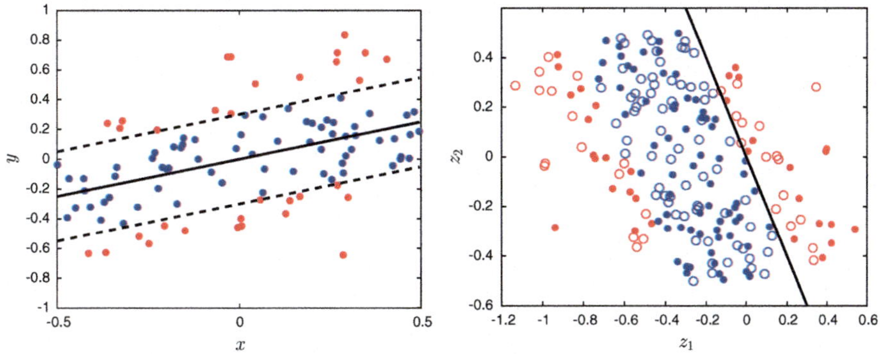

Fig. 5.5 Illustration of Lemma 5.2 for $d = 1$. Left: the blue points (•) are within the error threshold ϵ for the given linear model (plain line) and thus in $\mathcal{I}_1(\boldsymbol{\theta})$, while the red ones (•) are not. Right: the classification data set $\mathcal{Z}$ as in (5.22) plotted with filled disks (• and •) for $k \leq N$ and empty circles (∘ and ∘) for $k > N$. For the blue points of the left plot, both z_k (•) and z_{k+N} (∘) are on the same side of the hyperplane of normal $\boldsymbol{g}_{\boldsymbol{\theta}}$ (plain line in the right plot), whereas for the red points in the left plot, the corresponding z_k (•) and z_{k+N} (∘) lie on different sides of this hyperplane

position and the results of the previous sections do not directly apply.[5] Thus, the (omitted) proof includes a number of technical details and a careful analysis of the maximal number of ambiguous points based on a general position assumption on the original data. In particular, this number must not be proportional to N for the complexity to remain polynomial.

Theorem 5.5 *If the points* $\left\{\left[y_k\ \boldsymbol{x}_k^\top\right]^\top\right\}_{k=1}^N$ *are in general position, i.e., no hyperplane of* $\mathbb{R}^{d+1}$ *contains more than* $d+1$ *of these points, and subproblem* (5.9) *can be solved in* $\mathcal{O}(N^c)$ *time with a constant* $c \geq 1$ *independent of* N*, Algorithm 4 solves Problem 5.2 in*

$$\mathcal{O}(N^{c+d})$$

operations.

For instance, for the saturated ℓ_2-loss ($p = 2$), (5.9) is a simple least squares problem that can be solved in $\mathcal{O}(d^2 N)$ operations and Algorithm 4 has a complexity in the order of $\mathcal{O}(N^{d+1})$.

[5]For any set of $d-1$ points $\boldsymbol{x}_k$, there is a vector $\boldsymbol{\theta}$ such that $\boldsymbol{x}_k^\top \boldsymbol{\theta} = 0$. Thus, in $\mathbb{R}^{d+1}$, there is a hyperplane of normal $\left[0\ \boldsymbol{\theta}^\top\right]^\top$ passing through the origin and the corresponding $2d-2$ points of $\mathcal{Z}$, z_k, and z_{k+N}.

Algorithm 4 Exact $\ell_{p,\epsilon}$-regression

Require: a data set $\{(\boldsymbol{x}_k, y_k)\}_{k=1}^N$, a threshold $\epsilon > 0$.
Initialize $J^* \leftarrow \epsilon^p N$.
for all $\mathcal{S} \subset \mathcal{Z}$ with cardinality $|\mathcal{S}| = d$ **do** ▷ loop A
Compute the normal $\boldsymbol{g}$ to the hyperplane passing through $\mathcal{S} \cup \{\boldsymbol{0}\}$ with orientation such that $g_1 \geq 0$.
if $g_1 \neq 0$ **then**
Classify the points excluding those exactly on the hyperplane:

$$\forall k \in [2N], \quad q_k = \text{sign}_0(\boldsymbol{g}^\top \boldsymbol{z}_k),$$

where $\text{sign}_0(a)$ returns 0 if $a = 0$ and $\text{sign}(a)$ otherwise.
Set $\mathcal{I}_0 = \{k \in [2N] : q_k = 0\}$ and $n = |\mathcal{I}_0|$.
for all classification of the n points in $\mathcal{I}_0$, i.e., for all $\boldsymbol{c} \in \{-1, +1\}^n$ **do** ▷ loop B
Set the entries of $\boldsymbol{q}$ with index in $\mathcal{I}_0$ to $\boldsymbol{c}$.
Compute $\mathcal{I}_1^c = \{k \in [N] : q_k = q_{k+N} = -1\}$.
if $\epsilon^p(N - |\mathcal{I}_1^c|) < J^*$ **then**
Compute $\hat{\boldsymbol{\theta}}$ as in (5.9) with $\mathcal{I}_1^c$ replacing $\mathcal{I}_1(\boldsymbol{\theta}^*)$.
Compute the cost function value $J = \sum_{k=1}^N \ell_{p,\epsilon}(y_k - \boldsymbol{x}_k^\top \hat{\boldsymbol{\theta}})$
and update the best solution as $(J^*, \boldsymbol{\theta}^*) \leftarrow (J, \hat{\boldsymbol{\theta}})$ if $J < J^*$.
end if
end for
end if
end for
return $\boldsymbol{\theta}^*$.

5.4 Global Optimization with Branch-and-Bound

The applicability of the polynomial-time algorithms of the previous section is very limited by the dimension d, which appears as an exponent in their time complexities. In order to tackle slightly larger-dimensional problems, we will resort to the global optimization strategy, focusing on the real parameter vectors and implicitly embedding the discrete classification in the cost function, for instance as (5.2) was embedded in (5.4). Here, the gain in speed will be obtained thanks to heuristics. However, these heuristics are not of the same nature as those described in the next chapters. In particular, the heuristics only make us gain time, they do not incur a loss in accuracy (beyond the small tolerance used in the stopping criterion). Yet, they can be referred to as heuristics since they cannot be proved to yield a significant gain in speed in every circumstance: The average computational time observed in practice is decreased to a reasonable amount while the worst-case computational time remains exponential in the dimension D of the problem (here, D is the number of optimization variables). In particular, when discussing the computational complexity of the approach below, we will focus on the complexity of a single iteration of the algorithm, while the precise number of iterations depends on the data.

Branch-and-bound is a standard strategy for the global optimization of some cost function $J(\boldsymbol{\theta})$ of a vector of parameters $\boldsymbol{\theta} \in \mathbb{R}^D$ over a box $B_{\text{init}} = [\boldsymbol{u}_{\text{init}}, \boldsymbol{v}_{\text{init}}] \subset \mathbb{R}^D$. The main steps are summarized in Algorithm 5. It relies on computing upper and lower bounds ($\overline{J}$ and $\underline{J}$ in Algorithm 5) on the global optimum $\min_{\boldsymbol{\theta} \in B_{\text{init}}} J(\boldsymbol{\theta})$. Then, regions B of the search space in which the local lower bound $\underline{J}(B)$ is larger than the global upper bound $\overline{J}$ can be discarded, reducing the volume left to explore until the relative optimality gap, $(\overline{J} - \underline{J})/\overline{J}$, decreases below a predefined tolerance TOL.

Here, the considered regions are always boxes, i.e., hyperrectangles. Upper bounds $\overline{J}(B)$ can be easily computed by some local optimization or heuristic method for a problem of interest. Alternatively, $\overline{J}(B)$ can be computed merely as the cost function value at the box base point $\boldsymbol{u}$ or at a random point inside the box, while local optimization is only used periodically. On the other hand, lower bounds $\underline{J}(B)$ require a careful derivation, the efficiency of the approach relying mostly on the tightness of these bounds.

Algorithm 5 General branch-and-bound scheme (reprinted from [1], Copyright (2018), with permission from Elsevier).

Require: A data set $\{(\boldsymbol{x}_k, y_k)\}_{k=1}^N \subset \mathbb{R}^d \times \mathbb{R}$, initial box bounds $B_{\text{init}} = [\boldsymbol{u}_{\text{init}}, \boldsymbol{v}_{\text{init}}] \subset \mathbb{R}^D$ and $TOL > 0$. Optionally, an initial guess of $\boldsymbol{\theta} \in B_{\text{init}}$.
Initialize the global bounds $\underline{J} \leftarrow 0$, $\overline{J} \leftarrow +\infty$ or $\overline{J} \leftarrow J(\boldsymbol{\theta})$ if $\boldsymbol{\theta}$ is provided, and the list of boxes $\mathcal{B} \leftarrow \{B_{\text{init}}\}$.
while $(\overline{J} - \underline{J})/\overline{J} > TOL$ **do**
  Split the current box B into B^1 and B^2 such that $B = B^1 \cup B^2$.
  Compute upper bounds $\overline{J}(B^1)$ and $\overline{J}(B^2)$.
  Update $\overline{J} \leftarrow \min\{\overline{J}, \overline{J}(B^1), \overline{J}(B^2)\}$ and the best solution $\boldsymbol{\theta}^*$.
  Compute lower bounds $\underline{J}(B^1)$ and $\underline{J}(B^2)$.
  For $i = 1, 2$, append B^i to the list of active boxes $\mathcal{B}$ if $\underline{J}(B^i) \leq \overline{J}$.
  Remove B from the list of active boxes: $\mathcal{B} \leftarrow \mathcal{B} \setminus \{B\}$.
  Select the next box $B \leftarrow \operatorname{argmin}_{B \in \mathcal{B}} \underline{J}(B)$ and set $\underline{J} \leftarrow \underline{J}(B)$.
end while
return $\boldsymbol{\theta}^*$ and $\overline{J} = J(\boldsymbol{\theta}^*) \approx \min_{\boldsymbol{\theta} \in B_{\text{init}}} J(\boldsymbol{\theta})$.

5.4.1 Switching Regression

Consider the switching linear regression Problem (5.1) with the ℓ_2-loss, i.e., the minimization of a cost function

$$J(\boldsymbol{\theta}) = \frac{1}{N} \sum_{k=1}^{N} \min_{j \in [s]} (y_k - \boldsymbol{x}_k^\top \boldsymbol{\theta}_j)^2 \tag{5.24}$$

with $\boldsymbol{\theta} = \left[\boldsymbol{\theta}_1^\top \ \ldots \ \boldsymbol{\theta}_s^\top\right]^\top \in \mathbb{R}^D$ and $D = sd$. Note that for symmetry reasons, this cost function is invariant to permutations of the subvectors $\boldsymbol{\theta}_j$ in $\boldsymbol{\theta}$; hence, the minimizer is not unique. Such symmetries can be broken by arbitrarily imposing an ordering on the modes, for instance as

$$\theta_{j,1} \leq \theta_{j+1,1}, \quad j = 1, \dots, s-1, \tag{5.25}$$

where $\theta_{j,i}$ denotes the ith component of the jth parameter vector. Note that ties in the case $\theta_{j,1} = \theta_{j+1,1}$ can be broken by imposing similar constraints recursively on the remaining components. However, these additional constraints might be more difficult to take into account in the branch-and-bound approach, while they might also be of little use since the event corresponding to a tie in a global minimizer has zero measure with noisy data.

The symmetry-breaking constraints (5.25) can be simply imposed at the branching level by explicitly discarding regions of subboxes without feasible solutions. More precisely, we compute $B^1 = [\boldsymbol{u}^1, \boldsymbol{v}^1]$ and $B^2 = [\boldsymbol{u}^2, \boldsymbol{v}^2]$ from $B = [\boldsymbol{u}, \boldsymbol{v}]$ by first applying a standard split along the longest side of the box:

$$(j^*, i^*) = \operatorname*{argmax}_{j \in [s], i \in [d]} v_{j,i} - u_{j,i}, \tag{5.26}$$

and

$$\boldsymbol{u}^1 = \boldsymbol{u}, \quad v^1_{j,i} = \begin{cases} (u_{j,i} + v_{j,i})/2, & \text{if } (j,i) = (j^*, i^*), \\ v_{j,i}, & \text{otherwise,} \end{cases} \tag{5.27}$$

$$\boldsymbol{v}^2 = \boldsymbol{v}, \quad u^2_{j,i} = \begin{cases} (u_{j,i} + v_{j,i})/2, & \text{if } (j,i) = (j^*, i^*), \\ u_{j,i}, & \text{otherwise.} \end{cases} \tag{5.28}$$

Then, in the case $i^* = 1$, which is the only one concerned by (5.25), we correct the box bounds recursively for $j = j^* - 1, \dots, 1$ with

$$v^1_{j,1} = \min \left\{ v^1_{j,1}, v^1_{j+1,1} \right\} \tag{5.29}$$

and, for $j = j^* + 1, \dots, s$, with

$$u^2_{j,1} = \max \left\{ u^2_{j,1}, u^2_{j-1,1} \right\}. \tag{5.30}$$

Figure 5.6 illustrates the splitting rule.

5.4.1.1 Lower Bounds

When devising a lower bound, one always faces a trade-off between the speed at which the lower bound can be computed and its tightness. We describe below two lower bounds: one that can be very efficiently computed and one that is more tight.

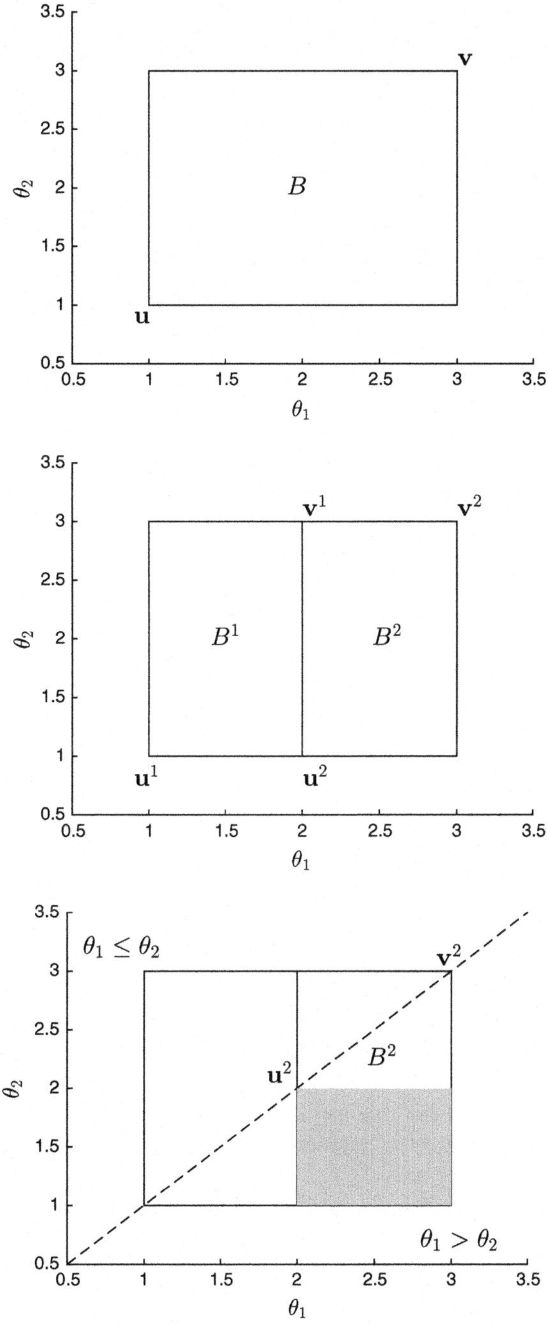

Fig. 5.6 Illustration of the splitting procedure when $s = 2$ and $d = 1$. *Top:* a box $B = [\boldsymbol{u}, \boldsymbol{v}]$ is a rectangular region of the plane of axis (θ_1, θ_2) with bottom-left and top-right corners at $\boldsymbol{u}$ and $\boldsymbol{v}$. *Middle:* B is split into $B^1 = [\boldsymbol{u}^1, \boldsymbol{v}^1]$ and $B^2 = [\boldsymbol{u}^2, \boldsymbol{v}^2]$ by application of (5.26)–(5.28). *Bottom:* B^2 is corrected as in (5.29)–(5.30) to remove the shaded area that does not contain any feasible solution according to (5.25), i.e., $\theta_1 > \theta_2$ for all $\boldsymbol{\theta}$ in the shaded area. Reprinted from [1], Copyright (2018), with permission from Elsevier

The first lower bound is based on a pointwise decomposition of the optimization problem with respect to the index k of data points. In particular, we use the fact that for any $k \in [N]$ and $j \in [s]$, the pointwise error of a parameter vector $\boldsymbol{\theta}_j$ at a given point $(\boldsymbol{x}_k, y_k)$,

$$e_k(\boldsymbol{\theta}_j) = y_k - \boldsymbol{x}_k^\top \boldsymbol{\theta}_j, \tag{5.31}$$

can be made smaller in magnitude if we are not trying to simultaneously minimize the errors at other points. Therefore, the global cost $J(\boldsymbol{\theta})$ must be at least as large as the sum of independently optimized pointwise errors. Formally, for any $k \in [N]$ and box $B_j = [\boldsymbol{u}_j, \boldsymbol{v}_j] \subset \mathbb{R}^d$, let

$$\begin{aligned} e_k^L(B_j) &= e_k(\boldsymbol{u}_j) - (\boldsymbol{v}_j - \boldsymbol{u}_j)^\top (\boldsymbol{x}_k)_- \\ e_k^U(B_j) &= e_k(\boldsymbol{u}_j) - (\boldsymbol{v}_j - \boldsymbol{u}_j)^\top (\boldsymbol{x}_k)_+, \end{aligned} \tag{5.32}$$

where $(\boldsymbol{x}_k)_+$ and $(\boldsymbol{x}_k)_-$ denote the positive and negative parts of $\boldsymbol{x}_k$, computed entrywise. These quantities constitute the bounds within which the error $e_k(\boldsymbol{\theta}_j)$ can be with $\boldsymbol{\theta}_j \in B_j$. Then, we can deduce the minimal value for the squared error over the box B_j as

$$\min_{\boldsymbol{\theta}_j \in B_j} e_k^2(\boldsymbol{\theta}_j) = \left(e_k^U(B_j)\right)_+^2 + \left(e_k^L(B_j)\right)_-^2 . \tag{5.33}$$

By applying a similar scheme to all data points, this gives a lower bound on the optimum over B expressed as

$$\min_{\boldsymbol{\theta} \in B} J(\boldsymbol{\theta}) \geq \underline{J}(B)$$

with

$$\underline{J}(B) = \sum_{k=1}^{N} \min_{j \in [s]} \left\{ \left(e_k^U(B_j)\right)_+^2 + \left(e_k^L(B_j)\right)_-^2 \right\} . \tag{5.34}$$

The next lower bound is based on a constant classification argument. The idea is that for sufficiently small boxes B, the parameter vectors $\boldsymbol{\theta}_j$ can be so constrained that the minimum error for a given data point is always obtained for the same mode. If many points are constantly assigned to mode j, then the cost over all these points is at least as large as the smallest error with which these points can be estimated by a single model, independently of the rest of the problem. Formally, let

$$q_k(\boldsymbol{\theta}) = \operatorname*{argmin}_{j \in [s]} (y_k - \boldsymbol{x}_k^\top \boldsymbol{\theta}_j)^2 \tag{5.35}$$

be the mode estimate for the kth point based on $\boldsymbol{\theta}$, which includes an arbitrary tie-breaking procedure: $q_k(\boldsymbol{\theta})$ is the minimum index of the mode with minimum error. Also define, for all $j \in [s]$,

$$\mathcal{I}_j(B) = \{k \in [N] : \forall \boldsymbol{\theta} \in B,\ q_k(\boldsymbol{\theta}) = j\} \tag{5.36}$$

as the subset of indexes of points assigned to mode j over the entire box B. Then,

$$\forall k \in \mathcal{I}_j(B),\ \forall \boldsymbol{\theta} \in B,\quad \min_{i \in [s]} e_k^2(\boldsymbol{\theta}_i) = e_k^2(\boldsymbol{\theta}_j) \tag{5.37}$$

and

$$\sum_{j=1}^{s} \sum_{k \in \mathcal{I}_j(B)} \min_{i \in [s]} e_k^2(\boldsymbol{\theta}_i) = \sum_{j=1}^{s} \sum_{k \in \mathcal{I}_j(B)} e_k^2(\boldsymbol{\theta}_j). \tag{5.38}$$

By introducing the set $\mathcal{I}_0(B) = [N] \setminus \bigcup_{j=1}^{s} \mathcal{I}_j(B)$ of remaining indexes for which the classification is not constant over B, all indexes $k \in [N]$ are in exactly one of the sets $\mathcal{I}_j(B), j = 0, \dots, s$. Thus, we can express the cost function, for all $\boldsymbol{\theta} \in B$, as

$$\begin{aligned}
J(\boldsymbol{\theta}) &= \sum_{k=1}^{N} \min_{j \in [s]} e_k^2(\boldsymbol{\theta}_j) \\
&= \sum_{j=0}^{s} \sum_{k \in \mathcal{I}_j(B)} \min_{i \in [s]} e_k^2(\boldsymbol{\theta}_i) \\
&= \sum_{k \in \mathcal{I}_0(B)} \min_{i \in [s]} e_k^2(\boldsymbol{\theta}_i) + \sum_{j=1}^{s} \sum_{k \in \mathcal{I}_j(B)} e_k^2(\boldsymbol{\theta}_j) \\
&\geq \min_{\boldsymbol{\theta} \in B} \sum_{k \in \mathcal{I}_0(B)} \min_{i \in [s]} e_k^2(\boldsymbol{\theta}_i) + \sum_{j=1}^{s} \min_{\boldsymbol{\theta}_j \in B_j} \sum_{k \in \mathcal{I}_j(B)} e_k^2(\boldsymbol{\theta}_j) \\
&\geq \sum_{k \in \mathcal{I}_0(B)} \min_{j \in [s]} \left\{ \left(e_k^U(B_j)\right)_+^2 + \left(e_k^L(B_j)\right)_-^2 \right\} + \sum_{j=1}^{s} \min_{\boldsymbol{\theta}_j \in B_j} \sum_{k \in \mathcal{I}_j(B)} e_k^2(\boldsymbol{\theta}_j),
\end{aligned} \tag{5.39}$$

where the last inequality is obtained by lower bounding the first sum in a manner similar to our first lower bound, i.e., using (5.33).

In order to use (5.39) as a lower bound in the branch-and-bound algorithm, we first need to determine the index sets $\mathcal{I}_j(B)$, $j = 1, \dots, s$. This can be done in a computationally efficient manner from the quantities (5.32) as

$$\begin{aligned}
\mathcal{I}_j(B) = \Big\{ k \in [N] : &\max\left\{e_k^U(B_j)^2, e_k^L(B_j)^2\right\} < \min_{i<j} \left(e_k^U(B_i)\right)_+^2 + \left(e_k^L(B_i)\right)_-^2, \\
&\max\left\{e_k^U(B_j)^2, e_k^L(B_j)^2\right\} \leq \min_{i>j} \left(e_k^U(B_i)\right)_+^2 + \left(e_k^L(B_i)\right)_-^2 \Big\}.
\end{aligned} \tag{5.40}$$

We omit the details and simply note that (5.40) considers the specific tie-breaking procedure of (5.35).

5.4.1.2 Overall Procedure

We are now almost ready to apply Algorithm 5 with lower bounds computed as in (5.39). The only remaining part to specify is how we compute the upper bounds. In practice, good results have been observed when using the k-LinReg algorithm of Sect. 6.1.3 every 1000 iterations and merely the cost function value $J(\boldsymbol{u})$ at the box base point $\boldsymbol{u}$ otherwise (see Sect. 5.5). This is computationally very efficient and still offers enough accuracy (remember that Algorithm 5 calls the upper bounding method many times from different initializations).

The computational complexity of the approach is, for each iteration, mostly governed by the computation of the lower bounds. For the one in (5.39), the most intensive task is to solve the box-constrained least squares problem

$$\min_{\boldsymbol{\theta}_j \in B_j} \sum_{k \in \mathcal{I}_j(B)} e_k^2(\boldsymbol{\theta}_j) \tag{5.41}$$

for each $j \in [s]$. Such problems are simple quadratic programs for which very efficient dedicated solvers can be found. In addition, it is not always necessary to solve the problems for all $j \in [s]$. Since all the values of (5.41) computed one by one are accumulated in the lower bound (5.39), we can stop as soon as the partial sum is large enough to exclude the box B from the branch-and-bound search, i.e., when it reaches the global upper bound $\overline{J}$.

5.4.2 Bounded-Error Estimation

We now detail how the branch-and-bound approach above can be adapted to deal with the bounded-error estimation Problem 5.2. Recall that in bounded-error estimation, we estimate a single parameter vector $\boldsymbol{\theta} \in \mathbb{R}^d$. Therefore, all the boxes B are now d-dimensional. We limit the presentation to the case $p = 2$, i.e., the minimization of a sum of saturated squared loss (see Table 2.1),

$$J(\boldsymbol{\theta}) = \sum_{k=1}^{N} \min\left\{(y_k - \boldsymbol{x}_k^\top \boldsymbol{\theta})^2,\ \epsilon^2\right\}, \tag{5.42}$$

though a similar approach could be derived for, e.g., $p = 0$. The idea, to some extent already used in Sect. 5.3.3, is that bounded-error estimation can be seen as a switching regression problem with two modes: one with a linear model approximating the points with error less than ϵ and another one with a constant error of ϵ.

5.4.2.1 Lower Bounds

A pointwise lower bound similar to (5.34) can be derived in the bounded-error estimation setting, leading to

$$\underline{J}(B) = \sum_{k=1}^{N} \min\left\{\left(e_k^U(B)\right)_+^2 + \left(e_k^L(B)\right)_-^2,\ \epsilon^2\right\} \leq \min_{\theta\in B} J(\theta).$$

A second lower bound based on a constant classification argument can also be derived. Here, we classify the points in two groups, those with squared error less than ϵ^2 and the others, and define the index sets

$$\mathcal{I}_1(B) = \{k \in [N] : \forall \theta \in B,\ (y_k - x_k^\top \theta)^2 \leq \epsilon^2\},$$

$$\mathcal{I}_2(B) = \{k \in [N] : \forall \theta \in B,\ (y_k - x_k^\top \theta)^2 > \epsilon^2\}$$

and $\mathcal{I}_0(B) = [N] \setminus (\mathcal{I}_1(B) \cup \mathcal{I}_2(B))$. These sets can be easily computed from the quantities (5.32) as

$$\mathcal{I}_1(B) = \left\{k \in [N] : \max_{\theta\in B} e_k^2(\theta) = \max\left\{e_k^L(B)^2, e_k^U(B)^2\right\} \leq \epsilon^2\right\}$$

and

$$\mathcal{I}_2(B) = \left\{k \in [N] : \min_{\theta\in B} e_k^2(\theta) = \left(e_k^U(B)\right)_+^2 + \left(e_k^L(B)\right)_-^2 > \epsilon^2\right\},$$

where we used (5.33). Clearly, for all points with index in $\mathcal{I}_1(B)$, $\min\left\{e_k^2(\theta),\ \epsilon^2\right\} = e_k^2(\theta)$, while for those with index in $\mathcal{I}_2(B)$, $\min\left\{e_k^2(\theta),\ \epsilon^2\right\} = \epsilon^2$. Thus, the cost function (5.42) can be lower bounded, for all $\theta \in B$, as

$$\begin{aligned} J(\theta) &= \sum_{j=0}^{2} \sum_{k\in\mathcal{I}_j(B)} \min\left\{e_k^2(\theta),\ \epsilon^2\right\} \\ &= \sum_{k\in\mathcal{I}_0(B)} \min\left\{e_k^2(\theta),\ \epsilon^2\right\} + \sum_{k\in\mathcal{I}_1(B)} e_k^2(\theta) + \epsilon^2|\mathcal{I}_2(B)| \\ &\geq \sum_{k\in\mathcal{I}_0(B)} \min_{j\in[s]} \left\{\left(e_k^U(B_j)\right)_+^2 + \left(e_k^L(B_j)\right)_-^2\right\} + \min_{\theta\in B} \sum_{k\in\mathcal{I}_1(B)} e_k^2(\theta) + \epsilon^2|\mathcal{I}_2(B)|, \end{aligned} \tag{5.43}$$

where the inequality is obtained as in (5.39) by using (5.33).

5.4.2.2 Overall Procedure

Algorithm 5 can be applied for bounded-error estimation with lower bounds computed as in (5.43). For the upper bounds, we recommend the computationally efficient heuristic of Sect. 6.2.2 used in combination with the cost function value $J(\boldsymbol{u})$ at the box base point $\boldsymbol{u}$.

As for switching regression, the computational complexity of the approach is, for each iteration, dominated by the (rather low) complexity of solving the box-constrained least squares problem

$$\min_{\boldsymbol{\theta}\in B} \sum_{k\in\mathcal{I}_1(B)} e_k^2(\boldsymbol{\theta}).$$

5.4.3 PWA Regression

Unfortunately, we are not aware of a technique to generalize the methods above to the PWA case. Indeed, the need to estimate the classifier g simultaneously complicates the procedure to the point where it becomes inefficient.

Instead, we here describe a global optimization strategy, also involving branch-and-bound, but with a quite different perspective. This approach does not directly estimate PWA models as presented in Sect. 4.1.3.2, but deals with so-called hinging hyperplanes, which form a subclass of continuous PWA models. In addition, the classification will be encoded in the problem with binary variables, resulting in Mixed-Integer Linear or Quadratic Programming (MILP/MIQP) problems. The interest of this approach is that a vast amount of work has been devoted to MILP/MIQP and generic solvers implementing efficient heuristics can be found. However, these remain highly limited by the number of integer variables, which will be here proportional to the number of data, N. Hence, this approach will be practical only for very small problem sizes, typically with $N < 200$.

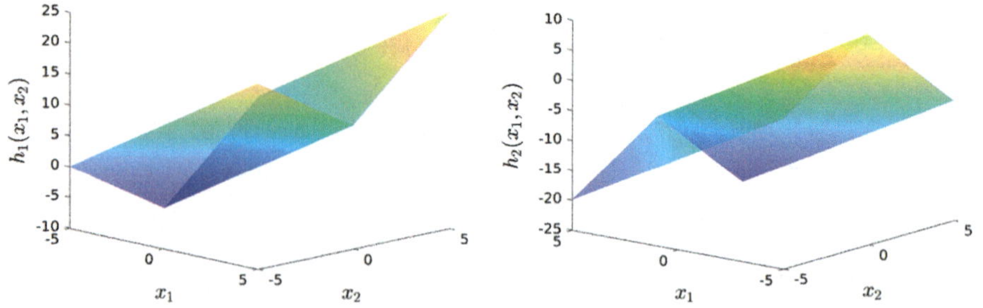

Fig. 5.7 Hinge functions $h_1(\boldsymbol{x}) = \max\{\boldsymbol{x}^\top\boldsymbol{\theta}_1^+, \boldsymbol{x}^\top\boldsymbol{\theta}_1^-\}$ and $h_2(\boldsymbol{x}) = -\max\{\boldsymbol{x}^\top\boldsymbol{\theta}_2^+, \boldsymbol{x}^\top\boldsymbol{\theta}_2^-\}$ for $\boldsymbol{\theta}_1^+ = \boldsymbol{\theta}_2^+ = [2\ 3]^\top$ and $\boldsymbol{\theta}_1^- = \boldsymbol{\theta}_2^- = [2\ -2]^\top$

A hinging hyperplane model,

$$f(\boldsymbol{x}) = \sum_{j=1}^{s'} h_j(\boldsymbol{x}), \tag{5.44}$$

is a sum of hinge functions

$$h_j(\boldsymbol{x}) = \pm \max\{\boldsymbol{x}^\top \boldsymbol{\theta}_j^+, \boldsymbol{x}^\top \boldsymbol{\theta}_j^-\},$$

which consist of two half-hyperplanes of parameter vectors $\boldsymbol{\theta}_j^+$ and $\boldsymbol{\theta}_j^-$. As illustrated in Fig. 5.7, in dimension 2, these functions take the form of an open book. The $\pm$ sign denotes a convex, $\max\{\boldsymbol{x}^\top \boldsymbol{\theta}_j^+, \boldsymbol{x}^\top \boldsymbol{\theta}_j^-\}$, or concave, $-\max\{\boldsymbol{x}^\top \boldsymbol{\theta}_j^+, \boldsymbol{x}^\top \boldsymbol{\theta}_j^-\}$, hinge function. The two hyperplanes, $\{\boldsymbol{x} \in \mathbb{R}^d : \boldsymbol{x}^\top \boldsymbol{\theta}_j^+\}$ and $\{\boldsymbol{x} \in \mathbb{R}^d : \boldsymbol{x}^\top \boldsymbol{\theta}_j^-\}$, are joined together at $\{\boldsymbol{x} \in \mathbb{R}^d : \boldsymbol{x}^\top(\boldsymbol{\theta}_j^+ - \boldsymbol{\theta}_j^-) = 0\}$ and the joint, $\boldsymbol{\theta}_j = \boldsymbol{\theta}_j^+ - \boldsymbol{\theta}_j^-$, is defined as the hinge for the hyperplanes. Note that this model implements a continuous function by construction.

In (5.44), the signs in front of the max operations are assumed to be given or fixed while the parameters $\boldsymbol{\theta}_j^+$ and $\boldsymbol{\theta}_j^-$ are estimated. More precisely, by using the fact that

$$\max\{\boldsymbol{x}^\top \boldsymbol{\theta}_j^+, \boldsymbol{x}^\top \boldsymbol{\theta}_j^-\} = \boldsymbol{x}^\top \boldsymbol{\theta}_j^- + \max\{\boldsymbol{x}^\top(\boldsymbol{\theta}_j^+ - \boldsymbol{\theta}_j^-),\ 0\},$$

we estimate the parameters $\{\boldsymbol{\theta}_j\}_{j=0}^s$ in the reformulation of the model (5.44) as

$$f(\boldsymbol{x}) = \boldsymbol{x}^\top \boldsymbol{\theta}_0 + \sum_{j=1}^{s} \sigma_j \cdot \max\left\{\boldsymbol{x}^\top \boldsymbol{\theta}_j,\ 0\right\}, \tag{5.45}$$

where $\boldsymbol{\theta}_0 = \sum_{j=1}^s \sigma_j \cdot \boldsymbol{\theta}_j^-$ and $\sigma_j \in \{-1, +1\}$ denotes the sign of the jth hinge function h_j. Let $\boldsymbol{\theta} = \left[\boldsymbol{\theta}_0^\top \ \dots \ \boldsymbol{\theta}_s^\top\right]^\top \in \mathbb{R}^D$ with $D = (s+1)d$ and assume that box bounds $\boldsymbol{\theta} \in B = [\boldsymbol{u}, \boldsymbol{v}]$ are available. Then, the problem can be written as

$$\min_{\boldsymbol{\theta} \in B} \frac{1}{N} \sum_{k=1}^{N} \ell\left(y_k - \boldsymbol{x}_k^\top \boldsymbol{\theta}_0 - \sum_{j=1}^{s} \sigma_j \cdot \max\left\{\boldsymbol{x}_k^\top \boldsymbol{\theta}_j,\ 0\right\}\right) \tag{5.46}$$

This problem can be reformulated as a MIQP when ℓ is the ℓ_2-loss function or a MILP when ℓ is the ℓ_1-loss. To see this, introduce, for all $j \in [s]$ and $k \in [N]$, the binary variables $\beta_{jk} \in \{0, 1\}$ encoding whether $\max\left\{\boldsymbol{x}_k^\top \boldsymbol{\theta}_j,\ 0\right\}$ is positive or not, i.e., with

$$\beta_{jk} = \mathbb{1}_{\boldsymbol{x}_k^\top \boldsymbol{\theta}_j > 0}, \tag{5.47}$$

we have

$$z_{jk} = \max\left\{\boldsymbol{x}_k^\top \boldsymbol{\theta}_j,\ 0\right\} = \beta_{jk}\, \boldsymbol{x}_k^\top \boldsymbol{\theta}_j. \tag{5.48}$$

The relations (5.47)–(5.48) together with the constraint $\boldsymbol{\theta}_j \in B_j = [\boldsymbol{u}_j, \boldsymbol{v}_j] \subset \mathbb{R}^d$ are equivalent for $\beta_{jk} \in \{0, 1\}$ to the linear inequalities

$$\begin{cases} 0 \leq z_{jk} \leq U_k(B_j)\beta_{jk} \\ (1-\beta_{jk})L_k(B_j) + z_{jk} \leq \boldsymbol{x}_k^\top \boldsymbol{\theta}_j \leq z_{jk}, \end{cases} \tag{5.49}$$

where

$$L_k(B_j) = \boldsymbol{x}_k^\top \boldsymbol{u}_j + (\boldsymbol{v}_j - \boldsymbol{u}_j)^\top (\boldsymbol{x}_k)_-$$

and

$$U_k(B_j) = \boldsymbol{x}_k^\top \boldsymbol{u}_j + (\boldsymbol{v}_j - \boldsymbol{u}_j)^\top (\boldsymbol{x}_k)_+$$

are lower and upper bounds on $\boldsymbol{x}_k^\top \boldsymbol{\theta}_j$ when $\boldsymbol{\theta}_j \in B_j$. Indeed, when $\boldsymbol{x}_k^\top \boldsymbol{\theta}_j > 0$, the last inequality in (5.49) implies $z_{jk} > 0$ and $z_{jk} \leq U_k(B_j)\beta_{jk}$ yields $\beta_{jk} = 1$, which in turn gives $z_{jk} = \boldsymbol{x}_k^T \boldsymbol{\theta}_j$. On the other hand, when $\boldsymbol{x}_k^\top \boldsymbol{\theta}_j < 0$, the positivity of z_{jk} and the second line of (5.49) imply $\beta_{jk} = 0$ and thus $z_{jk} = 0$. Note that the equivalence does not hold when $\boldsymbol{x}_k^\top \boldsymbol{\theta}_j = 0$, in which case β_{jk} is undetermined in (5.49). However, for all values of β_{jk}, $z_{jk} = 0$ and thus remains consistent with (5.48).

Therefore, Problem (5.46) can be equivalently formulated for the ℓ_2-loss as a MIQP,

$$\begin{aligned} \min_{\boldsymbol{\theta}\in B, \boldsymbol{\beta}\in\{0,1\}^{sN}, \boldsymbol{z}\in\mathbb{R}^{sN}} \ & \frac{1}{N}\sum_{k=1}^{N}\left(y_k - \boldsymbol{x}_k^\top \boldsymbol{\theta}_0 - \sum_{j=1}^{s}\sigma_j z_{jk}\right)^2 \\ \text{s.t. } & 0 \leq z_{jk} \leq U_k(B_j)\beta_{jk}, \quad j = 1, \dots, s, \ k = 1, \dots, N \\ & (1-\beta_{jk})L_k(B_j) + z_{jk} \leq \boldsymbol{x}_k^\top \boldsymbol{\theta}_j \leq z_{jk}, \quad j = 1, \dots, s, \ k = 1, \dots, N, \end{aligned} \tag{5.50}$$

with $\boldsymbol{\beta}$ and $\boldsymbol{z}$ containing all the β_{jk}'s and z_{jk}'s, a quadratic cost function and linear constraints.

Conversely, with the ℓ_1-loss, following the approach depicted in (2.31)–(2.32) yields a MILP:

$$\begin{aligned} \min_{\boldsymbol{\theta}\in B, \boldsymbol{\beta}\in\{0,1\}^{sN}, \boldsymbol{z}\in\mathbb{R}^{sN}, \boldsymbol{\xi}\in\mathbb{R}^N} \ & \frac{1}{N}\sum_{k=1}^{N}\xi_k \\ \text{s.t. } & 0 \leq z_{jk} \leq U_k(B_j)\beta_{jk}, \quad j = 1, \dots, s, \ k = 1, \dots, N, \\ & (1-\beta_{jk})L_k(B_j) + z_{jk} \leq \boldsymbol{x}_k^\top \boldsymbol{\theta}_j \leq z_{jk}, \quad j = 1, \dots, s, \ k = 1, \dots, N, \\ & -\xi_k \leq y_k - \boldsymbol{x}_k^\top \boldsymbol{\theta}_0 - \sum_{j=1}^{s}\sigma_j z_{jk} \leq \xi_k, \quad k = 1, \dots, N, \end{aligned} \tag{5.51}$$

Table 5.1 Average and empirical standard deviation of the computing time for the global optimization of a switching linear model with s modes with different dimensions d and number of data N. Results reprinted from [1], Copyright (2018), with permission from Elsevier

s	d	N	Time (s)
2	2	500	0.1 ± 0.1
		1 000	0.2 ± 0.1
		10 000	0.6 ± 0.4
	3	500	0.8 ± 1.3
		1 000	0.6 ± 0.3
		10 000	2.3 ± 1.5
	4	500	4.1 ± 2.9
		1 000	5.8 ± 6.9
		10 000	11.9 ± 11.9
	5	500	24.0 ± 20.7
		1 000	35.3 ± 29.9
		10 000	66.7 ± 20.1
3	2	500	1.3 ± 0.9
		1 000	1.8 ± 1.4
		10 000	3.8 ± 2.2
	3	500	22.8 ± 23.2
		1 000	50.7 ± 45.3
		10 000	72.4 ± 38.8
	4	500	783 ± 626
		1 000	1404 ± 977
		10 000	2061 ± 1239

with slack variables ξ_k estimating $\left|y_k - \boldsymbol{x}_k^\top \boldsymbol{\theta}_0 - \sum_{j=1}^s \sigma_j z_{jk}\right|$ at the optimum due to the last set of constraints.

Both the MIQP and MILP formulations involve sN binary variables β_{jk}. Therefore, they can only be solved in practice for small numbers of modes and of data, as will be seen in the next section.

5.5 The Need for Approximation Schemes/Heuristics

It stems quite clearly from the preceding sections that exact methods cannot be applied systematically. Whenever the dimension d or the number of modes s grows too large, these become far too expensive computationally.

To get some intuition about these limitations, we perform a few numerical experiments, first with the global optimization strategy of Sect. 5.4.1. Table 5.1 reports the computing time of Algorithm 5 for various number of data and dimensions. For each problem size, we consider the average and empirical standard deviation of the computing time over 10 trials, in which the regression vectors and the true parameter vectors are randomly drawn from a uniform distribution in $[-5, 5]^d$. The results in Table 5.1 show that switching regression problems with up to 10 000 points in dimension 4 can be solved in about one minute on a standard laptop. But, as expected, the computing time quickly increases with the dimension and the number of modes. Yet, these results support the idea that the complexity of this global optimization approach remains reasonably low with respect to the number of data. In particular, Table 5.1 suggests a complexity in N that is less than linear, indicating that the number of data does not critically influences the number of iterations and mostly affects the linear algebra and convex optimization operations.

Given the $\mathcal{NP}$-hardness of the problem, the high complexity with respect to s and d appears hardly overcomable by any exact or global optimization approach. Hence, most of the literature to be reviewed in the next chapters actually focuses on approximations, heuristics, and local optimization.

The same observations can be made regarding PWA regression. Table 5.2 reports the time needed to solve PWA regression problems with $s = 2$ with the polynomial-time algorithm of Sect. 5.3.1 and the mixed-integer quadratic programming approach of Sect. 5.4.3. Note that these times are obtained with parallel implementations running on a computer with 12 processing cores.

Notes

This chapter only provided a brief introduction to computational complexity, mainly sufficient to understand the results dedicated to hybrid system identification. For a full understanding of the technical issues involved, many textbooks are available, among which we recommend [2] for a focus on $\mathcal{NP}$-completeness. Alternatively, an introduction to these concepts with an automatic control perspective is given in [3]. Two major papers set the foundations of this field: [4] identified the first $\mathcal{NP}$-complete problem and [5] established a list of 21 $\mathcal{NP}$-complete problems (including the Partition Problem 5.4), from which reductions are still used to prove the $\mathcal{NP}$-hardness of many problems. Note that models of computation over the reals have also been considered and we refer to [6] for more details on this topic.

The complexity results ($\mathcal{NP}$-hardness and polynomial-time algorithm) for PWA regression were derived in [7], which also formalized the equivalence between linear classifiers and hyperplanes passing through subsets of points. Similar results for switching linear regression and the proof of Theorem 5.1 for the noiseless case are found in [8], while the proof of Theorem 5.2 for the noisy case has not been published elsewhere. The MIN PFS formulation of the general bounded-error problem is due to [9], in which a proof of $\mathcal{NP}$-hardness for MIN PFS can be found. For the greedy approach in which the models are estimated one by one, the $\mathcal{NP}$-hardness of the

Table 5.2 Time (in seconds) and number of least squares solutions (#LS) required to solve Problem 5.3 exactly with the polynomial-time algorithm compared with the time and number of quadratic programs (#QP) required by the MIQP approach. "n/a" appears when the algorithm did not terminate in less than 10 h

d	N	Polynomial-time algorithm		MIQP approach	
		Time	#LS	Time	#QP
1	100	0.03	400	16	5.10^3
	200	0.05	800	95	4.10^4
	500	0.11	2.10^3	n/a	n/a
	1000	0.16	4.10^3	n/a	n/a
	10000	3.6	4.10^4	n/a	n/a
	50000	110	2.10^5	n/a	n/a
2	100	0.5	4.10^4	10.5	3.10^3
	200	0.9	2.10^5	32	8.10^4
	500	5.6	1.10^6	n/a	n/a
	1000	30	4.10^6	n/a	n/a
3	100	8	2.10^6	15	5.10^3
	200	65	2.10^7	81	3.10^4
	500	1536	3.10^8	n/a	n/a
	1000	16870	2.10^9	n/a	n/a
4	50	23	7.10^6	2.2	2.10^3
	100	355	1.10^8	12	4.10^3
	200	6506	2.10^9	55	1.10^4

bounded-error estimation subproblem, i.e., Problem 5.2, is due to [10], while the polynomial-time algorithm is due to [11].

The global optimization methods based on branch-and-bound for switching regression and bounded-error estimation were derived in [1]. Hinging hyperplane models were introduced in [12] and their global optimization set as a MIQP problem as presented in this chapter was proposed in [13]. Other works on the global optimization of switching linear regression models with a mixed-integer programming approach also appeared in the operation research community; see, e.g., [14, 15].

References

1. Lauer, F.: Global optimization for low-dimensional switching linear regression and bounded-error estimation. Automatica **89**, 73–82 (2018)
2. Garey, M.R., Johnson, D.S.: Computers and Intractability: A Guide to the Theory of NP-Completeness. W.H Freeman and Co, New York (1979)
3. Blondel, V.D., Tsitsiklis, J.N.: A survey of computational complexity results in systems and control. Automatica **36**(9), 1249–1274 (2000)

4. Cook, S.A.: The complexity of theorem-proving procedures. In: Proceedings of the 3rd Annual ACM Symposium on Theory of Computing, Shaker Heights, OH, USA, pp. 151–158 (1971)
5. Karp, R.M.: Reducibility among combinatorial problems. Complexity of Computer Computations. The IBM Research Symposia Series, pp. 85–103. Plenum Press/Springer, Berlin (1972)
6. Blum, L., Cucker, F., Shub, M., Smale, S.: Complexity and Real Computation. Springer, Berlin (1998)
7. Lauer, F.: On the complexity of piecewise affine system identification. Automatica **62**, 148–153 (2015)
8. Lauer, F.: On the complexity of switching linear regression. Automatica **74**, 80–83 (2016)
9. Amaldi, E., Mattavelli, M.: The MIN PFS problem and piecewise linear model estimation. Discret. Appl. Math. **118**, 115–143 (2002)
10. Amaldi, E., Kann, V.: The complexity and approximability of finding maximum feasible subsystems of linear relations. Theor. Comput. Sci. **147**(1–2), 181–210 (1995)
11. Lauer, F.: On the exact minimization of saturated loss functions for robust regression and subspace estimation. Pattern Recognit. Lett. **112**, 317–323 (2018)
12. Breiman, L.: Hinging hyperplanes for regression, classification, and function approximation. IEEE Trans. Inf. Theory **39**(3), 999–1013 (1993)
13. Roll, J., Bemporad, A., Ljung, L.: Identification of piecewise affine systems via mixed-integer programming. Automatica **40**(1), 37–50 (2004)
14. Lau, K., Leung, P., Tse, K.: A mathematical programming approach to clusterwise regression model and its extensions. Eur. J. Op. Res. **116**, 640–652 (1999)
15. Carbonneau, R.A., Caporossi, G., Hansen, P.: Globally optimal clusterwise regression by mixed logical-quadratic programming. Eur. J. Op. Res. **212**, 213–222 (2011)

Chapter 6
Estimation of Switched Linear Models

In this chapter, we review various identification methods for switched *linear/affine* hybrid systems. Specifically, we focus on the Input–Output (I/O) model (4.3),

$$y_k = f_{q_k}(\boldsymbol{x}_k) + \varepsilon_k = \boldsymbol{x}_k^\top \boldsymbol{\theta}_{q_k} + \varepsilon_k, \tag{6.1}$$

where $y_k \in \mathbb{R}$ is the output at discrete-time k and $\boldsymbol{x}_k = \left[-y_{k-1} \ \dots \ -y_{k-n_a} \ u_{k-1} \ \dots \ u_{k-n_b}\right]^\top \in \mathbb{R}^d$ is the regression vector of dimension $d = n_a + n_b$ with the past outputs y_{k-i} and inputs u_{k-i} and the model orders n_a and n_b. Here, $q_k \in [s] = \{1, \dots, s\}$ is the discrete state or mode, s is the number of submodels, $\boldsymbol{\theta}_j \in \mathbb{R}^d$, $j = 1, \dots, s$, are the parameter vectors defining each submodel and $\varepsilon_k \in \mathbb{R}$ is a noise term. In the case of affine subsystems, the regression vector $\boldsymbol{x}_k$ is simply replaced by $\tilde{\boldsymbol{x}}_k = \left[\boldsymbol{x}_k^\top \ 1\right]^\top$.

In most of the methods, n_a and n_b are supposed to be known and the following problem is considered:

Problem 6.1 Given a data set $\mathcal{D} = \{(\boldsymbol{x}_k, y_k)\}_{k=1}^N$ generated by a linear switched system, estimate the number of submodels s, the submodel parameter vectors $\boldsymbol{\theta}_j$, $j = 1, \dots, s$, and the switching sequence $\{q_k\}_{k=1}^N$.

The recent history of switched system identification witnessed three major stages in the approach to the problem. The seminal work of Vidal and co-authors on the algebraic method targeted closed-form solutions obtainable in restricted conditions, such as the absence of noise in the data. Most methods developed in the following years aimed at relaxing these conditions. This also led to a relaxation of the goals in terms of the quality (or optimality) of the solution, due to the formulation of the problem as a nonconvex optimization program. Representative examples are the continuous optimization method and the k-LinReg algorithm, which provides a baseline method one can try first with a low computational cost. In the most recent years, a new trend appeared with the aim of circumventing the major drawbacks of nonconvex optimization, namely the lack of guarantees on the optimality of the

F. Lauer and G. Bloch, *Hybrid System Identification*,
Lecture Notes in Control and Information Sciences 478,
https://doi.org/10.1007/978-3-030-00193-3_6

solution. The methods following this new trend are thus based on convex optimization and convex relaxations of nonconvex problems, while the focus is now back on the conditions under which we can expect the algorithm to yield the optimal solution. As these methods often rely on sparsity arguments, either in terms of the errors or of the parameter variations, results from the compressed sensing literature can be used for this purpose. Finally, it seems that these approaches offer a trade-off between high goals and not too restrictive conditions, thus arriving at a midpoint between the two previously taken paths of research.

This chapter presents some approaches at a rather high level and may omit certain details of the algorithmic implementations. Though they might be essential for good practical results, these details do not provide much insights into hybrid system identification. In addition, they are often related to an arbitrary choice of optimization strategy while other choices might lead to similar results for a given optimization problem formulation. For the fully detailed algorithms, we refer the reader to the original papers listed in the bibliographic notes section.

Chapter Outline

The chapter is divided into two main sections: Sect. 6.1 for the case where the number of modes is fixed and Sect. 6.2 for the case where it is not, thus visiting the two sides of the trade-off between the number of modes and the error of the model, discussed in Sect. 4.2.2.

More precisely, Sect. 6.1 describes the following practical methods: the seminal algebraic approach in Sect. 6.1.1, the continuous optimization approach in Sect. 6.1.2, with its optimization issues and choice of loss function for the robustness to outliers, and the block-coordinate descent approach in Sect. 6.1.3 with the k-LinReg algorithm. Then, Sect. 6.2 presents the first method implementing the bounded-error strategy with a greedy approach in Sect. 6.2.1, a block-coordinate descent approach for the bounded-error estimation problem in Sect. 6.2.2, and several sparsity-based approaches in Sect. 6.2.3. The latter includes the error sparsification method and methods based on the sparsification of an overparametrized model associating one parameter vector to each data point.

6.1 Fixed Number of Modes

We start with the description of methods that work with a fixed number of modes, s.

6.1.1 Algebraic Method

The algebraic approach starts from the so-called Hybrid Decoupling Constraint (HDC), a relation which, under the assumption of noiseless data and for a given number s of modes, always holds irrespective of the switching sequence. The HDC is expressed as

$$\forall k, \quad \prod_{j=1}^{s} \left(y_k - \boldsymbol{x}_k^\top \boldsymbol{\theta}_j \right) = 0. \tag{6.2}$$

Indeed, at each instant k, a particular mode $j = q_k$ is active, the corresponding error is zero and so is the product of all the subsystem errors. These polynomial equations decouple the two subproblems of estimating the parameters $\boldsymbol{\theta}_j$ and the switching sequence $\{q_k\}$. Solving these equations for $\boldsymbol{\theta}_j$ allows one to recover the system parameters without having to estimate simultaneously the mode q_k.

This HDC can be rewritten as

$$\forall k, \quad \prod_{j=1}^{s} \left(\boldsymbol{\beta}_j^\top \boldsymbol{z}_k \right) = 0, \tag{6.3}$$

where $\boldsymbol{\beta}_j = \left[1 \ \boldsymbol{\theta}_j^\top \right]^\top \in \mathbb{R}^{d+1}$ and $\boldsymbol{z}_k = \left[y_k \ -\boldsymbol{x}_k^\top \right]^\top \in \mathbb{R}^{d+1}$. Thus, we have a set of N equations corresponding to the N data points in order to determine the model parameter vectors $\boldsymbol{\beta}_j$.

For any vector $\boldsymbol{z} = \left[z_1 \ \dots \ z_i \ \dots \ z_{d+1} \right]^\top$, $P_s(\boldsymbol{z}) = \prod_{j=1}^{s} \left(\boldsymbol{\beta}_j^\top \boldsymbol{z} \right)$ is an homogeneous polynomial of degree s in $d+1$ variables, which can be written in a linearly parametrized form by introducing the Veronese map of degree s, $\boldsymbol{\nu}_s : \mathbb{R}^{d+1} \longrightarrow \mathbb{R}^{M_s}$. This map returns vectors $\boldsymbol{\nu}_s(\boldsymbol{z})$ including all possible monomials of total degree s, i.e., $z_1^{n_1} \dots z_i^{n_i} \dots z_{d+1}^{n_{d+1}}$ with $0 \leq n_i \leq s$ and $\sum_{i=1}^{d+1} n_i = s$, which are in number

$$M_s = \binom{s+d}{s} = \frac{(s+d)!}{s!d!}.$$

Specifically, this leads to

$$P_s(\boldsymbol{z}) = \prod_{j=1}^{s} \left(\boldsymbol{\beta}_j^\top \boldsymbol{z} \right) = \boldsymbol{h}^\top \boldsymbol{\nu}_s(\boldsymbol{z}), \tag{6.4}$$

where $\boldsymbol{h}$ is a vector of coefficients depending on the $\boldsymbol{\beta}_j$'s.

For example, in a case where $s = 2$ and $d = 2$, we have

$$\begin{aligned} P_2(\boldsymbol{z}) &= \prod_{j=1}^{2} \left(\boldsymbol{\beta}_j^\top \boldsymbol{z} \right) = \left(\boldsymbol{\beta}_1^\top \boldsymbol{z} \right) \left(\boldsymbol{\beta}_2^\top \boldsymbol{z} \right) \\ &= (\beta_{11} z_1 + \beta_{12} z_2 + \beta_{13} z_3)(\beta_{21} z_1 + \beta_{22} z_2 + \beta_{23} z_3) \\ &= \beta_{11}\beta_{21} z_1^2 + (\beta_{11}\beta_{22} + \beta_{12}\beta_{21}) z_1 z_2 + (\beta_{11}\beta_{23} + \beta_{13}\beta_{21}) z_1 z_3 \\ &\quad + \beta_{12}\beta_{22} z_2^2 + (\beta_{12}\beta_{23} + \beta_{13}\beta_{22}) z_2 z_3 + \beta_{13}\beta_{23} z_3^2. \end{aligned}$$

Then, $M_2 = 6$,

$$\boldsymbol{h} = \begin{bmatrix} \beta_{11}\beta_{21} \\ \beta_{11}\beta_{22} + \beta_{12}\beta_{21} \\ \beta_{11}\beta_{23} + \beta_{13}\beta_{21} \\ \beta_{12}\beta_{22} \\ \beta_{12}\beta_{23} + \beta_{13}\beta_{22} \\ \beta_{13}\beta_{23} \end{bmatrix} \quad \text{and} \quad \boldsymbol{v}_2(\boldsymbol{z}) = \begin{bmatrix} z_1^2 \\ z_1 z_2 \\ z_1 z_3 \\ z_2^2 \\ z_2 z_3 \\ z_3^2 \end{bmatrix}.$$

Given the linear form (6.4), the vector $\boldsymbol{h} \in \mathbb{R}^{M_s}$ can be determined from the data by solving the HDC (6.3), which now amounts to the system of linear equations

$$\begin{bmatrix} \boldsymbol{v}_s(\boldsymbol{z}_1)^\top \\ \boldsymbol{v}_s(\boldsymbol{z}_2)^\top \\ \vdots \\ \boldsymbol{v}_s(\boldsymbol{z}_N)^\top \end{bmatrix} \boldsymbol{h} = \boldsymbol{L}_s \boldsymbol{h} = \boldsymbol{0}, \tag{6.5}$$

with $\boldsymbol{L}_s \in \mathbb{R}^{N \times M_s}$. Then, from the gradient of $P_s(\boldsymbol{z})$ with respect to $\boldsymbol{z}$,

$$\nabla P_s(\boldsymbol{z}) = \frac{\partial P_s(\boldsymbol{z})}{\partial \boldsymbol{z}} = \sum_{j=1}^{s} \left(\prod_{l \neq j} \boldsymbol{\beta}_l^\top \boldsymbol{z} \right) \boldsymbol{\beta}_j, \tag{6.6}$$

the model parameter vectors $\boldsymbol{\beta}_j$ can be calculated as

$$\boldsymbol{\beta}_j = \frac{\nabla P_s(\boldsymbol{z}_j^*)}{\boldsymbol{a}_d^\top \nabla P_s(\boldsymbol{z}_j^*)}, \tag{6.7}$$

where $\boldsymbol{a}_d = \begin{bmatrix} 1 \; 0 \dots 0 \end{bmatrix}^\top \in \mathbb{R}^{d+1}$ (recall that $\beta_{j1} = 1$) and $\boldsymbol{z}_j^*$ is a point belonging to the mode j.

For example, in the case where $s = 2$, we have

$$\nabla P_2(\boldsymbol{z}) = (\boldsymbol{\beta}_2^\top \boldsymbol{z})\boldsymbol{\beta}_1 + (\boldsymbol{\beta}_1^\top \boldsymbol{z})\boldsymbol{\beta}_2.$$

Then, for a point $\boldsymbol{z}_1^*$ of the 1st mode, $\boldsymbol{\beta}_1^\top \boldsymbol{z}_1^* = 0$ and

$$\boldsymbol{\beta}_1 = \frac{\nabla P_2(\boldsymbol{z}_1^*)}{\boldsymbol{a}_d^\top \nabla P_2(\boldsymbol{z}_1^*)} = \frac{(\boldsymbol{\beta}_2^\top \boldsymbol{z}_1^*)\boldsymbol{\beta}_1}{\boldsymbol{\beta}_2^\top \boldsymbol{z}_1^*}.$$

To determine s parameter vectors by (6.7), one needs a point $\boldsymbol{z}_j^*$ for each mode, which is not a trivial problem in itself. However, such points can be chosen at the intersections of a line $\{\boldsymbol{z} : \boldsymbol{z} = \boldsymbol{z}_0 + \alpha \boldsymbol{z}', \ \alpha \in \mathbb{R}\}$ with all the hyperplanes $\{\boldsymbol{z} : \boldsymbol{\beta}_j^\top \boldsymbol{z} = 0\}$, $j = 1, \dots, s$. Constraining the points to lie on a line allows us to express them with a single variable α whose values can be determined by solving the polynomial equation $P_s(z) = P_s(z_0 + \alpha z') = 0$ encoding the intersection with the hyperplanes. Specifically, for any $z_0 \neq \mathbf{0}$ and $z' \neq \gamma z_0$ such that $\boldsymbol{\beta}_j^\top z' \neq 0$, $j = 1, \ldots, s$, in order for the line to intersect with all hyperplanes, the points of interest are given by $z_j^* = z_0 + \alpha_j z'$, where the α_j's are the s roots of the polynomial $P_s(z_0 + \alpha z')$ in α. However, note that finding these roots can become highly complex when the degree of the polynomial, here the number of modes s, is larger than 4.

After the estimation of the parameters, the switching sequence $\{q_k\}_{k=1}^N$ is estimated by (4.20), i.e.,

$$\hat{q}_k = \underset{j \in [s]}{\operatorname{argmin}} \left(y_k - \boldsymbol{x}_k^\top \boldsymbol{\theta}_j\right)^2 = \underset{j \in [s]}{\operatorname{argmin}} \left(\boldsymbol{\beta}_j^\top z_k\right)^2 .$$

The approach provides the exact solution in the noiseless case where the HDC (6.2) is valid. Note that, in this case, a rank condition allows us to determine the number of submodels s. If $\boldsymbol{L}_i \in \mathbb{R}^{n \times M_i}$ is a matrix built as $\boldsymbol{L}_s$ in (6.5), but with $s = i$ and only a large enough number of points n, then s can be estimated as the smallest i such that $\operatorname{rank}(L_i) = M_i - 1$. And in the case of noisy data, s is estimated by inspecting the singular values of the (full rank) matrix $\boldsymbol{L}_i$. As presented above, the approach assumes that the orders n_a and n_b of the ARX submodels are known, an assumption that can, however, be relaxed. Moreover, the method can be extended to MISO and MIMO systems. Note finally that a recursive version of the algebraic procedure has been proposed. It is presented in Chap. 8.

The algorithmic complexity of the algebraic approach is dominated by the computation of a solution to the homogeneous system (6.5), i.e., a nonzero vector $\boldsymbol{h}$ in the null space of $\boldsymbol{L}_s$. Such a vector can be taken as a right-singular vector of $\boldsymbol{L}_s$ associated with a zero singular value (see Appendix B.2.8). In the presence of noise, the system $\boldsymbol{L}_s \boldsymbol{h} = \mathbf{0}$ has no solution $\boldsymbol{h} \neq \mathbf{0}$ and it is only solved in a least squares sense as

$$\min_{\|\boldsymbol{h}\|_2 = 1} \|\boldsymbol{L}_s \boldsymbol{h}\|_2,$$

again via the computation of the singular value decomposition of $\boldsymbol{L}_s$. This makes the method computationally very efficient with respect to the number of data N. However, the complexity increases with the dimension M_s and thus quickly with the number of modes s and the dimension d.

6.1.2 Continuous Optimization Approach

Recall that the switched system identification problem can be formulated in very general terms as the mixed-integer programming problem (4.19), where both the classification and regression subproblems are solved simultaneously. In order to deal with numerous data while maintaining the dimension of the problem as low as possible and avoiding integer variables, the Minimum-of-Errors (ME) reformulation (4.21) can be considered. For linear submodels, this reads

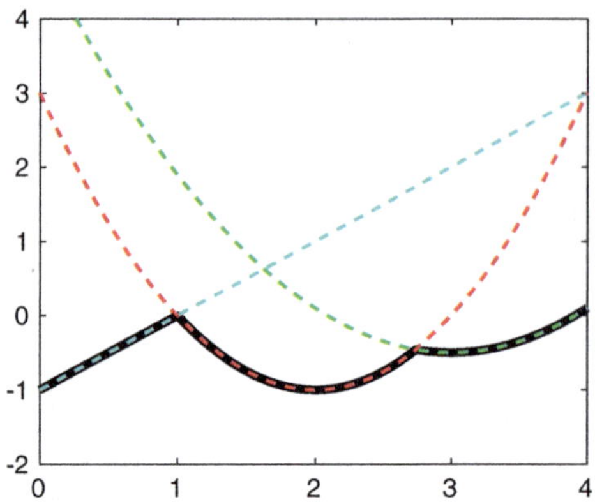

Fig. 6.1 Illustration of the continuity of the pointwise minimum (thick black line) of several continuous functions (dashed color lines)

$$\min_{\boldsymbol{\theta}\in\mathbb{R}^{sd}} J^{ME}(\boldsymbol{\theta}), \tag{6.8}$$

$$\text{with} \quad J^{ME}(\boldsymbol{\theta}) = \frac{1}{N}\sum_{k=1}^{N} \min_{j\in[s]} \ell(y_k - \boldsymbol{x}_k^\top \boldsymbol{\theta}_j),$$

with the vector of variables $\boldsymbol{\theta} = \left[\boldsymbol{\theta}_1^\top \ \dots \ \boldsymbol{\theta}_s^\top\right]^\top$ containing all the parameter vectors $\boldsymbol{\theta}_j$. Problem (6.8) is a nonconvex but continuous optimization problem with a continuous objective function of real variables. To see this, note that the cost function can be rewritten as

$$J^{ME}(\boldsymbol{\theta}) = \frac{1}{N}\sum_{k=1}^{N} \ell_{\min}(\boldsymbol{x}_k, y_k, \boldsymbol{\theta}),$$

where the loss function $\ell_{\min}$ is defined as the pointwise minimum of a set of s loss functions $\ell_j(\boldsymbol{x}_k, y_k, \boldsymbol{\theta}) = \ell(y_k - \boldsymbol{x}_k^\top\boldsymbol{\theta}_j)$, $j = 1, \dots, s$. Since the pointwise minimum of a set of continuous functions is continuous (see Fig. 6.1), J^{ME} is continuous (though nondifferentiable) with respect to the variables $\boldsymbol{\theta}$.

Though continuous, problem (6.8) is nonsmooth, which damages the convergence of many local optimization algorithms that one could apply otherwise. To circumvent this issue, a differentiable alternative to (6.8), known as the poduct-of-erors (PE), can be considered. In this formulation, the nonsmooth pointwise minimum is replaced with a product of errors:

$$\min_{\boldsymbol{\theta}\in\mathbb{R}^{sd}} J^{PE}(\boldsymbol{\theta}), \tag{6.9}$$

$$\text{with} \quad J^{PE}(\boldsymbol{\theta}) = \frac{1}{N}\sum_{k=1}^{N}\prod_{j=1}^{s} \ell\left(y_k - \boldsymbol{x}_k^\top\boldsymbol{\theta}_j\right).$$

In the noiseless case, both problems (ME and PE) share the same global minimum (zero) and minimizers. However, in noisy cases, the PE formulation yields less accurate estimates. This issue is intrinsically implied by the use of product terms as with the algebraic approach (see below for a discussion on the connections between the two approaches). However, experiments indicate that the PE approach can be more robust to noise than the algebraic method. This is also partly due to the possibility of explicitly choosing the loss function ℓ in the PE framework, as elaborated on below.

The number of variables involved in (6.8) and (6.9) is small and fixed to the number of model parameters, $s \times d$, for any number of data N. This is the major advantage of the ME/PE formulations over the MINLP formulation in (4.19): Generic solvers for continuous optimization problems can be applied to find a satisfactory solution in reasonable time, despite the $\mathcal{NP}$-hard nature of the problem. The trick is that the dimension d is typically small for linear dynamical subsystems, since it corresponds to the sum of the input and output orders: $d = n_a + n_b$. Thus, a number of iterations exponential in sd can still be reasonable when compared with the exponential complexity in the number of data N for problem (4.19). Note, however, that the cost functions in (6.8) and (6.9) require to compute a sum over N terms, hence leading to a linear complexity with respect to N.

6.1.2.1 Optimization Issues

Problems (6.8) and (6.9) are continuous optimization problems, but still nonconvex ones which may have many local minima. The multilevel coordinate search (MCS) algorithm is a generic solver for black-box global optimization of unconstrained continuous objective functions that can be used to solve these nonconvex optimization problems. Here, "black-box" refers to the fact that the objective function is only considered through sampled values and that no information on the derivatives is used (though derivatives are internally estimated by the algorithm from the function values). The MCS algorithm combines global and local optimization steps, where the global component ensures that the search space is sufficiently explored and the local component accelerates the convergence once a point in the basin of attraction of the solution is found. Of course, though convergence toward the global optimum can be guaranteed in a finite number of steps, the precise number of steps required for a particular problem instance is unknown and can be quite large. In practice, the algorithm stops after a predetermined number of iterations or cost function evaluations and returns a local minimizer with no certificate of global optimality. In the end, the quality of the estimation ultimately depends on the amount of time allocated to the algorithm.

The following Sect. 6.1.3 discusses another algorithm for the optimization of the ME objective function in (6.8). This algorithm will instead employ a local-only strategy in order to maximize the computational efficiency.

6.1.2.2 Loss Functions and Robustness to Outliers

The formulation in (6.8) is very general in that it allows for different loss functions ℓ. The choice typically boils down to a trade-off between loss function convexity and robustness to outliers. Convexity of the loss function is not a major issue, since the overall objective functions, (6.8) and (6.9), are intrinsically nonconvex. But nonconvex loss functions might still add a number of local minima to these objective functions. On the other hand, robustness to outliers can be a desired feature even for outlier-free data sets. In particular, for the PE approach, it also serves as a means to reject points of other modes. Indeed, in the PE cost function (6.9), though a small noise term corresponds to a small error term with respect to the correct submodel, it is always subject to multiplication by large error terms with respect to the submodels of the other modes. This multiplicative effect makes the PE estimator less accurate than the ME estimator in large noise regimes. However, this effect can be lessen by using a robust loss function which bounds the influence of a single data point.

6.1.2.3 Relationship with the Algebraic Approach

For the squared loss function (2.28), the objective function J^{PE} can be rewritten as

$$J^{PE}(\boldsymbol{\theta}) = \frac{1}{N}\sum_{k=1}^{N}\prod_{j=1}^{s}(y_k - \boldsymbol{x}_k^\top \boldsymbol{\theta}_j)^2.$$

In the noiseless case, the global optimum of this objective function is zero and, since all terms in the products are positive, it coincides with the solution of the following system of equations:

$$\prod_{j=1}^{s}(y_k - \boldsymbol{x}_k^\top \boldsymbol{\theta}_j) = 0, \quad k = 1, \dots, N. \tag{6.10}$$

This system is very close to the one solved in the algebraic approach of Sect. 6.1.1, where each row is a hybrid decoupling polynomial. Therefore, the PE estimator can be seen as a generalization of the algebraic approach to other loss functions, which may include robust ones as discussed above.

Another slight difference is that the algebraic approach focuses on the linear system obtained by expanding the products and a change of variables, whereas the PE approach directly minimizes the product with respect to the original parameters of the model. As a result, the algebraic method can be seen as a relaxation of the nonlinear (and nonconvex) optimization problem (6.9). To see where the relaxation takes place, note that the algebraic method does not constrain the solution in the variables $\boldsymbol{h}$ to be an expansion of the original product. In this respect, the estimated variables $\boldsymbol{h}$ may not lead to a factored form as in (6.10).

6.1.3 Block-Coordinate Descent Approach

We now turn to a direct approach to the minimization of the mixed-integer programming problem (4.19) in the case of a squared loss function (2.28) and linear submodels f_j:

$$\min_{\{\theta_j \in \mathbb{R}^d\}_{j=1}^s, \{\beta_{jk}\}} \frac{1}{N} \sum_{k=1}^{N} \sum_{j=1}^{s} \beta_{jk} \left(y_k - \boldsymbol{x}_k^\top \boldsymbol{\theta}_j\right)^2 \tag{6.11}$$

$$\text{s.t.} \quad \beta_{jk} \in \{0, 1\}, \; j = 1, \ldots, s, \; k = 1, \ldots, N,$$

$$\sum_{j=1}^{s} \beta_{jk} = 1, \; k = 1, \ldots, N,$$

where the β_{jk} are binary variables encoding the assignment of the kth point to the jth submodel.

In particular, the k-LinReg algorithm builds on the relationship between problem (6.11) and the classical clustering problem described in Sect. 3.2. These problems share the common difficulty of simultaneously computing a classification of the data points and a model of each group of points (the group centers in clustering). In the clustering literature, the baseline method typically used to solve such problems is the k-means algorithm, which alternates between assignments of data points to groups and updates of the model parameters. Applying a similar strategy in the context of switched regression leads to the k-LinReg algorithm, which is depicted in Algorithm 6, where we let $\boldsymbol{X} = \begin{bmatrix}\boldsymbol{x}_1 \ldots \boldsymbol{x}_N\end{bmatrix}^\top$ be the regression matrix and $\boldsymbol{y} = \begin{bmatrix}y_1 \ldots y_N\end{bmatrix}^\top$ be the target output vector.

When the number of points assigned to a mode becomes too small, Algorithm 6 simply returns the current estimate. This usually indicates that the algorithm converges to an unsatisfactory solution. However, refined strategies based on randomly drawing a new value for the defective parameter vector or dropping it to obtain a model with fewer modes could also be considered.

With an optimization viewpoint, Algorithm 6 can be interpreted as a block-coordinate descent algorithm, where the cost function in (6.11) is alternatively optimized over two sets of variables: the β_{jk}'s and the $\boldsymbol{\theta}_j$'s. Such strategies can be shown to converge toward a local minimum. Specifically, by considering the equivalent formulation in (6.8), we have the following result.

Proposition 6.1 *Algorithm 6 monotonically decreases the cost function* $J^{ME}(\boldsymbol{\theta}^i)$ *in* (6.8), *in the sense that* $\forall i \geq 0, \; J^{ME}(\boldsymbol{\theta}^{i+1}) \leq J^{ME}(\boldsymbol{\theta}^i)$.

While convergence toward a local optimum cannot be used to guarantee the accuracy of the resulting model, the main strength of this approach lies at another level. Indeed, all the computations in Algorithm 6 are simple ones that can be efficiently performed for large data sets in large dimensions. Drawing on this efficiency, a typical strategy to improve the accuracy is to restart the algorithm from many different initializations and retain the best model, i.e., the one leading to the smallest cost function value in (6.8).

Algorithm 6 k-LinReg

Require: the data set $(\boldsymbol{X}, \boldsymbol{y}) \in \mathbb{R}^{N\times d} \times \mathbb{R}^N$, the number of modes s and an initial vector $\boldsymbol{\theta}^0 = \begin{bmatrix}\boldsymbol{\theta}_1^{0\top} \dots \boldsymbol{\theta}_s^{0\top}\end{bmatrix}^\top$.

Initialize $i \leftarrow 0$.

repeat

Classify the data points according to

$$q_k^i = \operatorname*{argmin}_{j\in[s]}(y_k - \boldsymbol{x}_k^\top \boldsymbol{\theta}_j^i)^2, \qquad k = 1, \dots, N. \tag{6.12}$$

for $j = 1$ to s **do**

if $|\{k \,:\, q_k^i = j\}| < d$ **then return** $\hat{\boldsymbol{\theta}} = \boldsymbol{\theta}^i$ and $J^{ME}(\hat{\boldsymbol{\theta}})$.

Build the matrix $\boldsymbol{X}_j^i$, containing all the kth rows of $\boldsymbol{X}$ for which $q_k^i = j$, and the target vector $\boldsymbol{y}_j^i$ with the corresponding components from $\boldsymbol{y}$.

Update the model parameters for mode j with

$$\boldsymbol{\theta}_j^{i+1} = \operatorname*{argmin}_{\boldsymbol{\theta}_j\in\mathbb{R}^d} \|\boldsymbol{y}_j^i - \boldsymbol{X}_j^i\boldsymbol{\theta}_j\|_2^2. \tag{6.13}$$

As seen in Sect. 2.1.3.2, the solution to this least squares problem is given if $\boldsymbol{X}_j^i$ is full rank by

$$\boldsymbol{\theta}_j^{i+1} = (\boldsymbol{X}_j^{i\top}\boldsymbol{X}_j^i)^{-1}\boldsymbol{X}_j^{i\top}\boldsymbol{y}_j^i.$$

end for

Increase the counter $i \leftarrow i + 1$.

until convergence, e.g., until

$$\|\boldsymbol{\theta}^{i+1} - \boldsymbol{\theta}^i\|_2 \leq \epsilon,$$

or no more changes occur in the classification.

return $\hat{\boldsymbol{\theta}} = \boldsymbol{\theta}^{i+1}$ and $J^{ME}(\hat{\boldsymbol{\theta}})$.

The global convergence of the strategy based on multiple restarts is hard to theoretically analyze. However, a large set of experiments with random data sets might help us to detect recurrent patterns and grasp some of the convergence properties of the approach. Such an empirical analysis led to the results in Fig. 6.2. From these, the optimal number of restarts required to obtain the global solution with probability at least $1 - \delta$ could be estimated as

$$\left\lceil \frac{\log\delta}{\log\left(1 - \hat{P}_{success}(s, d, N)\right)} \right\rceil,$$

where $\hat{P}_{success}(s, d, N)$ is an estimate of the probability of finding the global optimum with a single run of Algorithm 6 given by

$$\hat{P}_{success}(s, d, N) = (1.02 - 0.023s)\left(1 - \exp\left(\frac{-(N - 1.93\times 2^s d - 37)}{52\sqrt{2^s d} - 220}\right)\right), \tag{6.14}$$

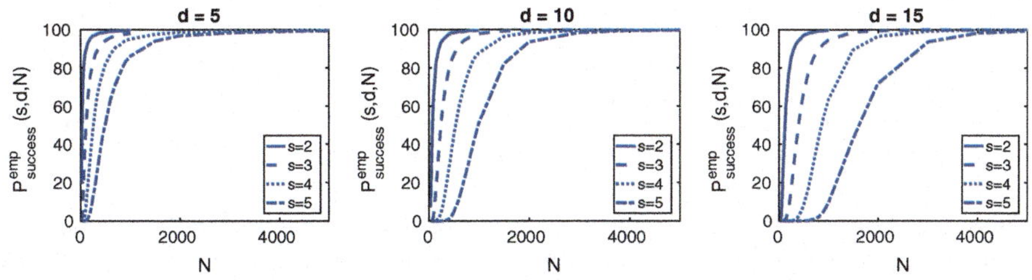

Fig. 6.2 Probability of success of the k-LinReg Algorithm 6 with a random initialization as empirically estimated over a random sample of data sets. Each plot corresponds to a different dimension d, and each curve corresponds to a different number of modes s

if $N > 1.93 \times 2^s d - 37$ and $\hat{P}_{success}(s, d, N) = 0$ otherwise.

The formula (6.14) also suggests that with a large enough data set, a single run of Algorithm 6 should find the global optimum with high probability.

6.2 Free Number of Modes

The methods described below follow the bounded-error viewpoint of Sect. 4.2.4 and aim at finding a hybrid model with sufficient accuracy on the given data rather than one with a given number of modes.

6.2.1 Bounded-Error Approach

The bounded-error approach aims at finding a hybrid model f with the minimal number $\hat{s}$ of submodels that allows the error on all the training data, $(\boldsymbol{x}_k, y_k)$, to be bounded by a predefined threshold ϵ, i.e.,

$$|y_k - f(\boldsymbol{x}_k)| \leq \epsilon, \quad k = 1, \dots, N. \tag{6.15}$$

For linear submodels, these constraints can be rewritten as a set of N *linear complementary inequalities*,

$$\begin{cases} \boldsymbol{x}_k^\top \boldsymbol{\theta}_{q_k} \leq y_k + \epsilon \\ \boldsymbol{x}_k^\top \boldsymbol{\theta}_{q_k} \geq y_k - \epsilon \end{cases}, \quad k = 1, \dots, N, \tag{6.16}$$

with respect to the parameter vectors $\boldsymbol{\theta}_j$. Here, *complementary* means that a pair of inequalities has to be satisfied simultaneously in order to satisfy one inequality of (6.15).

Consider now this system of inequalities expressed with respect to a single linear model, i.e.,

$$\begin{cases} \boldsymbol{x}_k^\top \boldsymbol{\theta} \leq y_k + \epsilon \\ \boldsymbol{x}_k^\top \boldsymbol{\theta} \geq y_k - \epsilon \end{cases}, \quad k = 1, \ldots, N. \tag{6.17}$$

Then, the problem of estimating both the number of submodels and their parameter vectors can be formulated as the problem of finding, from the (a priori infeasible) system of linear complementary inequalities (6.17), a partition in a minimum number of feasible subsystems (the MIN PFS problem). This MIN PFS problem is not directly tractable, and a standard approach tackles it via a greedy procedure based on iteratively solving a sequence of Maximum Feasible Subsystem (MAX FS) problems to estimate the submodels one by one. The basic idea, summarized in Algorithm 7, is to look for the maximum number of inequalities (and thus data points) in (6.15) that can be satisfied by a single linear submodel (the MAX FS problem). Then, these inequalities are removed and a new MAX FS iteration operates on the set of remaining points, while considering a new submodel. The algorithm terminates in a finite number of steps when all the inequalities are satisfied.

The hyperparameter ϵ can be interpreted as a regularization parameter, i.e., ϵ is used to tune the trade-off between the model complexity (the number of modes $\hat{s}$) and the accuracy. However, if prior knowledge on s is available or a particular hybrid structure is desired, the tuning of the error threshold ϵ may become intricate to obtain $\hat{s} = s$.

The MAX FS subproblem described above is a modification of a standard MAX FS problem dealing with (noncomplementary) linear inequalities. However, one must keep in mind that this subproblem is of combinatorial nature: if at iteration j, there are $|\sum_j|$ remaining inequalities, the feasibility of all the possible subsets of size $|\sum_j| - n$ must be tested for an increasing sequence of n until one subset is found to yield a feasible system. It can indeed be shown to be $\mathcal{NP}$-hard, making the overall procedure in need for approximation methods.

Note that Algorithm 7 can be slightly modified in order to take this suboptimality into account. In particular, each parameter vector estimated, $\hat{\boldsymbol{\theta}}_j$, is tested against all previous subsystems $\sum_n$, $n = 1, \ldots, j-1$. If, for a given n, the number of satisfied

---

Algorithm 7 Bounded-error approach

Require: the data set $\{(\boldsymbol{x}_k, y_k)\}_{k=1}^N$ and the error tolerance ϵ.
  Initialize $j \leftarrow 0$ and $I = \{1, \ldots, N\}$.
  repeat
    $j \leftarrow j + 1$.
    Estimate $\hat{\boldsymbol{\theta}}_j$ by solving the MAX FS problem for the system of complementary inequalities $\sum_j = \{|y_k - \boldsymbol{x}_k^\top \boldsymbol{\theta}| \leq \epsilon \ : \ k \in I\}$.
    Let $G_j = \{k \in I \ : \ |y_k - \boldsymbol{x}_k^\top \hat{\boldsymbol{\theta}}_j| \leq \epsilon\}$ and $I \leftarrow I \setminus G_j$.
  until $I = \emptyset$.
  return the number of modes $\hat{s} = j$ and the estimated groups and parameter vectors $\{G_j, \hat{\boldsymbol{\theta}}_j\}_{j=1}^{\hat{s}}$.

---

inequalities is larger than the cardinality of G_n, then this means that $\hat{\boldsymbol{\theta}}_j$ should have been found at the nth iteration and the algorithm is reset with $\hat{\boldsymbol{\theta}}_n = \hat{\boldsymbol{\theta}}_j$, $G_n = G_j$ and $j = n$.

The original approach to the MAX FS problem is a randomized algorithm, in which at each iteration i a data point is randomly selected and the parameter vector $\boldsymbol{\theta}^i$ is updated so as to decrease the error on this point. But since a single linear model cannot a priori approximate all data points well, we need to decrease the influence of those that apparently could not be and that should be left aside of the estimation of the current $\hat{\boldsymbol{\theta}}_j$ (and retained for the estimation of the next mode). To do so, the amount of update, $\boldsymbol{\theta}^i - \boldsymbol{\theta}^{i-1}$, applied to $\boldsymbol{\theta}^{i-1}$ to obtain $\boldsymbol{\theta}^i$ can be weighted by an exponentially decreasing function of the error $|y_k - \boldsymbol{x}_k^\top \boldsymbol{\theta}|$. During this process, the number of complementary inequalities satisfied by all $\boldsymbol{\theta}^i$ is computed and the estimate with the maximal number is returned.

Such an algorithm can be described as a stochastic gradient descent minimizing the cost function

$$J_{\epsilon,\beta}(\boldsymbol{\theta}) = \frac{1}{N} \sum_{k=1}^{N} \ell_{\epsilon,\beta}^{\exp} \left(y_k - \boldsymbol{x}_k^\top \boldsymbol{\theta} \right)$$

based on the loss function

$$\ell_{\epsilon,\beta}^{\exp}(e) = \mathbb{1}_{|e|>\epsilon} \exp\left(\frac{-|e|}{\beta} \right). \tag{6.18}$$

plotted in Fig. 6.3. This loss function is tailored to be insensitive to errors less than ϵ in magnitude and more sensitive to errors that are only slightly larger than this threshold than to very large ones.

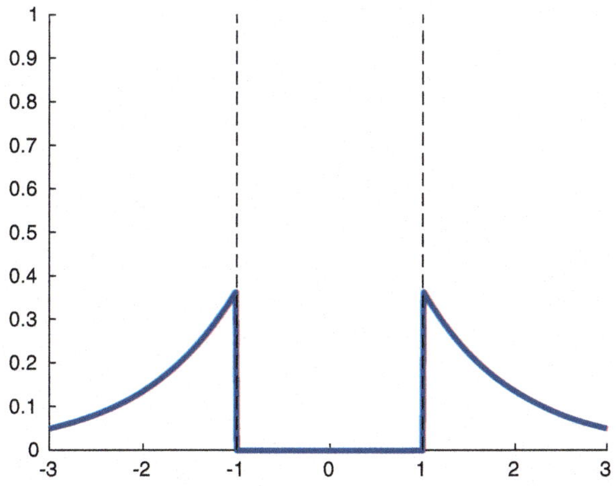

Fig. 6.3 Exponentially decreasing ϵ-insensitive loss function (6.18) (plotted with $\epsilon = 1$ and $\beta = 1$)

Stochastic gradient descent is a computationally efficient and popular optimization method in machine learning for the minimization of convex and nonconvex cost functions. At every iteration i, a random index k is drawn and the parameters are drawn in the direction opposite to the gradient of the kth term in the cost function:

$$\boldsymbol{\theta}^i \leftarrow \boldsymbol{\theta}^{i-1} - \mu_i \boldsymbol{g}_k(\boldsymbol{\theta}^{i-1})$$

where $\mu_i \in (0, 1]$ is a learning rate (either constant or decreasing in i) and the gradient is computed as

$$\boldsymbol{g}_k(\boldsymbol{\theta}) = \frac{\partial \ell^{\exp}_{\epsilon,\beta}(y_k - \boldsymbol{x}_k^\top \boldsymbol{\theta})}{\partial \boldsymbol{\theta}} = \begin{cases} \frac{1}{\beta}\boldsymbol{x}_k e^{\frac{|y_k - \boldsymbol{x}_k^\top \boldsymbol{\theta}|}{\beta}}, & \text{if } y_k - \boldsymbol{x}_k^\top \boldsymbol{\theta} > \epsilon \\ \frac{-1}{\beta}\boldsymbol{x}_k e^{\frac{|y_k - \boldsymbol{x}_k^\top \boldsymbol{\theta}|}{\beta}}, & \text{if } y_k - \boldsymbol{x}_k^\top \boldsymbol{\theta} < -\epsilon \\ \boldsymbol{0}, & \text{otherwise.} \end{cases}$$

In the original approach, this scheme is modified to make the parameter β a decreasing function of i. The idea is that by starting with a large enough β, all data points have almost the same weight, and then the estimation focuses more and more on the points close to be well approximated by the current model.

6.2.2 Block-Coordinate Descent Approach

Recall the formulation of bounded-error regression given in Problem 5.2. We will now describe a simple heuristic approach to this problem in the flavor of the k-LinReg algorithm of Sect. 6.1.3, i.e., we will alternate between the estimation of the model parameters and the classification of the data. More precisely, we focus on the formulation in (5.7) for $p = 2$, which involves the determination of the index set

$$\mathcal{I}_1(\boldsymbol{\theta}) = \{k \in [N] : |y_k - \boldsymbol{x}_k^\top \boldsymbol{\theta}| \leq \epsilon\},$$

i.e., the classification of the data into two groups: one with points that can be well approximated with $\boldsymbol{\theta}$ and the others.

Starting at iteration $i = 0$ from an initial guess $\boldsymbol{\theta}^0$, the algorithm alternates between a classification step and a least squares regression as follows.

1. Identify the set of points satisfying the bounded-error criterion:

$$\mathcal{I}_1(i) = \{k \in [N] : (y_k - \boldsymbol{x}_k^\top \boldsymbol{\theta}^i)^2 \leq \epsilon^2\}. \tag{6.19}$$

2. Update the parameter vector with

$$\boldsymbol{\theta}^{i+1} = \operatorname*{argmin}_{\boldsymbol{\theta} \in \mathbb{R}^d} \sum_{k \in \mathcal{I}_1(i)} (y_k - \boldsymbol{x}_k^\top \boldsymbol{\theta})^2. \tag{6.20}$$

3. Set $i \leftarrow i + 1$ and repeat from step (1) until convergence.

This algorithm could be interpreted as a block-coordinate descent since it alternates between the optimization with respect to the integer variables encoding $\mathcal{I}_1$ and with respect to the model parameters.

This algorithm can be shown to converge to a local minimum of Problem 5.2 for $p = 2$. On the computational side, it involves only simple tasks, and basically amounts to solving a least squares problem at each iteration (see Sect. 2.1.2). This makes it a good candidate for computing upper bounds in the global optimization procedure of Sect. 5.4.2. Alternatively, it can be used on its own by considering multiple restarts from different initializations $\boldsymbol{\theta}^0$, as for the k-LinReg algorithm of Sect. 6.1.3.

6.2.3 Error Sparsification Method

The error sparsification method estimates the submodels one by one, and thus the number of submodels, as above, but with a different method for the estimation of each submodel. In particular, here, the method is based on a sparsity argument.

Specifically, let us consider the noiseless case. Then, for the estimation of the first submodel, the error vector $\boldsymbol{e} = [e_1 \dots e_N]^\top$, with entries $e_k = y_k - \boldsymbol{x}_k^\top \boldsymbol{\theta}$, is sparse, i.e., with many zero entries, if $\boldsymbol{\theta}$ is one of the true submodel parameter vectors (see Fig. 6.4). Moreover, there exists a parameter vector, corresponding to the mode dominating the data set, which maximizes the number of zero entries of $\boldsymbol{e}$. In other words, it is the solution to the following sparse optimization problem,

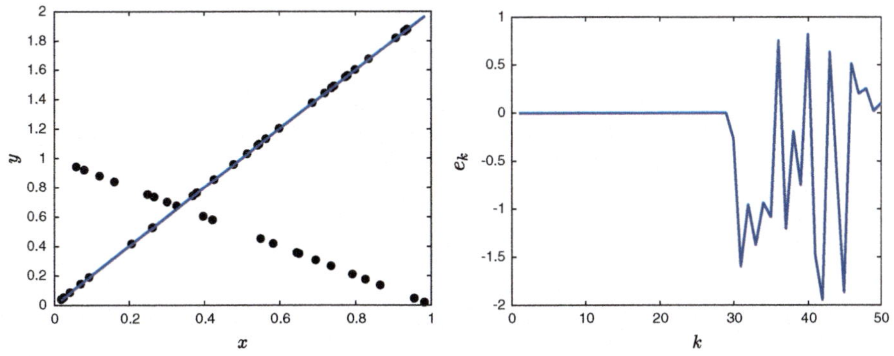

Fig. 6.4 Basic idea of the error sparsification approach: When the model coincides with one of the true submodels (plain line) and the data is noiseless (left), the error vector $\boldsymbol{e}$ is sparse (right). For clarity, the data points have been ordered with respect to their true mode, while in general the zeros in $\boldsymbol{e}$ can be found at any index

$$\min_{e \in \mathbb{R}^N, \theta \in \mathbb{R}^d} \|e\|_0 \quad \text{s.t. } e = y - X\theta, \tag{6.21}$$

where $y = \begin{bmatrix} y_1 & \dots & y_N \end{bmatrix}^\top$, $X = \begin{bmatrix} x_1 & \dots & x_N \end{bmatrix}^\top$ and the ℓ_0-pseudo-norm $\|e\|_0$ denotes the number of nonzero entries of e as in (B.2). After finding the first submodel, all data points corresponding to $e_k = 0$ are removed from the data set to estimate the next submodel. This procedure is repeated until the data set is empty.

The problem (6.21) is a hard nonconvex optimization problem in general, with a combinatorial component. A straightforward approach would be to start with $e = 0$ (or $e = y$) and increase (or decrease) the number of nonzero entries in e until a parameter θ can be found to satisfy the constraints. But this naive algorithm is not practical, since for each value of $\|e\|_0$ one also has to select $\|e\|_0$ indexes among N for the nonzero entries in e.

However, sparse optimization problems such as (6.21) have been largely studied over the last decade and a number of practical heuristics have been proposed. In particular, a popular approach is to consider ℓ_1-norm based convex relaxations. For the optimization program (6.21), this leads to

$$\min_{\theta \in \mathbb{R}^d} \|W_x(y - X\theta)\|_1, \tag{6.22}$$

where W_x is a diagonal weighting matrix whose precise definition will come from the analysis below.

Being convex, the problem (6.26) is much more amenable than the original (6.21). In addition, solving the relaxation (6.26) can lead to the optimal solution of (6.21) in situations characterized by the following theorem.

Theorem 6.1 *If there is a vector θ achieving an error $e = y - X\theta$ such that*

$$\|e\|_0 \leq \frac{1}{2}\left(1 + \frac{1}{m(X)}\right) \tag{6.23}$$

with

$$m(X) = \max_{1 \leq i < k \leq N} \frac{|M_{ik}|}{\sqrt{(1 - M_{ii})(1 - M_{kk})}}, \tag{6.24}$$

where M_{ik} is the (i, k)-entry of $M = X(X^\top X)^{-1}X^\top$, then θ is the unique solution to (6.22).

Problems (6.21) and (6.22) can be analyzed in the more general framework of *sparse recovery*. As such, Theorem 6.1 is a direct consequence of a standard result in this field. A sparse recovery problem is that of finding the sparsest solution to an underdetermined system of linear equalities. Given a matrix $A \in \mathbb{R}^{n \times N}$ and a vector b, this can be written as

$$(\textit{Sparse recovery}) \qquad \min_{z \in \mathbb{R}^n} \|z\|_0 \quad \text{s.t. } Az = b. \tag{6.25}$$

This problem and its ℓ_1-relaxation have been the focus of much attention over the last decade and many results are available. Among these, we here consider one that is based on the *mutual coherence* of the matrix A, defined as

$$\mu(A) = \max_{1 \le i < k \le N} \frac{|A_i^\top A_k|}{\|A_i\|_2 \|A_k\|_2},$$

where A_k denotes the kth column of A.

Theorem 6.2 (Exact recovery by ℓ_1 -minimization) *If there is a vector $z \in \mathbb{R}^N$ such that $Az = b$ and*

$$\|z\|_0 \le \frac{1}{2}\left(1 + \frac{1}{\mu(A)}\right),$$

then z is the unique solution to both $\min_{z \in \mathbb{R}^n} \|Wz\|_1$, s.t. $Az = b$ and (6.25), with $W = \operatorname{diag}\left(\left[\|A_1\|_2 \dots \|A_N\|_2\right]\right)$.

In order to apply this result to (6.22), we first restate it as

$$\begin{aligned} \min_{e \in \mathbb{R}^N, \theta \in \mathbb{R}^d} \ & \|W_x e\|_1 \\ \text{s.t. } & Pe = Py, \\ & e = y - X\theta \end{aligned} \tag{6.26}$$

with the projection matrix $P = I_N - X(X^\top X)^{-1} X^\top$ and the weighting matrix $W_x = \operatorname{diag}\left(\left[\|P_1\|_2 \dots \|P_k\|_2 \dots \|P_N\|_2\right]\right)$, where P_k refers to the kth column of P. Indeed, due to $PX = 0$, the first constraint in (6.26) is always satisfied when the second one is and adding it does not change the problem. For the sake of the analysis, we will consider solving (6.26) in two steps: first solve

$$\begin{aligned} \min_{e \in \mathbb{R}^N} \ & \|W_x e\|_1 \\ \text{s.t. } & Pe = Py, \end{aligned} \tag{6.27}$$

for e and then solve $X\theta + e = y$ for θ.

Then, it is easy to recognize (6.27) as a sparse recovery problem (6.25) with $A = P$ and $b = Py$. Therefore, the proof of Theorem 6.1 boils down to showing that $\mu(P) = m(X)$.[1] To see this, note that M is a symmetric projection matrix and

[1] Actually, we also need to prove that $X\theta + e = y$ is feasible when e is the solution to (6.27) to justify the decomposition of the problem (6.26) in two steps. The condition in Theorem 6.1 also ensures this, since by Theorem 6.2 any e that satisfies (6.23) can be recovered as the unique solution to (6.27) and so must be as the one in Theorem 6.1.

thus that $\boldsymbol{M}^\top \boldsymbol{M} = \boldsymbol{M}$, which implies $M_{ik} = \boldsymbol{M}_i^\top \boldsymbol{M}_k$, with $\boldsymbol{M}_i$ and $\boldsymbol{M}_k$ the ith and kth columns of $\boldsymbol{M}$. Then, let $\boldsymbol{u}_i$ denote the ith vector of the canonical basis of $\mathbb{R}^N$ and write

$$\begin{aligned}\mu(\boldsymbol{P}) &= \mu(\boldsymbol{I} - \boldsymbol{M}) \\ &= \max_{1\leq i<k\leq N} \frac{|(\boldsymbol{u}_i - \boldsymbol{M}_i)^\top(\boldsymbol{u}_k - \boldsymbol{M}_k)|}{\sqrt{(\boldsymbol{u}_i - \boldsymbol{M}_i)^\top(\boldsymbol{u}_i - \boldsymbol{M}_i)(\boldsymbol{u}_k - \boldsymbol{M}_k)^\top(\boldsymbol{u}_k - \boldsymbol{M}_k)}} \\ &= \max_{1\leq i<k\leq N} \frac{|-M_{ik} - M_{ki} + M_{ik}|}{\sqrt{(1 - 2M_{ii} + M_{ii})(1 - 2M_{kk} + M_k)}} \\ &= m(\boldsymbol{X}).\end{aligned} \tag{6.28}$$

6.2.3.1 Condition on the Distribution of the Mode

The analysis above leads to a condition on the distribution of the mode: For the method to be guaranteed to recover the true parameter vector of a mode, enough data must be generated in this mode. We refer to the mode from which the largest number of data have been generated as the *dominating mode*. It is clear that the sparsest error vector $\boldsymbol{e}$ is obtained when $\boldsymbol{\theta}$ is the parameter vector of this mode.

Let N_j denote the number of data generated from mode j and assume that this is the dominating mode. Further assume for simplicity that the modes are separated in the sense that the data is such that $y_k - \boldsymbol{x}_k^\top \boldsymbol{\theta} \neq 0$ for all $\boldsymbol{\theta}$ not equal to the true parameter vector that generated $(\boldsymbol{x}_k, y_k)$. Then, the sparsest $\boldsymbol{e}$ has $\|\boldsymbol{e}\|_0 = N - N_j$ nonzero entries and the sparsity condition in Theorem 6.1 can be rewritten as the following condition on N_j:

$$N_j \geq N - \frac{1}{2}\left(1 + \frac{1}{m(\boldsymbol{X})}\right).$$

It can be further shown that this condition requires, in particular, $N_j \geq N/2$, i.e., that at least half of the data come from the dominating mode. This might not be satisfied in many practical situations, but this is only a sufficient condition, not a necessary one. In practice, the ℓ_1-relaxation approach to sparse recovery can be more robust and improved as described next.

6.2.3.2 Increasing the Sparsity

The sparsity of the solution to the ℓ_1-relaxation can be increased (and the condition on the number of data generated by the dominating mode relaxed) by using reweighting techniques. One of these techniques associates every variable involved in the ℓ_0-norm, here the e_k's, with smooth weights computed as

$$w_k = \frac{1}{|y_k - \boldsymbol{x}_k^\top \boldsymbol{\theta}| + \delta}$$

for some small $\delta > 0$ avoiding divisions by zero. Then, with the weight matrix $\boldsymbol{W} = \text{diag}(w_1, \ldots, w_N)$ initialized to the identity matrix, a sequence of problems

$$\min_{\boldsymbol{\theta} \in \mathbb{R}^d} \|\boldsymbol{W}(\boldsymbol{y} - \boldsymbol{X}\boldsymbol{\theta})\|_1. \tag{6.29}$$

is solved with increased sparsity at each step.

Another technique considers hard thresholding and a selective procedure. At each step, the variable with largest absolute value is removed from the computation of the ℓ_1-norm, thus letting it be free, and putting more weight on the remaining variables. This can be implemented as above with the update of a single weight at each step:

$$w_{k^*} = 0, \quad \text{for } k^* = \operatorname*{argmax}_{k \in [N]} |y_k - \boldsymbol{x}_k^\top \boldsymbol{\theta}|.$$

6.2.3.3 Noisy Data

With noisy data, the error vector cannot be sparse anymore. Instead, it is "compressible" with the magnitude of most entries below a threshold ϵ. The ℓ_1-relaxation remains applicable in this setting. However, exact recovery guarantees do not apply and, instead, stability results can be obtained. These state that the obtained error vector should be close to the one we could get in the noiseless setting, at a distance increasing with the noise level.

Therefore, with noise, our ability to correctly estimate the mode by thresholding the entries in the error vector depends on the noise level and the separability of the modes. Indeed, if the magnitude of the noise is small compared to the smallest error $|y_k - \boldsymbol{x}_k^\top \boldsymbol{\theta}_j|$ obtained with a parameter vector of another mode $j \neq q_k$, we can expect the entries in $\boldsymbol{e}$ to be in two distinct groups in accordance to the membership of the data points to the dominating mode. At the opposite, if the noise level is high and the true parameter vectors do not differ much between the modes, the entries of $\boldsymbol{e}$ would blend together, thus making the dominating mode indistinguishable from the others.

6.2.4 Parameter Sparsification Method

As seen above, looking for a sparse solution is a strong assumption that can have tremendous consequences in terms of how easily we can expect to find that solution. While the previous approach considered sparse error vectors, sparsity can also be found at a different level in switched system identification. Consider an over-parametrized model with one parameter vector for each data point, or, equivalently, a model with a single parameter *function* that returns a parameter vector for every data point. Then, the fact that the system has only a few modes translates into the assumption that a sparse solution could be found in terms of the variations of this parameter function.

Let $\boldsymbol{\theta} : [N] \rightarrow \mathbb{R}^d$ denote the parameter function and

$$f(\boldsymbol{x}_k) = \boldsymbol{x}_k^\top \boldsymbol{\theta}(k)$$

be the overparametrized model. Note that this model will only be used for estimation purposes and does not constitute a predictive model in its own: Its value $f(\boldsymbol{x})$ is only defined for $\boldsymbol{x}$ in the set $\{\boldsymbol{x}_k\}_{k=1}^N$ (f might as well be written as a function of $k \in [N]$). In particular, we will try to obtain a parameter function with only sparse variations and thus only a few different parameter values. The sparsity of the variations can be measured by the ℓ_0-pseudo-norm[2] (see (B.2)) of a strictly lower triangular matrix $\boldsymbol{\Delta}$ of entries given by

$$\forall (k, l) \in [N]^2, \quad \Delta_{kl} = \mathbb{1}_{k<l} \|\boldsymbol{\theta}(k) - \boldsymbol{\theta}(l)\|_p$$

for any choice of ℓ_p-norm $\|\cdot\|_p$.

Thus, the general idea decomposes in two main steps. In the first step, we minimize the number of different parameter vectors while maintaining the error below a given threshold:

$$\begin{aligned} \min_{\boldsymbol{\theta}:[N]\rightarrow\mathbb{R}^d, \boldsymbol{e}\in\mathbb{R}^N} \quad & \|\boldsymbol{\Delta}\|_0 \\ \text{s.t. } & y_k - \boldsymbol{x}_k^\top \boldsymbol{\theta}(k) = e_k, \ k = 1, \dots, N, \\ & \|\boldsymbol{e}\|_{p'} \leq \epsilon, \end{aligned} \tag{6.30}$$

where the error is measured by some $\ell_{p'}$-norm. Of course, this combinatorial problem is rarely considered in practice and it is customary to rely on an ℓ_1-relaxation and the reweighting techniques introduced in Sect. 6.2.3 to obtain sparse solutions after a sequence of convex optimizations.

Then, the second step clusters the data and estimates the modes with a rule assigning data points with similar parameter vectors to the same mode. The most simple of those rules considers that parameters are similar when they are equal and is such that

$$\forall (k, l) \in [N]^2, \quad q_k = q_l \Leftrightarrow \boldsymbol{\theta}(k) = \boldsymbol{\theta}(l). \tag{6.31}$$

After this step, the number of modes s is estimated as the number of different parameter vectors and the parameters of each mode can be re-estimated with any of the methods of Chap. 2.

Other rules can be used, since this step amounts to a clustering problem (as presented in Sect. 3.2) for which many methods are available. Specifically, the task

---

[2] We naturally extend the definition of ℓ_p-(pseudo-)norms to matrices by applying them to the vectorized forms of the matrices. Thus, $\|\boldsymbol{\Delta}\|_0$ denotes the number of nonzero entries in $\boldsymbol{\Delta}$.

is to recover s homogeneous groups of parameter vectors among $\{\boldsymbol{\theta}(k)\}_{k=1}^N$. The idea of clustering local parameter vectors to estimate the modes will be discussed once again in Sect. 7.3.1 for PWA system identification.

An important issue of the approach just described is the following. Though a solution to (6.30) must have the smallest number of parameter vectors yielding the desired accuracy, $\|\boldsymbol{\Delta}\|_0$ is not precisely equal to the number of different parameter vectors. It is indeed equal to the number of different pairs of parameter vectors, which is related but also much larger. As a result the optimal matrix $\boldsymbol{\Delta}$ might not be very sparse and heuristics such as those based on convex relaxations might not lead to the sparsest solution. The problem comes from the fact that $\boldsymbol{\Delta}$ takes the variations of $\boldsymbol{\theta}$ over all pairs of indexes (k, l) into account. To circumvent this issue, one needs to be more specific when defining what are the variations of the parameter function that matter in the setting under consideration, as detailed next.

6.2.4.1 Segmentation of Data Ordered in Time

If the data points can be ordered in a sequence such that the modes can be defined as segments of that sequence, then the approach can easily and drastically be improved. Such a setting occurs for instance with time series data or in the context of system identification with data originating from a single trajectory of the system and a minimum dwell time in each mode. In this case, the problem amounts to a segmentation one and mode switchings only occur between two consecutive time steps.

Under such an assumption, variations of the parameter function between two nonconsecutive indexes do not increase the number of modes and shall not be penalized. This leads to the redefinition of $\boldsymbol{\Delta}$ as a matrix with a single diagonal (below the main one) of nonzero entries:

$$\forall (k, l) \in [N]^2, \quad \boldsymbol{\Delta}_{kl} = \mathbb{1}_{k=l+1} \|\boldsymbol{\theta}(k) - \boldsymbol{\theta}(l)\|_p.$$

In this case, the relaxed (and reweighted) version of the sparse optimization problem (6.30) reads

$$\begin{aligned} \min_{\{\boldsymbol{\theta}(k)\in\mathbb{R}^d\}_{k=1}^N, \boldsymbol{e}\in\mathbb{R}^N, \boldsymbol{z}\in\mathbb{R}^{N-1}} \quad & \|\boldsymbol{W}\boldsymbol{z}\|_1 \\ \text{s.t.} \quad & \|\boldsymbol{\theta}(k+1) - \boldsymbol{\theta}(k)\|_p \leq z_k, \ k = 1, \dots, N-1 \\ & y_k - \boldsymbol{x}_k^\top \boldsymbol{\theta}(k) = e_k, \quad k = 1, \dots, N, \\ & \|\boldsymbol{e}\|_{p'} \leq \epsilon, \end{aligned} \tag{6.32}$$

where $\boldsymbol{W}$ is the diagonal matrix of weights that can be computed at each iteration to increase the sparsity of the solution (see the discussion in Sect. 6.2.3.2).

Note that such segmentation problems can also be viewed as one-dimensional PWA regression problems over the time-axis and solved with exact methods as in Example 5.1 for a fixed and small s.

6.2.4.2 General Unordered Data

In the much more challenging and general case where the data points cannot be ordered to yield a segmentation problem, an alternative to (6.30) is to fall back to an iterative procedure in the flavor of the bounded-error approach (see Sects. 4.2.4.1 and 6.2.1).

Specifically, we now aim at the estimation of a parameter function with minimal variations between its values and a "mean" parameter vector $\tilde{\boldsymbol{\theta}}$ modeling the current/dominating mode. In terms of the matrix $\boldsymbol{\Delta}$, this leads to

$$\forall (k,l) \in [N]^2, \quad \boldsymbol{\Delta}_{kl} = \mathbb{1}_{k=l} \|\boldsymbol{\theta}(k) - \tilde{\boldsymbol{\theta}}\|_p$$

and the relaxed optimization problem

$$\begin{aligned} \min_{\{\boldsymbol{\theta}(k)\in\mathbb{R}^d\}_{k=1}^N, \tilde{\boldsymbol{\theta}}\in\mathbb{R}^d, \boldsymbol{e}\in\mathbb{R}^N, \boldsymbol{z}\in\mathbb{R}^N} \quad & \|\boldsymbol{W}\boldsymbol{z}\|_1 \\ \text{s.t.}\ & \|\boldsymbol{\theta}(k) - \tilde{\boldsymbol{\theta}}\|_p \leq z_k,\ k = 1,\dots,N \\ & y_k - \boldsymbol{x}_k^\top \boldsymbol{\theta}(k) = e_k,\ k = 1,\dots,N, \\ & \|\boldsymbol{e}\|_{p'} \leq \epsilon. \end{aligned} \tag{6.33}$$

This problem is solved iteratively while, at each step j, recording $\tilde{\boldsymbol{\theta}}$ as the parameter vector of mode j and removing the data points for which it provides a satisfactory approximation, for instance, such that $y_k - \boldsymbol{x}_k^\top \tilde{\boldsymbol{\theta}} \leq \epsilon$.

6.2.4.3 Unconstrained Formulations and Sum-of-Norms Optimization

Before concluding the section, let us highlight that the approach above can also be considered with unconstrained formulations. For instance, for the general case of (6.30), this leads to

$$\min_{\{\boldsymbol{\theta}(k)\in\mathbb{R}^d\}_{k=1}^N} \sum_{k=1}^N \ell\left(y_k - \boldsymbol{x}_k^\top \boldsymbol{\theta}(k)\right) + \lambda \sum_{1\leq k<l\leq N} \|\boldsymbol{\theta}(k) - \boldsymbol{\theta}(l)\|_p, \tag{6.34}$$

where ℓ is a loss function and λ is the regularization parameter that controls the trade-off between the model accuracy and the variation of its parameters. These formulations are used in what are known as *sum-of-norms* methods due to the form of the regularization term.

The main difference here is that the error is not constrained below a predefined threshold but minimized together with some increasing function of the number of modes. The error thus obtained now depends on the value of the hyperparameter λ whose influence is opposite to that of ϵ in (6.30): For a sufficiently large λ, the variation term will be zero and a single mode will be found, while a sufficiently small λ yields N different modes and zero error.

For the segmentation of ordered data as in (6.32), the sum-of-norms formulation (6.34) becomes

$$\min_{\{\boldsymbol{\theta}(k)\in\mathbb{R}^d\}_{k=1}^N} \sum_{k=1}^{N} \ell\left(y_k - \boldsymbol{x}_k^\top \boldsymbol{\theta}(k)\right) + \lambda \sum_{k=2}^{N} \|\boldsymbol{\theta}(k) - \boldsymbol{\theta}(k-1)\|_p. \tag{6.35}$$

For the general and unordered case considered in (6.33), it instead takes the form:

$$\min_{\{\boldsymbol{\theta}(k)\in\mathbb{R}^d\}_{k=1}^N, \tilde{\boldsymbol{\theta}}\in\mathbb{R}^d} \sum_{k=1}^{N} \ell\left(y_k - \boldsymbol{x}_k^\top \boldsymbol{\theta}(k)\right) + \lambda \sum_{k=1}^{N} \|\boldsymbol{\theta}(k) - \tilde{\boldsymbol{\theta}}\|_p, \tag{6.36}$$

which has to be solved once for each mode.

Note that the regularization terms can also be iteratively reweighted in these unconstrained formulations to increase the sparsity of the solution. This is done by weighting each term in the sum of norms by a decreasing function of the norm obtained at the previous iteration.

6.2.4.4 Choice of Norms and Optimization

A common choice for p' in the error term of the constrained formulations (6.32)–(6.33) is $p' = 2$, which leads to constraints on the square root of the mean squared error.

Regarding the choice of ℓ_p-norm used for the control of the variations, first note that it is crucial not to use *squared* ℓ_p-norms. Indeed, using, for instance, the *squared* ℓ_2-norm in (6.36) amounts to minimizing

$$\sum_{k=1}^{N} \|\boldsymbol{\theta}(k) - \tilde{\boldsymbol{\theta}}\|_2^2 = \|\boldsymbol{\theta}_{1:N} - \mathbf{1}_N \otimes \tilde{\boldsymbol{\theta}}\|_2^2,$$

with a vector $\boldsymbol{\theta}_{1:N} \in \mathbb{R}^{dN}$ concatenating all $\boldsymbol{\theta}(k)$'s. This would lead to a smooth optimization problem, but not to sparse solutions. Without squaring the norms, another problem arises for $p = 1$. Despite leading to sparse solutions, the sparsity would be measured in terms of the number of nonzero entries in $(\boldsymbol{\theta}_{1:N} - \mathbf{1}_N \otimes \tilde{\boldsymbol{\theta}}) \in \mathbb{R}^{dN}$ and solutions with few switchings between completely different $\boldsymbol{\theta}(k)$'s could not be favored over others with many switchings of a single entry inside $\boldsymbol{\theta}(k)$. Thus, the method would not be able to differentiate between a hybrid model with a small number of very different modes and one with many very similar modes.

Therefore, standard choices of ℓ_p-norm to measure the variations are based on $p = 2$ and $p = \infty$, which allow the sparsity to be measured at the level of complete vectors, i.e., to be related to the number of nonzero vectors. Similar characteristics and choices will apply in Sect. 7.2 to the regularization scheme in (7.7) also based on a vector-level notion of sparsity.

The choice of p and p' in the various problems depicted above also directly influences the type of the resulting optimization problem. The ℓ_∞-norm leads to linear programming formulations and thus computational efficiency, compared to the ℓ_2-norm which introduces second-order cone constraints that are convex, but much more demanding for the solver.

Finally, note that contrary to the error sparsification method (Sect. 6.2.3), the conditions under which the relaxed problems truly yield the sparsest solutions have not been determined yet.

Notes

The algebraic approach was pioneered by [1] and further developed in [2] with respect to the estimation of the model orders and the limitation on the number s of modes due to the polynomial equation of degree s solved to find a point z_j^* of each mode. Then, it has been extended in [3] to the noisy case, in [4] to a recursive algorithm and in [5] to the MIMO case, to name only a few. In the computer vision community, it is a well-known subspace clustering method known as Generalized Principal Component Analysis (GPCA) [6, 7]. Other authors also worked on making the method more robust to noise [8, 9].

The continuous optimization approach was proposed in [10] and an experimental study of the properties of various loss functions in the product-of-errors framework was conducted in [11]. The MCS solver used in these papers is due to [12]. A regularized version of the Minimum-of-Errors formulation was considered in [13], where a difference of convex functions programming algorithm was also derived as an alternative to the MCS solver for larger dimensional systems.

The k-LinReg algorithm and the analysis of its probability of success are due to [14].

The bounded-error approach described in Sect. 6.2.1 was developed in [15–17] and heavily relies on the greedy algorithm for MIN PFS and randomized heuristics for the MAX FS subproblem proposed in [18]. The $\mathcal{NP}$-hardness of MAX FS was proved by the authors of the latter in [19]. The block-coordinate descent method for bounded-error estimation was developed in [20] as a heuristic for computing upper bounds in a global optimization procedure.

By following an overall similar strategy in which the submodels are estimated one by one until a satisfactory level of error is reached, the error sparsification method of Sect. 6.2.3 was proposed in [21], where the Theorem 6.1 is taken from.

The parameter sparsification approaches were developed in [22–25]. In particular, the unconstrained sum-of-norms formulations are due to [24, 25], while [22, 23] considered the constrained versions. A solution based on a greedy sliding window for the segmentation case can also be found in [26].

These methods based on sparsity arguments take inspiration from the compressed sensing literature [27, 28], which had a deep impact in many fields from signal processing to machine learning and computer vision. A good overview of this field is given in the textbook [29]. Theorem 6.2 was proved in [30]. Motivated by the error sparsification approach, an analysis of robust estimation based on these principles was proposed in [31].

Reweighting techniques used to improve the sparsity of the solution to ℓ_1-relaxations were developed in [32] with smooth weighting functions and in [33] with selective/hard weights.

Software

MATLAB code for the algebraic method of Sect. 6.1.1 is available at http://vision.jhu.edu/gpca.htm. More precisely, this offers an implementation of GPCA with Polynomial Differentiation and spectral clustering for subspace classification, which is the core of the algebraic method. MATLAB code for the continuous optimization framework is available at https://members.loria.fr/FLauer/files/COFSR/COFSR.zip. It relies heavily on the code of the MCS solver from http://www.mat.univie.ac.at/~neum/software/mcs/. The k-LinReg software implements the block-coordinate descent method of Sect. 6.1.3. It is available in MATLAB code or a parallel implementation in C for Linux at https://members.loria.fr/FLauer/files/klinreg/. All sparsity-based methods amount to solving convex optimization problems that can be efficiently modeled with the CVX toolbox [34, 35] available at http://cvxr.com/cvx. For ℓ_1-minimization problems, dedicated solvers have been developed, see, e.g., NESTA [36] at https://statweb.stanford.edu/~candes/nesta/ or ℓ_1-magic at https://statweb.stanford.edu/~candes/l1magic/.

References

1. Vidal, R., Soatto, S., Ma, Y., Sastry, S.: An algebraic geometric approach to the identification of a class of linear hybrid systems. In: Proceedings of the 42nd IEEE Conference on Decision and Control (CDC), Maui, HI, USA, pp. 167–172 (2003)
2. Vidal, R.: Identification of PWARX hybrid models with unknown and possibly different orders. In: Proceedings of the 2004 American Control Conference (ACC), Boston, MA, USA, pp. 547–552 (2004)
3. Ma, Y., Vidal, R.: Identification of deterministic switched ARX systems via identification of algebraic varieties. In: Proceedings of the 8th International Workshop on Hybrid Systems: Computation and Control (HSCC), Zürich, Switzerland. Lecture Notes in Computer Science, vol. 3414, pp. 449–465 (2005)
4. Vidal, R.: Recursive identification of switched ARX systems. Automatica **44**(9), 2274–2287 (2008)
5. Bako, L., Vidal, R.: Algebraic identification of switched MIMO SARX models. In: Proceedings of the 11th International Workshop on Hybrid Systems: Computation and Control (HSCC), St. Louis, MO, USA. Lecture Notes in Computer Science, vol. 4981, pp. 43–57 (2008)
6. Vidal, R., Ma, Y., Sastry, S.: Generalized principal component analysis (GPCA). IEEE Trans. Pattern Anal. Mach. Intell. **27**(12), 1945–1959 (2005)

7. Vidal, R., Ma, Y., Sastry, S.: Generalized Principal Component Analysis. Springer, Berlin (2016)
8. Ozay, N., Lagoa, C.M., Sznaier, M.: Robust identification of switched affine systems via moments-based convex optimization. In: Proceedings of the 48th IEEE Conference on Decision and Control (CDC) - 28th Chinese Control Conference, Shanghai, China, pp. 4686–4691 (2009)
9. Ozay, N., Lagoa, C.M., Sznaier, M.: Set membership identification of switched linear systems with known number of subsystems. Automatica **51**, 180–191 (2015)
10. Lauer, F., Bloch, G., Vidal, R.: A continuous optimization framework for hybrid system identification. Automatica **47**(3), 608–613 (2011)
11. Lauer, F., Vidal, R., Bloch, G.: A product-of-errors framework for linear hybrid system identification. In: Proceedings of the 15th IFAC Symposium on System Identification (SYSID), Saint-Malo, France, IFAC Proceedings Volumes, vol. 42(10), pp. 563–568 (2009)
12. Huyer, W., Neumaier, A.: Global optimization by multilevel coordinate search. J. Glob. Optim. **14**(4), 331–355 (1999)
13. Pham Dinh, T., Le Thi, H.A., Le, H.M., Lauer, F.: A difference of convex functions algorithm for switched linear regression. IEEE Trans. Autom. Control **59**(8), 2277–2282 (2014)
14. Lauer, F.: Estimating the probability of success of a simple algorithm for switched linear regression. Nonlinear Anal. Hybrid Syst. **8**, 31–47 (2013)
15. Bemporad, A., Garulli, A., Paoletti, S., Vicino, A.: A greedy approach to identification of piecewise affine models. In: Proceedings of the 6th International Workshop on Hybrid Systems: Computation and Control (HSCC), Prague, Czech Republic. Lecture Notes in Computer Science, vol. 2623, pp. 97–112 (2003)
16. Bemporad, A., Garulli, A., Paoletti, S., Vicino, A.: Set membership identification of piecewise affine models. In: Proceedings of the 13th IFAC Symposium on System Identification, Rotterdam, The Netherlands, IFAC Proceedings Volumes, vol. 36(16), pp. 1789–1794 (2003)
17. Bemporad, A., Garulli, A., Paoletti, S., Vicino, A.: A bounded-error approach to piecewise affine system identification. IEEE Trans. Autom. Control **50**(10), 1567–1580 (2005)
18. Amaldi, E., Mattavelli, M.: The MIN PFS problem and piecewise linear model estimation. Discret. Appl. Math. **118**, 115–143 (2002)
19. Amaldi, E., Kann, V.: The complexity and approximability of finding maximum feasible subsystems of linear relations. Theor. Comput. Sci. **147**(1–2), 181–210 (1995)
20. Lauer, F.: Global optimization for low-dimensional switching linear regression and bounded-error estimation. Automatica **89**, 73–82 (2018b)
21. Bako, L.: Identification of switched linear systems via sparse optimization. Automatica **47**(4), 668–677 (2011)
22. Ozay, N., Sznaier, M., Lagoa, C., Camps, O.: A sparsification approach to set membership identification of a class of affine hybrid systems. In: Proceedings of the 47th IEEE Conference on Decision and Control (CDC), Cancun, Mexico, pp. 123–130 (2008)
23. Ozay, N., Sznaier, M., Lagoa, C.M., Camps, O.I.: A sparsification approach to set membership identification of switched affine systems. IEEE Trans. Autom. Control **57**(3), 634–648 (2012)
24. Ohlsson, H., Ljung, L., Boyd, S.: Segmentation of ARX-models using sum-of-norms regularization. Automatica **46**(6), 1107–1111 (2010)
25. Ohlsson, H., Ljung, L.: Identification of switched linear regression models using sum-of-norms regularization. Automatica **49**(4), 1045–1050 (2013)
26. Gionis, A., Mannila, H.: Segmentation algorithms for time series and sequence data. In: Tutorial at the 5th SIAM International Conference on Data Mining, Newport Beach, CA, USA (2005)
27. Donoho, D.L.: Compressed sensing. IEEE Trans. Inf. Theory **52**(4), 1289–1306 (2006)
28. Candès, E.J.: Compressive sampling. In: Proceedings of the International Congress of Mathematicians, Madrid, Spain, vol. 3, pp. 1433–1452 (2006)
29. Foucart, S., Rauhut, H.: A Mathematical Introduction to Compressive Sensing. Birkhäuser, New York (2013)
30. Donoho, D.L., Elad, M.: Optimally sparse representation in general (nonorthogonal) dictionaries via ℓ_1 minimization. Proc. Natl. Acad. Sci. **100**(5), 2197–2202 (2003)

31. Bako, L., Ohlsson, H.: Analysis of a nonsmooth optimization approach to robust estimation. Automatica **66**, 132–145 (2016)
32. Candès, E.J., Wakin, M.B., Boyd, S.P.: Enhancing sparsity by reweighted ℓ_1 minimization. J. Fourier Anal. Appl. **14**(5–6), 877–905 (2008)
33. Le, V.L., Lauer, F., Bloch, G.: Selective ℓ_1 minimization for sparse recovery. IEEE Trans. Autom. Control **59**(11), 3008–3013 (2014)
34. Grant, M., Boyd, S.: CVX: Matlab software for disciplined convex programming, version 2.1. (2017). http://cvxr.com/cvx
35. Grant, M.C., Boyd, S.P.: Graph implementations for nonsmooth convex programs. In: Blondel, V., Boyd, S.P., Kimura, H. (eds.) Recent Advances in Learning and Control. Lecture Notes in Control and Information Sciences, vol. 2, pp. 95–110. Springer, Berlin (2008). http://stanford.edu/boyd/papers/pdf/graph_dcp.pdf
36. Becker, S., Bobin, J., Candès, E.J.: NESTA: a fast and accurate first-order method for sparse recovery. SIAM J. Imaging Sci. **4**(1), 1–39 (2011)

Chapter 7
Estimation of Piecewise Affine Models

This chapter presents practical methods for the identification of single-input, single-output (SISO) piecewise ARX (PWARX) models. These models, introduced in Sect. 4.1.3.2 (see (4.11), (4.12)), are of the form (4.3)

$$y_k = \boldsymbol{x}_k^\top \boldsymbol{\theta}_{g(\boldsymbol{x}_k)} + \varepsilon_k, \tag{7.1}$$

where $y_k \in \mathbb{R}$ is the output, $\varepsilon_k \in \mathbb{R}$ is a noise term, and $\boldsymbol{x}_k = \left[-y_{k-1} \ \dots \ -y_{k-n_a} \ u_{k-1} \ \dots \ u_{k-n_b}\right]^\top \in \mathcal{X} \subseteq \mathbb{R}^d$ is the regression vector of dimension $d = n_a + n_b$ with the past outputs y_{k-i} and inputs u_{k-i} and the model orders n_a and n_b. Here, $\boldsymbol{\theta}_j \in \mathbb{R}^{n_a+n_b}$, $j = 1, \dots, s$, are the parameter vectors defining each one of the s submodels and the function

$$g : \mathcal{X} \to \mathcal{Q} = [s] = \{1, \dots, s\}$$

is a classifier partitioning the regression space $\mathcal{X}$ into regions $\mathcal{X}_j$, in which a particular submodel is active; i.e., the discrete state $q_k \in [s]$ is given by $g(\boldsymbol{x}_k)$.

Recall that for such systems, the identification problem involves two intertwined tasks: the estimation of the parameter vectors $\{\boldsymbol{\theta}_j\}_{j=1}^s$ and the estimation of the classifier g partitioning the regression space into s modes. As in Chap. 6, we will focus on regression and classification issues and merely consider the $\mathbf{x}_k$'s as regression vectors in $\mathbb{R}^d$, ignoring the structure related to dynamical systems if any. Therefore, we aim at estimating a piecewise affine (PWA) function from a data set $\mathcal{D} = \{(\boldsymbol{x}_k, y_k)\}_{k=1}^N \subset \mathbb{R}^d \times \mathbb{R}$, under the assumption that it has been generated as in (7.1).

Specifically, we discuss three pathways for obtaining a PWA model. The first one relies on the application of a switching regression method to obtain preliminary estimates of the parameter vectors. Thus, one can benefit from the large body of work in that field and the methods already described in Chap. 6. However, this also has

F. Lauer and G. Bloch, *Hybrid System Identification*,
Lecture Notes in Control and Information Sciences 478,
https://doi.org/10.1007/978-3-030-00193-3_7

the obvious drawback of not exploiting all the available information, namely that the modes are supposed to be separated in the regression space.

The second pathway also aims at taking advantage of the availability of well-known methods, and more precisely, those devised for nonlinear regression. Indeed, contrary to an arbitrarily switched model, a PWA model is a well-defined nonlinear function. However, classical approaches for nonlinear regression have difficulties in dealing with data generated by a PWA system, since they rely on a smoothness assumption and are typically designed to avoid PWA-like solutions. Nonetheless, such difficulties can be circumvented by imposing the smoothness assumption on the model in a local manner.

Finally, we consider approaches truly dedicated to PWA estimation. These come in different flavors depending on whether the number of modes is known/fixed or a desired amount of error is tolerated. Yet, most of them can be seen as instances of a general framework, in which the classification of the data into modes results from the clustering of a large number (usually N) of local models. This can be interpreted as a two-step process in which an overparametrized model involving many local ones is first estimated before seeing its complexity reduced by regrouping the local models into a few significantly different modes. In this framework, the various methods mostly differ in the way they implement the first step: by looking at a small neighborhood for the estimation of a single local model at a time or by simultaneously estimating all local models at once.

Chapter Outline

Section 7.1 starts with the general approach to PWA model estimation based on the application of switching regression methods detailed in Chap. 6. Then, Sect. 7.2 discusses how to estimate PWA models with nonlinear regression techniques highly connected to the nonparametric methods of Sect. 2.4.2. Finally, Sect. 7.3 presents methods dedicated to PWA models and based on the estimation of a large number of local models.

7.1 From Switched Affine Models to PWA Models

The most difficult subtask in PWA system identification is to estimate the mode of all data points, from which both the partitioning function g and the models f_j can be easily estimated. This point of view leads to the overall procedure depicted in Algorithm 8.

Algorithm 8 Two-stage approach to PWA system identification

1. Estimate the modes $\{\hat{q}_k\}_{k=1}^N$ (and $\hat{s} = \max_{k\in[N]} \hat{q}_k$)
2. a. Estimate the classifier g from $\{(\boldsymbol{x}_k, \hat{q}_k)\}_{k=1}^N$.
   b. Estimate the submodels f_j from the local data sets $\{(\boldsymbol{x}_k, y_k) : \hat{q}_k = j\}$, $j = 1, \dots, \hat{s}$.

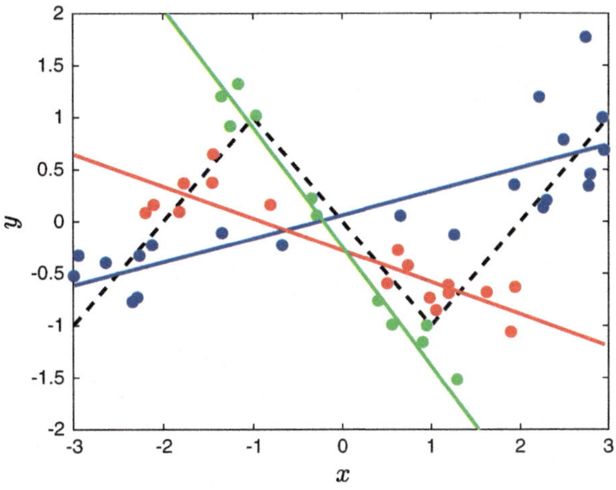

Fig. 7.1 Plot shows a one-dimensional PWA regression data set for which a global solution to the switching regression Problem 5.1 (three models plotted in plain lines and the classification corresponding to the point colors) does not provide a satisfactory solution for PWA regression (whose optimal solution is plotted in dashed line)

This two-stage approach first estimates the mode. Then the partition of the regression space is obtained by solving a discrimination problem (see Chap. 3) on the data labeled by the estimated mode and the submodels by solving independent regression problems. The distinction between the two stages is made such that the second stage applies standard learning techniques while the first one requires methods dedicated to PWA models.

As an intermediate level of dedication, the methods described in Chap. 6 can be used in the first stage. These provide baseline estimates of the submodels but also of the mode to which every data point should be assigned. However, when applying such a procedure to estimate a PWA model, additional difficulties can be encountered. Firstly, the classifier g learned from a finite sample cannot be 100% accurate and classification errors will occur in prediction for regression vectors close to the mode boundaries. For discontinuous models possibly including large jumps in output value at these boundaries, this may lead to high prediction errors. Secondly, the piecewise constraint is introduced in the model only at the end of the procedure. As a result, the discrete state and the parameters are first estimated without this constraint, which gives rise to nonoptimal local solutions that can be difficult to correct in the subsequent steps.

With respect to this issue, illustrated by Fig. 7.1, the methods presented in the rest of the chapter may perform better by taking into account the final form of the model from the start.

7.2 From Nonlinear Models to PWA Models

A PWA model is basically a nonlinear function with a particular structure. Thus, we can consider the optimal PWA model as the nonlinear target function of a nonlinear regression problem, in which one estimates a single function f. The aim in doing so is to set the PWA system identification problem in a classical learning framework to see if we could benefit from the large body of work on nonlinear regression.

In particular, the following discusses the standard approach based on regularized learning and depicted in Sects. 2.4.2.2–2.4.2.3.

7.2.1 Limitation of the Classical Regularization Schemes

The classical regularizer (2.70) based on the RKHS norm $\|f\|_{\mathcal{H}}$ is a global measure of the function smoothness and is particularly suitable for cases without prior information on the shape of the target function in order to avoid overfitting. However, minimizing $\|f\|_{\mathcal{H}}$ influences the shape of f over the entire regression space $\mathcal{X}$. For instance, it similarly penalizes a noisy function with many oscillations and a very smooth function with large jumps at a few locations. Thus, in the case of piecewise smooth target functions, minimizing $\|f\|_{\mathcal{H}}$ yields globally smoother functions and discards optimal solutions without distinguishing them from noisy functions. See Example 7.2 below for an illustration of this limitation.

7.2.2 Local Regularization

We now introduce local regularization terms which can distinguish between globally nonsmooth functions and smooth functions with large derivatives at sparse locations. For this, we define a local regularization functional of order (n, p) over the set $\mathcal{C}^n$ of functions of class C^n (i.e., the functions for which all the partial derivatives of order up to n exist and are continuous) as

$$\forall f \in \mathcal{C}^n,\ \forall \boldsymbol{x} \in \mathcal{X}, \qquad R_{n,p}(\boldsymbol{x}, f) = \|D^{(n)} f(\boldsymbol{x})\|_p, \tag{7.2}$$

where p is a parameter that selects a particular norm and $D^{(n)}$ is a differential operator of order n defined as[1]

$$D^{(n)} = \nabla D^{(n-1)} = \left[\frac{\partial}{\partial x_1} D^{(n-1)} \ \dots \ \frac{\partial}{\partial x_d} D^{(n-1)}\right]^\top, \tag{7.3}$$

with $D^{(0)}$ being the identity operator. In particular, if the function $f : \mathbb{R}^d \to \mathbb{R}$ is of class C^n, then, for $n \geq 1$, $D^{(1)} f = \nabla f$ is the gradient of f, for $n \geq 2$, $D^{(2)} f = \text{vec}(\nabla^2 f)$ is the vector representation (see (B.12)) of the Hessian and $D^{(n)} f$ is a function of $\mathbb{R}^d$ to $\mathbb{R}^{d^n}$ that computes all the nth-order derivatives of f.

[1] Note that the notation $D^{(n)}$ is not connected with the Jacobian operator D.

Given a function class $\mathcal{F} \subseteq C^n$, the learning problem with a local regularizer as in (7.2) reads

$$\min_{f \in \mathcal{F}} \sum_{k=1}^{N} \ell(y_k - f(\boldsymbol{x}_k)) + \lambda \sum_{i=1}^{M} R_{n,p}(z_i, f), \tag{7.4}$$

where local regularization terms are minimized for M sample points $z_i \in \mathcal{X}$ in order to globally regularize f over $\mathcal{X}$. Various sampling strategies can be considered here. The z_i's can be chosen on a grid in order to obtain a sufficient coverage of $\mathcal{X}$ or equal to the data points, $\boldsymbol{x}_k$, in order to obtain a representative sample of the data distribution (the latter is used in all experiments below). Another choice combining the two features is to sample z_i as perturbed versions of the $\boldsymbol{x}_k$.

7.2.3 Learning Smooth Models of PWA Functions by Convex Optimization

We now consider the squared loss function and models f built as kernel expansions over the data set as in (2.72)[2]:

$$f(\boldsymbol{x}) = \sum_{k=1}^{N} \alpha_k K(\boldsymbol{x}_k, \boldsymbol{x}). \tag{7.5}$$

By introducing the vector of parameters $\boldsymbol{\alpha} = \begin{bmatrix}\alpha_1 & \dots & \alpha_N\end{bmatrix}^\top$, the kernel matrix $\boldsymbol{K}$ as in (2.65) and the target vector $\boldsymbol{y} = \begin{bmatrix}y_1 & \dots & y_N\end{bmatrix}^\top$, the data term in (7.4) can be written as

$$\frac{1}{2}\|\boldsymbol{y} - \boldsymbol{K}\boldsymbol{\alpha}\|_2^2.$$

Observe that f in (7.5) is a linear combination of kernel functions with weights $\boldsymbol{\alpha}$. Thus, any derivative of f is also linear with respect to the weights $\boldsymbol{\alpha}$ and can be expressed at a point z_i as $\boldsymbol{d}_{il}^\top \boldsymbol{\alpha}$ with a vector $\boldsymbol{d}_{il}^\top \in \mathbb{R}^N$ containing the coefficients of the linear combination of the weights for that particular derivative (see Example 7.1 below). Hence, we can rewrite $D^{(n)} f(z_i)$ as

$$D^{(n)} f(z_i) = \begin{bmatrix} \boldsymbol{d}_{i1}^\top \\ \vdots \\ \boldsymbol{d}_{id^n}^\top \end{bmatrix} \boldsymbol{\alpha} = \boldsymbol{D}_i \boldsymbol{\alpha}, \tag{7.6}$$

[2]Note that here form (7.5) is arbitrarily fixed in advance and not the solution of a particular regularized learning problem as for (2.72).

with the matrix $\boldsymbol{D}_i$ containing the coefficients for all the d^n derivatives involved in $D^{(n)} f(z_i)$. This yields the finite-dimensional optimization problem

$$\min_{\boldsymbol{\alpha}\in\mathbb{R}^N} \frac{1}{2}\|\boldsymbol{y} - \boldsymbol{K}\boldsymbol{\alpha}\|_2^2 + \lambda \sum_{i=1}^{M} \|\boldsymbol{D}_i\boldsymbol{\alpha}\|_p, \tag{7.7}$$

where the convexity of the data term results from the choice of a convex loss function and where the regularization term is a sum of norms of linear functions, hence convex. In particular, problem (7.7) can be set as a quadratic program for $p = \infty$ or as a second-order cone program for $p = 2$.

In (7.7), we minimize the ℓ_1-norm

$$\sum_{i=1}^{M} \|\boldsymbol{D}_i\boldsymbol{\alpha}\|_p = \sum_{i=1}^{M} \|D^{(n)} f(z_i)\|_p = \left\| \begin{bmatrix} \|D^{(n)} f(z_1)\|_p \\ \vdots \\ \|D^{(n)} f(z_M)\|_p \end{bmatrix} \right\|_1 .$$

This leads to a sparse solution in which the vectors of derivatives, $D^{(n)} f(z_i)$, will be drawn to zero at points z_i where their norm is small, while large derivatives will be left at few points. Note that it is crucial here not to use *squared* ℓ_p-norms of the $\boldsymbol{D}_i\boldsymbol{\alpha}$'s in (7.7) which would not favor sparse derivatives, as already emphasized in Sect. 6.2.4.4.

Standard choices for p are $p = 2$ or $p = \infty$ rather than $p = 1$ which would lead to sparse vectors of derivatives but in terms of their components. As a result, the regularization term would not be able to distinguish between a single jump with a nonzero vector $D^{(n)} f(z_i)$ and many axis-aligned jumps with multiple sparse vectors $D^{(n)} f(z_i)$.

PWA functions have zero second-order derivatives almost everywhere. Thus, they can be learned by setting $n = 2$ in (7.6)–(7.7). For $p = 2$, this amounts to penalizing the Frobenius norm (B.5) of the Hessian matrix computed at all sample points z_i. Note that in this approach, no error is implied by the computation of derivatives from the data. We only use the derivatives of a smooth model (7.5), which can be exactly and analytically computed for all $z_i \in \mathcal{X}$.

Example 7.1 (*Derivatives of a kernel expansion with the Gaussian kernel*) Consider the case $n = 1$ for which we have to compute the first-order derivatives of f in (7.5). Focusing on the Gaussian kernel (2.69) and denoting by $x_{k,l}$ (respectively, $z_{i,l}$) the lth component of the kth data point $\boldsymbol{x}_k$ (respectively, ith point z_i), we have, for any $i \in [M]$ and $l \in [d]$,

$$
\begin{aligned}
\left.\frac{\partial f(\boldsymbol{x})}{\partial x_l}\right|_{\boldsymbol{x}=\boldsymbol{z}_i} &= \sum_{k=1}^{N} \alpha_k \left.\frac{\partial K(\boldsymbol{x}_k, \boldsymbol{x})}{\partial x_l}\right|_{\boldsymbol{x}=\boldsymbol{z}_i} \\
&= \sum_{k=1}^{N} \alpha_k K(\boldsymbol{x}_k, \boldsymbol{z}_i) \frac{x_{k,l} - z_{i,l}}{\sigma^2} \\
&= \boldsymbol{d}_{il}^\top \boldsymbol{\alpha}
\end{aligned}
$$

with

$$
\boldsymbol{d}_{il}^\top = \frac{1}{\sigma^2} \left[K(\boldsymbol{x}_1, \boldsymbol{z}_i)(x_{1,l} - z_{i,l}) \ \dots \ K(\boldsymbol{x}_N, \boldsymbol{z}_i)(x_{N,l} - z_{i,l}) \right]
$$

to be used in (7.6) for the construction of $\boldsymbol{D}_i$. When $n = 2$, the second-order derivatives can be computed similarly through the derivatives of the kernel function.

Example 7.2 Figure 7.2 illustrates the application of the method for learning a PWA target function over $\mathcal{X} = [-10, 10]$ with a large jump at $x = 1$ from a noisy data set (plot a). Using a Gaussian kernel of width $\sigma = 0.5$ and with $\lambda = 0.1$ yields both an accurate estimation of the jump at $x = 1$ and a very smooth model at other locations (plot b). On the other hand, the classical regularization scheme (2.70) used by kernel ridge regression with λ tuned so that the model accurately fits the jump yields a model with many oscillations that also tends to fit the noise (plot c). In addition, increasing the regularization parameter λ reduces the effect of noise, but also dramatically decreases the accuracy of the model, in particular around the jump of the target function (plot d).

7.2.4 Recovering PWA Models

A simple procedure can be used to recover a PWA model from the solution of (7.7). In few words, this relies on the clustering of vectors built from the gradient of f computed at the data points. In this procedure, a preliminary step, which allows for the points in different "pieces" to be correctly clustered, is the removal of the points close to the boundaries between the pieces. This can be implemented by thresholding the norm of derivatives, $\|D^{(2)} f(\boldsymbol{x}_k)\|_p$, since the smoothness assumptions on f in (7.7) and the submodels f_j imply a zero norm inside each region $\mathcal{X}_j$. Once the data is clustered, the mode estimates $\hat{q}_k$ are obtained and we can proceed as in Step 2 of Algorithm 8 to estimate the PWA model.

7.3 From Local Models to PWA Models

We now turn to methods that are truly dedicated to the estimation of PWA models. However, due to the difficulty of estimating the intertwined submodels and classifier

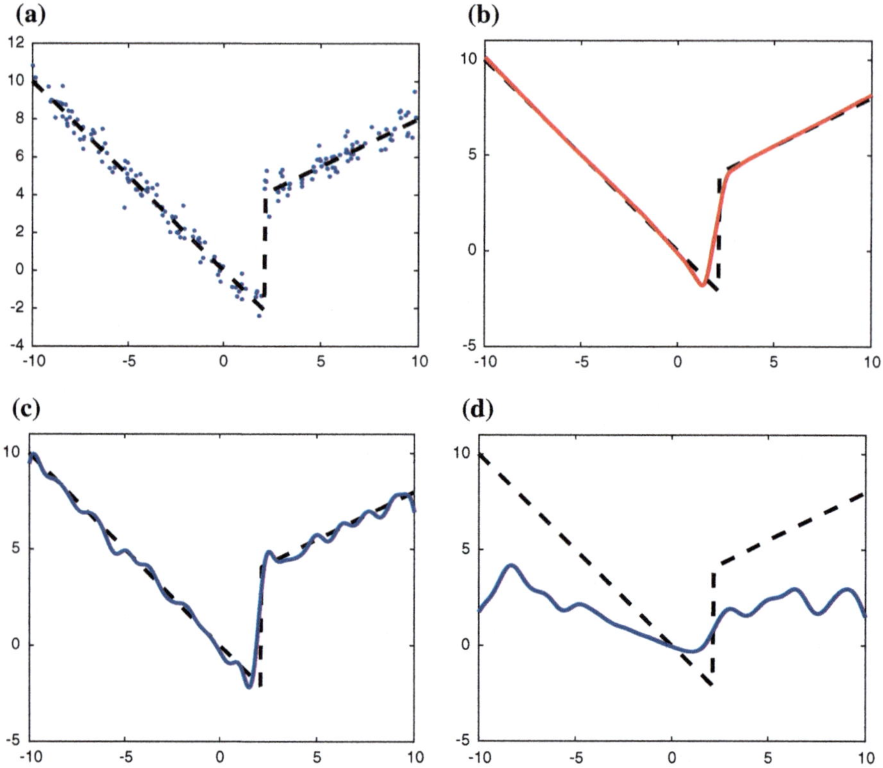

Fig. 7.2 Learning a PWA target function (dash line) from a noisy data set (**a**): model obtained with local regularization (**b**) and models obtained with global regularization for a small (**c**) or large (**d**) regularization parameter λ

simultaneously, most of these approaches follow the two-stage scheme of Algorithm 8. But, contrary to the procedure described in Sect. 7.1, they introduce the structure of the PWA model in the first stage for the estimation of the mode. This is usually done by decomposing this stage in two steps: one that estimates local models and one that clusters these into modes. This leads to the overall procedure of Algorithm 9.

Algorithm 9 Overview of methods based on local models for PWA model estimation

1. Estimate the modes $\{\hat{q}_k\}_{k=1}^N$ (and $\hat{s} = \max_{k\in[N]} \hat{q}_k$)

   a. Estimate one local model f_k for the approximation of each data point $(\boldsymbol{x}_k, y_k)$.
   b. Cluster the local models into modes, i.e., estimate $\{\hat{q}_k\}_{k=1}^N$.

2. a. Estimate the classifier g from $\{(\boldsymbol{x}_k, \hat{q}_k)\}_{k=1}^N$.
   b. Estimate the submodels f_j from the local data sets $\{(\boldsymbol{x}_k, y_k) : \hat{q}_k = j\}$, $j = 1, \dots, \hat{s}$.

The rationale behind this approach is the following. Given an overparametrization that considers one local model f_k for each data point, the data points generated by the same mode should be accurately approximated by similar local models. However, without further constraints/assumptions, the local models can arbitrarily well estimate the noise and end up being all different. Two major strategies can be considered at this point. In the first one, the locality of a local model is extended in the estimation phase 1.(a) to a neighborhood around a data point rather than a single point, thus averaging the effect of the noise. This is reasonable because a small neighborhood of a point in one of the regions $\mathcal{X}_j$ of $\mathcal{X}$ is likely to be contained in that same region. In the second approach, the variations between local models are controlled during their estimation in order to limit the number of different local models obtained.

The following details these two approaches for linear local models

$$f_k(\boldsymbol{x}) = \boldsymbol{x}^\top \boldsymbol{\theta}(k) \tag{7.8}$$

with parameter vectors $\boldsymbol{\theta}(k) \in \mathbb{R}^d$.

7.3.1 Fitting Local Models over Neighborhoods

Here, we focus on Step 1.(a) of Algorithm 9 and consider the estimation of local models over neighborhoods $\mathcal{N}(\boldsymbol{x}_k)$ of the regression vectors $\boldsymbol{x}_k$. Note that we restrict these neighborhoods to subsets of $\mathcal{D}_x = \{\boldsymbol{x}_k\}_{k=1}^N$ where information is available via $\{y_k\}_{k=1}^N$ and not actual (continuous) neighborhoods in $\mathcal{X}$. Such neighborhoods can be constructed in different manners, by focusing either on the distance to the neighbors or their number. Since we will need to ensure that the neighborhoods contain enough data points, here we define $\mathcal{N}(\boldsymbol{x}_k)$ as the set of n nearest neighbors from $\boldsymbol{x}_k$ in $\mathcal{D}_x$.

Then, given a choice of loss function ℓ, the local models can be estimated as

$$\boldsymbol{\theta}(k) = \operatorname*{argmin}_{\boldsymbol{\theta} \in \mathbb{R}^d} \sum_{\boldsymbol{x}_i \in \mathcal{N}(\boldsymbol{x}_k)} \ell(y_i - \boldsymbol{x}_i^\top \boldsymbol{\theta}), \qquad k = 1, \dots, N. \tag{7.9}$$

For the squared loss (2.28), this can be done efficiently. Indeed, this amounts to N independent and small least squares problems over n points in dimension d, whose solutions can be shown as in Sect. 2.1.2 to take the form (2.33), i.e.,

$$\boldsymbol{\theta}(k) = (\boldsymbol{X}_k^\top \boldsymbol{X}_k)^{-1} \boldsymbol{X}_k^\top \boldsymbol{y}_k,$$

where $\boldsymbol{X}_k$ is the submatrix of $\boldsymbol{X}$ containing only the points in $\mathcal{N}(\boldsymbol{x}_k)$ and $\boldsymbol{y}_k$ is a subvector of $\boldsymbol{y}$ containing the corresponding output values. Note that this formula only makes sense when $\boldsymbol{X}_k$ is full rank, which requires $n \geq d$.

This approach can also be made more robust by taking into account the distance to the neighbors in a weighted least squares formulation. Given a positive proximity function $\rho(\boldsymbol{x}_k, \boldsymbol{x}_i)$ decreasing with the distance $\|\boldsymbol{x}_k - \boldsymbol{x}_i\|_2$, e.g.,

$\rho(\boldsymbol{x}_k, \boldsymbol{x}_i) = 1/\|\boldsymbol{x}_k - \boldsymbol{x}_i\|_2$, one then computes the parameter vectors with

$$\boldsymbol{\theta}(k) = \underset{\boldsymbol{\theta}\in\mathbb{R}^d}{\operatorname{argmin}} \sum_{\boldsymbol{x}_i\in\mathcal{N}(\boldsymbol{x}_k)} \rho(\boldsymbol{x}_k, \boldsymbol{x}_i)(y_i - \boldsymbol{x}_i^\top\boldsymbol{\theta})^2, \qquad k = 1, \dots, N,$$

instead of (7.9). Such weighted least squares problem can also be solved explicitly. Let us write the cost function of the kth problem in matrix form as

$$J_k^{WLS}(\boldsymbol{\theta}) = (\boldsymbol{y}_k - \boldsymbol{X}_k\boldsymbol{\theta})^\top \boldsymbol{R}_k(\boldsymbol{y}_k - \boldsymbol{X}_k\boldsymbol{\theta})$$

with a diagonal matrix $\boldsymbol{R}_k$ of entries $\rho(\boldsymbol{x}_k, \boldsymbol{x}_i)$ on the diagonal. Then, the solution is the vector $\boldsymbol{\theta}$ at which the gradient

$$\nabla J_k^{WLS}(\boldsymbol{\theta}) = -2\boldsymbol{X}_k^\top \boldsymbol{R}_k^\top \boldsymbol{y}_k + 2\boldsymbol{X}_k^\top \boldsymbol{R}_k\boldsymbol{X}_k\boldsymbol{\theta}$$

is zero. If $\boldsymbol{X}_k^\top \boldsymbol{R}_k\boldsymbol{X}_k$ is invertible, this leads to

$$\boldsymbol{\theta}(k) = (\boldsymbol{X}_k^\top \boldsymbol{R}_k\boldsymbol{X}_k)^{-1}\boldsymbol{X}_k^\top \boldsymbol{R}_k^\top \boldsymbol{y}_k.$$

In practice, this strategy can be very efficiently implemented but requires careful tuning of the number n of neighbors. Indeed, the optimal choice depends on the (typically unknown) noise level. Without noise, n could be set to its minimum value, d, in order to limit the number of neighborhoods crossing the mode boundaries. One the other hand, larger amounts of noise call for larger values of n, and a trade-off has to be found between the number of neighborhoods including points of different modes and the impact of the noise on the estimation of the parameter vectors.

7.3.2 Controlling the Variations of the Local Models

We now consider the alternative approach for the estimation of the local models (Step 1.(a) of Algorithm 9), in which their variations are controlled during their estimation. This means that we now aim at the simultaneous estimation of the collection of N local models.

For this purpose, we will consider the method of Sect. 6.2.4.3 based on a sum-of-norms regularization term for SARX system identification. Recall that this approach was based on the minimization of the variations of a parameter function also assigning a parameter vector to every data point and that a main issue was how to compute these variations. For data points ordered in time, the data set could be segmented in a few modes by computing variations between two time steps. For a PWA model, the variations should only appear at the mode boundaries. This can be taken into account by computing variations of the parameter function between points that are close in the regression space $\mathcal{X}$. Indeed, points that are close in $\mathcal{X}$ are likely to be in the same region $\mathcal{X}_j$.

This leads to the (convex) optimization problem

$$\min_{\{\boldsymbol{\theta}(k)\in\mathbb{R}^d\}_{k=1}^N} \sum_{k=1}^N \ell\left(y_k - \boldsymbol{x}_k^\top \boldsymbol{\theta}(k)\right) + \lambda \sum_{1\le k<l\le N} \rho(\boldsymbol{x}_k, \boldsymbol{x}_l)\|\boldsymbol{\theta}(k) - \boldsymbol{\theta}(l)\|_p, \tag{7.10}$$

where λ is a regularization parameter to control the trade-off between the accuracy and the variations of the model parameters and $\rho(\boldsymbol{x}_k, \boldsymbol{x}_l)$ is a proximity function to evaluate the similarity between two regressors. It may be a binary function that evaluates to 1 for two points that are considered as neighbors and to 0 otherwise, for instance based on n-nearest neighbors, or computed as

$$\rho(\boldsymbol{x}_k, \boldsymbol{x}_l) = \mathbb{1}_{\|x_l - x_k\|_2 \le \epsilon}$$

for some threshold $\epsilon > 0$. But, it can also be a continuous and decreasing function of the distance between the points.

Similarly as in Sect. 6.2.3.2, to get more zeros in the regularization term and further limit the number of different parameter vectors, an iteratively reweighted technique can be used as

$$\min_{\{\boldsymbol{\theta}(k)\in\mathbb{R}^d\}_{k=1}^N} \sum_{k=1}^N \ell\left(y_k - \boldsymbol{x}_k^\top \boldsymbol{\theta}(k)\right) + \lambda \sum_{1\le k<l\le N} w_{kl}^{(i)} \rho(\boldsymbol{x}_k, \boldsymbol{x}_l)\|\boldsymbol{\theta}(k) - \boldsymbol{\theta}(l)\|_p, \tag{7.11}$$

where $w_{kl}^{(i)}$ is the weight at the ith iteration, with $w_{kl}^{(0)} = 1$.

As for Sect. 6.2.4.3, the computational burden of this approach depends on the choice of loss function ℓ for the data term and ℓ_p-norm for the regularization term. With the squared loss (2.28), (7.10)–(7.11) can be cast as second-order cone programs when choosing the ℓ_2-norm ($p = 2$). However, the number of conic constraints will be quadratic in N, which limits the practicality of the approach on large sets. This increase of complexity when compared to the linear number of conic constraints in Sect. 6.2.4.3 is due to the difficulty of measuring the variations of the parameter function over a multi-dimensional space $\mathcal{X}$ instead of just the axis of time.

With a proper choice of the parameter λ, this method can yield a very small number of different parameter vectors, or at least a set of parameter vectors tightly clustered around a few. This makes the next clustering step easier and possibly even unnecessary. Ideally, the number of modes s could be estimated by the number of different parameter vectors and the mode estimates directly given by the values of the $\boldsymbol{\theta}(k)$ if for all $k \in [N]$, $\boldsymbol{\theta}(k) \in \{\boldsymbol{\theta}_j\}_{j=1}^s$ for some small $s \ll N$.

7.3.3 Clustering the Local Models

Clustering of the local models constitutes Step 1.(b) of the global procedure in Algorithm 9. It can be performed by standard methods discussed in Sect. 3.2, but can also be dedicated and improved for PWA model estimation.

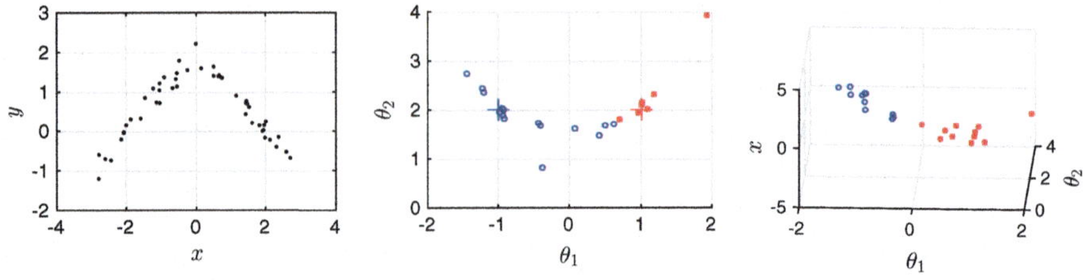

Fig. 7.3 *Left:* data set from Example 7.3 for PWA regression with two modes. *Middle:* local parameter vectors in $\mathbb{R}^2$ estimated from local neighborhoods and clustered into two modes ($\circ$ and $*$) with 22% of classification errors. The true parameter vectors appear as large $+$ marks. *Right:* feature vectors $\boldsymbol{\varphi}_k$ in $\mathbb{R}^3$ computed by (7.12) and easily clustered into two modes without errors

The general idea is to estimate the mode q_k by clustering local models in parameter space, i.e., to cluster the parameter vectors $\boldsymbol{\theta}(k)$ of the linear models in (7.8). But additional information can be included by applying clustering to recover groups of feature vectors $\boldsymbol{\varphi}_k$, computed for instance as

$$\boldsymbol{\varphi}_k = \begin{bmatrix} \boldsymbol{\theta}(k) \\ \boldsymbol{x}_k \end{bmatrix}, \qquad k = 1, \dots, N.$$

With such feature vectors, the location of the local models is taken into account in addition to their parameters, thus yielding a more spatially coherent clustering. For the method of Sect. 7.3.1 based on local neighborhoods, this technique is typically made more robust by using the center of the neighborhood instead:

$$\boldsymbol{\varphi}_k = \begin{bmatrix} \boldsymbol{\theta}(k) \\ \frac{1}{n}\sum_{\boldsymbol{x}_i \in \mathcal{N}(\boldsymbol{x}_k)} \boldsymbol{x}_i \end{bmatrix}, \qquad k = 1, \dots, N. \tag{7.12}$$

Another piece of information that we can use is contained in y_k and we can also define two kinds of feature vectors depending on whether we include y_k in $\boldsymbol{\varphi}_k$ or the average output value over the neighborhood, $\frac{1}{n}\sum_{\boldsymbol{x}_i \in \mathcal{N}(\boldsymbol{x}_k)} y_i$.

Example 7.3 (Influence of the feature space components on the clustering of local models) Figure 7.3 shows how the straightforward clustering of the parameter vectors can be improved by using additional information on the data points. In particular, directly clustering the parameter vectors given by (7.9) with $n = 7$ yields many classification errors due to the noise in the data, which makes the local parameter vectors distributed in an unstructured elongated point cloud of $\mathbb{R}^2$. On the other hand, by including information on the location $\boldsymbol{x}_k$ in $\boldsymbol{\varphi}_k$ as in (7.12), two clusters clearly emerge in $\mathbb{R}^3$ and all classification errors are avoided.

Notes

The estimation of PWA models via switching regression as discussed in Sect. 7.1 was considered in, e.g., [1, 2]. An analysis of the behavior of the error sparsification method of Sect. 6.2.3 for PWA model estimation was conducted in [3].

The method of Sect. 7.2 based on the estimation of a smooth kernel model is described in [4].

The clustering-based method due to [5] was one of the seminal works on PWA system identification. It suggested the general two-stage approach of Algorithm 8 and also paved the way for the approaches based on the estimation of one local model for every data point. Among these, the sum-of-norms approach originally proposed for the segmentation of time series in [6] was extended to PWA systems in [7, 8], as described in Sect. 7.3.1. The minimization of a trade-off between a data fitting term and regularization term controling the complexity of the overall model is also at the core of the data compression approach of [9, 10], in which the complexity is measured via the Delaunay triangulation associated to the data points.

The clustering-based method of [5] has also been modified and extended in many ways. Variations around that theme include mainly methods focusing on the clustering step [11–13]. This clustering step was also directly tackled in [14] from the data without the estimation of local models. In [15], this is performed with a variant of the k-means algorithm in the flavor of k-LinReg: The iterative algorithm alternates between assignments of the points to the modes and parameter updates. An extension of the clustering-based approach to PWA output error models was proposed in [16] based on the instrumental variable method [17].

Software

The Hybrid Identification Toolbox (HIT) for MATLAB implements the clustering-based method for regression with PWA maps and the identification of PWARX models. It is available at http://sisdin.unipv.it/lab/personale/pers_hp/ferrari/HIT_toolbox.php.

References

1. Bemporad, A., Garulli, A., Paoletti, S., Vicino, A.: A bounded-error approach to piecewise affine system identification. IEEE Trans. Autom. Control **50**(10), 1567–1580 (2005)
2. Juloski, A.Lj., Weiland, S., Heemels, W.P.M.H.: A Bayesian approach to identification of hybrid systems. IEEE Trans. Autom. Control **50**(10), 1520–1533 (2005)
3. Bako, L., Boukharouba, K., Lecoeuche, S.: Recovering piecewise affine maps by sparse optimization. In: Proceedings of the 16th IFAC Symposium on System Identification (SYSID), Brussels, Belgium, IFAC Proceedings Volumes **45**(16), 356–361 (2012)
4. Lauer, F., Le, V.L., Bloch, G.: Learning smooth models of nonsmooth functions via convex optimization. In: Proceedings of the IEEE International Workshop on Machine Learning for Signal Processing (MLSP), Santander, Spain, pp. 1–6 (2012)
5. Ferrari-Trecate, G., Muselli, M., Liberati, D., Morari, M.: A clustering technique for the identification of piecewise affine systems. Automatica **39**(2), 205–217 (2003)

6. Ohlsson, H., Ljung, L., Boyd, S.: Segmentation of ARX-models using sum-of-norms regularization. Automatica **46**(6), 1107–1111 (2010)
7. Ohlsson, H., Ljung, L.: Identification of piecewise affine systems using sum-of-norms regularization. In: Proceedings of the 18th IFAC World Congress, Milano, Italy, IFAC Proceedings Volumes, vol. 44(1), pp. 6640–6645 (2011)
8. Ohlsson, H., Ljung, L.: Identification of switched linear regression models using sum-of-norms regularization. Automatica **49**(4), 1045–1050 (2013)
9. Maruta, I., Sugie, T.: Identification of PWA models via data compression based on ℓ_1 optimization. In: Proceedings of the 50th IEEE Conference on Decision and Control and European Control Conference (CDC-ECC), Orlando, FL, USA, pp. 2800–2805 (2011)
10. Maruta, I., Sugie, T.: Identification of PWA models via optimal data compression. In: Proceedings of the 16th Symposium on System Identification, Brussels, Belgium, IFAC Proceedings Volumes, vol. 45(16), pp. 368–373 (2012)
11. Gegundez, M.E., Aroba, J., Bravo, J.M.: Identification of piecewise affine systems by means of fuzzy clustering and competitive learning. Eng. Appl. Artif. Intell. **21**(8), 1321–1329 (2008)
12. Baptista, R.D.S., Ishihara, J.Y., Borges, G.A.: A split and merge algorithm for identification of piecewise affine systems. In: Proceedings of the 2011 American Control Conference, San Francisco, CA, USA, pp. 2018–2023 (2011)
13. Lassoued, Z., Abderrahim, K.: An experimental validation of a novel clustering approach to PWARX identification. Eng. Appl. Artif. Intell. **28**, 201–209 (2014)
14. Nakada, H., Takaba, K., Katayama, T.: Identification of piecewise affine systems based on statistical clustering technique. Automatica **41**(5), 905–913 (2005)
15. Boukharouba, K., Bako, L., Lecoeuche, S.: Identification of piecewise affine systems based on Dempster-Shafer theory. In: Proceedings of the 15th IFAC Symposium on System Identification (SYSID), Saint-Malo, France, IFAC Proceedings Volumes, vol. 42(10), pp. 1662–1667 (2009)
16. Canty, N., O'Mahony, T., Cychowski, M.T.: An output error algorithm for piecewise affine system identification. Control Eng. Pract. **20**(4), 444–452 (2012)
17. Söderström, T.D., Stoica, P.: Instrumental Variable Methods for System Identification. Lecture Notes in Control and Information Sciences, vol. 57. Springer, Berlin (1983)

Chapter 8
Recursive and State-Space Identification of Hybrid Systems

In batch algorithms presented so far, the model is identified from all the collected input–output data. Thus, batch methods are not suitable for real-time applications in which the data need to be processed on-line. Contrarily, recursive identification algorithms seek a model on-line, i.e., adjusted by taking into account only the new current measurement (see Sect. 2.3). For hybrid models, such algorithms can also be an interesting alternative to the batch ones from a computational point of view, since the complexity of most batch algorithms increases rapidly with the number of data points. Moreover, the data used for off-line identification may not cover all the modes of the system. This problem can be solved by on-line methods that can add new modes during the process. In addition, segmentation of time series, which is in itself a whole domain, can be understood as detecting and isolating the time segments of a signal with different behaviors. These behaviors can be expressed in form of models or parameters in models, thus the link to hybrid systems. Such a problem has been tackled in the preceding chapters by batch approaches, particularly in Sect. 6.2.4.1. Therefore, there are three main reasons to develop recursive approaches for hybrid systems identification: availability of an on-line updated model, reduction of the computational cost of the batch approaches, and on-line segmentation of time series.

In the preceding chapters, ARX input–output (I/O) hybrid models were considered. Although the dynamical aspect of the submodels was included by incorporating delayed measurements in the regression vector, the problem stayed fundamentally a *regression* problem dealing with static and dynamical data in the same way. Note nevertheless the previously mentioned extension of the sum-of-norms to time series processing.

Dealing specifically with dynamical systems is the automatic control domain, which encompasses both state-space (SS) model formulations and real-time processing. Thus, in this more control-oriented chapter, we regroup the recursive identification of switched and piecewise linear I/O models and the batch identification of SS hybrid models. As indicated in the introduction (Chap. 1), in what follows, the regression vector $\boldsymbol{x}$ of I/O models must not be confused with the state vector $\mathbf{x}$ of SS models.

F. Lauer and G. Bloch, *Hybrid System Identification*,
Lecture Notes in Control and Information Sciences 478,
https://doi.org/10.1007/978-3-030-00193-3_8

Chapter Outline

This chapter presents separately the recursive identification of I/O hybrid models in Sect. 8.1 and the identification of SS hybrid models in Sect. 8.2.

Section 8.1.1 details the approaches which alternate between the assignation of the current data point to the closest mode and the update of the corresponding model, for switched and piecewise linear/affine systems. Then, Sect. 8.1.2 presents two different approaches to track the output signal: the extension of the algebraic approach of Sect. 6.1.1 and the methods stemming from the field of adaptive filtering.

For the identification of linear SS hybrid models, after the presentation of the problem and the associated difficulties in Sect. 8.2.1, the various batch methods are described in Sect. 8.2.2.

8.1 Input–Output Models

Some "historical" batch approaches for the identification of hybrid I/O systems can be quite easily extended to recursive ones. In fact, two major ways emerge. The first one runs in parallel several recursive identifiers, from a known or assumed number of submodels, assigning the current data to the "closest" model. In the second one, only one model is recursively identified to track the behavior of the output signal. This can yield directly the model parameters, as in the natural extension of the algebraic approach (see Sect. 6.1.1) to a recursive version. Otherwise, techniques of detection of abrupt changes are applied, at each time instant or backward, after the time series has been processed. In the latter case, one resorts to techniques stemming from the field of adaptive filtering and identification of time-varying systems, outside the hybrid system identification framework presented so far.

8.1.1 Parallel Identifiers

The idea developed here is to run recursively as many identification algorithms as the number of submodels. In this spirit, two ways can be derived to identify linear/affine ARX hybrid models (4.3), $y_k = \boldsymbol{x}_k^\top \boldsymbol{\theta}_{q_k} + \varepsilon_k$, with known orders n_a and n_b. In the first and simple approach, the parameters of the currently recognized submodel are directly updated. The second one estimates the probability density functions (pdf) of the parameters instead of a simple value.

8.1.1.1 On-line Mode Detection and Parameter Update

Switched ARX Model

Starting from some initial values for the parameter vectors of the s different submodels, the procedure alternates between data assignment to submodels and parameter update. At each time instant, the discrete state is determined as the index of the closest submodel in terms of prediction error (or posterior error). Given the estimated discrete state, the associated parameter vector is updated with an adaptive linear identifier.

More precisely, a recursive least squares (RLS) algorithm (2.60) can be used. If the mode j is active at time k, i.e., $q_k = j$, only the parameter vector $\hat{\boldsymbol{\theta}}_j$ is updated as

$$\begin{cases} \hat{\boldsymbol{\theta}}_{jk} = \hat{\boldsymbol{\theta}}_{j(k-1)} + \boldsymbol{l}_{jk}\left(y_k - \boldsymbol{x}_k^\top \hat{\boldsymbol{\theta}}_{j(k-1)}\right) \\ \boldsymbol{l}_{jk} = \dfrac{\boldsymbol{P}_{j(k-1)}\boldsymbol{x}_k}{\gamma + \boldsymbol{x}_k^\top \boldsymbol{P}_{j(k-1)}\boldsymbol{x}_k} \\ \boldsymbol{P}_{jk} = \dfrac{1}{\gamma}\left(\boldsymbol{I} - \boldsymbol{l}_{jk}\boldsymbol{x}_k^\top\right)\boldsymbol{P}_{j(k-1)}, \end{cases} \tag{8.1}$$

where $\gamma \in (0, 1)$ is the forgetting factor, while the others are kept unchanged: $\hat{\boldsymbol{\theta}}_{ik} = \hat{\boldsymbol{\theta}}_{i(k-1)}$, $\boldsymbol{l}_{ik} = \boldsymbol{l}_{i(k-1)}$ and $\boldsymbol{P}_{ik} = \boldsymbol{P}_{i(k-1)}$, $i \neq j$. To initialize the parallel estimators, for $j = 1, ..., s$, $\hat{\boldsymbol{\theta}}_{j0}$ can be randomly drawn and $\boldsymbol{P}_{j0}$ is set to $p_0\boldsymbol{I}$, with $p_0 \gg 1$.

The unknown discrete state q_k is estimated as the index of the submodel that has most presumably generated the data pair $(\boldsymbol{x}_k, y_k)$ in the sense of a decision criterion d_{jk}:

$$\hat{q}_k = \operatorname*{argmin}_{j\in[s]} d_{jk}. \tag{8.2}$$

The decision criterion can be a normalized function of the fitting prior error $e_{jk} = y_k - \boldsymbol{x}_k^\top \hat{\boldsymbol{\theta}}_{j(k-1)} = \boldsymbol{z}_k^\top \hat{\boldsymbol{\beta}}_{j(k-1)}$, with $\boldsymbol{z}_k = \begin{bmatrix} y_k & -\boldsymbol{x}_k^\top \end{bmatrix}^\top$ and $\hat{\boldsymbol{\beta}}_{j(k-1)} = \begin{bmatrix} 1 & \hat{\boldsymbol{\theta}}_{j(k-1)}^\top \end{bmatrix}^\top$:

$$d_{jk}^{prior} = \frac{\left|e_{jk}\right|}{\|\hat{\boldsymbol{\beta}}_{j(k-1)}\|_2} = \frac{\left|e_{jk}\right|}{\sqrt{1 + \hat{\boldsymbol{\theta}}_{j(k-1)}^\top \hat{\boldsymbol{\theta}}_{j(k-1)}}},$$

or of the posterior error $\eta_{jk} = y_k - \boldsymbol{x}_k^\top \hat{\boldsymbol{\theta}}_{jk} = \boldsymbol{z}_k^\top \hat{\boldsymbol{\beta}}_{jk}$, with $\boldsymbol{z}_k = \begin{bmatrix} y_k & -\boldsymbol{x}_k^\top \end{bmatrix}^\top$ and $\hat{\boldsymbol{\beta}}_{jk} = \begin{bmatrix} 1 & \hat{\boldsymbol{\theta}}_{jk}^\top \end{bmatrix}^\top$:

$$d_{jk}^{post} = \frac{\left|\eta_{jk}\right|}{\|\hat{\boldsymbol{\beta}}_{jk}\|_2} = \frac{\gamma\left|e_{jk}\right|}{\left(\gamma + \boldsymbol{x}_k^\top \boldsymbol{P}_{j(k-1)}\boldsymbol{x}_k\right)\sqrt{1 + \|\hat{\boldsymbol{\theta}}_{j(k-1)} + e_{jk}\boldsymbol{l}_{j(k-1)}\|_2^2}}.$$

This posterior error criterion can be obtained without explicitly calculating $\hat{\boldsymbol{\theta}}_{jk}$, thus $\hat{\boldsymbol{\beta}}_{jk}$. Indeed, from (8.1), we have:

$$\begin{aligned}\eta_{jk} &= y_k - \boldsymbol{x}_k^\top \hat{\boldsymbol{\theta}}_{jk} = y_k - \boldsymbol{x}_k^\top \left(\hat{\boldsymbol{\theta}}_{j(k-1)} + \boldsymbol{l}_{jk} e_{jk}\right) = e_{jk} - \boldsymbol{x}_k^\top \boldsymbol{l}_{jk} e_{jk} \\ &= \left(1 - \frac{\boldsymbol{x}_k^\top \boldsymbol{P}_{j(k-1)} \boldsymbol{x}_k}{\gamma + \boldsymbol{x}_k^\top \boldsymbol{P}_{j(k-1)} \boldsymbol{x}_k}\right) e_{jk} = \left(\frac{\gamma}{\gamma + \boldsymbol{x}_k^\top \boldsymbol{P}_{j(k-1)} \boldsymbol{x}_k}\right) e_{jk}.\end{aligned}$$

It can be noted that the true number s of submodels does not need to be known. The algorithm is able to operate with an upper bound $\bar{s} \ll N$ of initially presumed submodels. Once the algorithm has converged, the $\bar{s} - s$ submodels to which a significant number of data have not been affected can be recognized, and thus eliminated.

Switched OE Model

The switched output error (SOE) model can be written $y_k = \hat{y}_k + \varepsilon_k = \boldsymbol{x}_k^\top \boldsymbol{\theta}_{q_k} + \varepsilon_k$, with $\boldsymbol{x}_k = \left[-\hat{y}_{k-1} \ \dots \ -\hat{y}_{k-n_a} \ u_{k-1} \ \dots \ u_{k-n_b}\right]^\top$ (4.8) and $\boldsymbol{\theta}_{q_k} = \left[a_{q_k 1} \ \dots \ a_{q_k n_a} \ b_{q_k 1} \ \dots \ b_{q_k n_b}\right]$ (see 4.5)).

For SOE recursive identification, in the preceding algorithm of mode detection and parameter update, the regression vector $\boldsymbol{x}_k$ is replaced by

$$\hat{\boldsymbol{x}}_k = \left[-\hat{y}_{k-1|k-1} \ \dots \ -\hat{y}_{k-n_a|k-n_a} \ u_{k-1} \ \dots \ u_{k-n_b}\right]^\top,$$

where $\hat{y}_{k-i|k-i}$ is the output at time $k-i$ estimated from the parameter vector $\hat{\boldsymbol{\theta}}_{\hat{q}_{k-i}}$.

A convergence problem can arise from the nonlinear nature of the optimization problem. The initial values of the parameter vectors can be randomly drawn or set to those given from the identification of a SARX model, but this can lead to biased estimates, due to possible convergence to local optima. The use of a low-pass prefilter to generate the estimated outputs improves the convergence to accurate estimates.

Piecewise Affine (PWA) Model

The preceding identification algorithm for switched (SARX) models can be extended to piecewise affine (PWARX) models based on a Voronoï partition of the regression space $\mathcal{X}$. Given a set of s centers, $\{\boldsymbol{c}_j\}_{j=1}^s \subset \mathcal{X}$, the regions are defined as in (4.16) by

$$\mathcal{X}_j = \left\{\boldsymbol{x} \in \mathcal{X} \ : \ j = \underset{i \in [s]}{\operatorname{argmin}} \|\boldsymbol{x} - \boldsymbol{c}_i\|_2\right\}, \quad j = 1, \dots, s.$$

The current mode is recursively determined as

$$\hat{q}_k = \underset{j \in [s]}{\operatorname{argmin}} \, d_{jk}^{prior} + \alpha \|\boldsymbol{x}_k - \hat{\boldsymbol{c}}_{j(k-1)}\|_2, \tag{8.3}$$

where the second term encodes the switching mechanism of the PWARX model, and α is a weighting factor.

If the estimated current mode is i, the center $\hat{\boldsymbol{c}}_i$ is estimated as the mean of the regression vectors $\boldsymbol{x}_k$ that are presumed to be relevant to this mode, while the other centers remain unchanged:

$$\hat{\boldsymbol{c}}_{ik} = \begin{cases} \gamma\hat{\boldsymbol{c}}_{i(k-1)} + (1-\gamma)\boldsymbol{x}_k, & \text{if } \hat{q}_k = i \\ \hat{\boldsymbol{c}}_{i(k-1)}, & \text{otherwise.} \end{cases}$$

When dealing with SARX models, only one hyperparameter, the forgetting factor $\gamma \in (0, 1)$, needs to be tuned, influencing the trade-off between the convergence speed and the smoothing of the estimates. For PWARX identification, the same forgetting factor is used, but there is an additional hyperparameter α (see (8.3)). Moreover, the initialization of the algorithm, i.e., fixing $\hat{\boldsymbol{\theta}}_{j0}$, $\boldsymbol{P}_{j0}$ and, in case of PWARX models, $\hat{\boldsymbol{c}}_{j0}$, for $j = 1, ..., s$, influences the convergence. Despite the lack of theoretical convergence analysis, good results are observed in practice.

The submodel recognition can also include the noise variance σ_ε^2, assumed to be known, and the covariance matrices $\boldsymbol{R}_j$ of the clusters $\mathcal{C}_j = \{\boldsymbol{x}_k \in \mathcal{X}_j\}$, $j = 1, \ldots, s$, initialized to $\boldsymbol{R}_{j0} = \boldsymbol{I}$:

$$\hat{q}_k = \operatorname*{argmin}_{j\in[s]} \frac{(y_k - \boldsymbol{x}_k^\top \hat{\boldsymbol{\theta}}_{j(k-1)})^2}{\sigma_\varepsilon^2} + \left(\boldsymbol{x}_k - \hat{\boldsymbol{c}}_{j(k-1)}\right)^\top \boldsymbol{R}_{j(k-1)}^{-1} \left(\boldsymbol{x}_k - \hat{\boldsymbol{c}}_{j(k-1)}\right).$$

If the conditional pdf $p(\boldsymbol{x}_k \mid \boldsymbol{x}_k \in \mathcal{C}_j)$ is a Gaussian function centered at $\boldsymbol{c}_j$ for the cluster $\mathcal{C}_j$ with covariance matrix $\boldsymbol{R}_j$, the criterion maximized over $[s]$ in that equation is the conditional posterior pdf $p((\boldsymbol{x}_k, \boldsymbol{y}_k) \mid \boldsymbol{x}_k \in \mathcal{C}_j)$.

The parameter vector can be updated as in (8.1). Then, the number $c_{\hat{q}_k}$ of data points in the cluster $\mathcal{C}_{\hat{q}_k}$ is incremented: $c_{\hat{q}_k k} = c_{\hat{q}_k(k-1)} + 1$. With

$$\boldsymbol{\delta} = \frac{1}{c_{\hat{q}_k k}} \left(\boldsymbol{x}_k - \hat{\boldsymbol{c}}_{\hat{q}_k(k-1)}\right),$$

the center is updated by

$$\hat{\boldsymbol{c}}_{\hat{q}_k k} = \hat{\boldsymbol{c}}_{\hat{q}_k(k-1)} + \boldsymbol{\delta},$$

and the corresponding covariance matrix by

$$\boldsymbol{R}_{\hat{q}_k k} = \frac{c_{\hat{q}_k k} - 1}{c_{\hat{q}_k k} - 2}\boldsymbol{R}_{\hat{q}_k(k-1)} + \boldsymbol{\delta}\boldsymbol{\delta}^\top + \frac{1}{c_{\hat{q}_k k} - 1}\left(\boldsymbol{x}_k - \hat{\boldsymbol{c}}_{\hat{q}_k k)}\right)\left(\boldsymbol{x}_k - \hat{\boldsymbol{c}}_{\hat{q}_k k)}\right)^\top,$$

while the other parameter vectors, cluster cardinalities, centers, and covariance matrices are kept unchanged.

8.1.1.2 Bayesian Approach

We now consider a probabilistic setting in which all the parameter vectors, $\boldsymbol{\theta}_j \in \Theta_j \subseteq \mathbb{R}^d$, of the ARX submodels, $y_k = \boldsymbol{x}_k^\top \boldsymbol{\theta}_{q_k} + \varepsilon_k$, are mutually independent random vectors. In this Bayesian approach, each parameter vector is described by a pdf, $p_{\boldsymbol{\theta}_j}(\boldsymbol{\theta})$, with a known prior pdf $p^0_{\boldsymbol{\theta}_j}(\boldsymbol{\theta})$. The number s of submodels, the orders n_a and n_b of the regressor $\boldsymbol{x}_k$ are assumed to be known and the noise is a sequence of independent and identically distributed random variables, with an a priori known pdf p_ε. In this setting, the aim is to estimate the mode sequence $\{q_k\}_{k=1}^N$ and the pdfs of the parameter vectors given the data set, i.e.,

$$p_{\boldsymbol{\theta}_j}(\boldsymbol{\theta} \mid \{(\boldsymbol{x}_k, y_k)\}_{k=1}^N), \quad j = 1, \dots, s,$$

from which various point estimates can be computed, such as through the expected value:

$$\hat{\boldsymbol{\theta}}_j = \mathbb{E}[\boldsymbol{\theta}_j] = \int_{\Theta_j} \boldsymbol{\theta}\, p_{\boldsymbol{\theta}_j}(\boldsymbol{\theta} \mid \{(\boldsymbol{x}_k, y_k)\}_{k=1}^N)\, d\boldsymbol{\theta}.$$

This batch problem can be dealt with in a greedy manner by the recursive identification of the pdfs, and we denote by $p_{\boldsymbol{\theta}_j}(\boldsymbol{\theta}; k)$ the pdf of $\boldsymbol{\theta}_j$ at time k. Similarly to (8.2), but now in a probabilistic framework, at each time k, the data pair $(\boldsymbol{x}_k, y_k)$ is assigned to the mode with the highest likelihood:

$$\hat{q}_k = \operatorname*{argmax}_{j \in [s]} p((\boldsymbol{x}_k, y_k) \mid q_k = j),$$

where

$$p((\boldsymbol{x}_k, y_k) \mid q_k = j) = \int_{\Theta_j} p((\boldsymbol{x}_k, y_k) \mid \boldsymbol{\theta})\, p_{\boldsymbol{\theta}_j}(\boldsymbol{\theta}; k-1)\, d\boldsymbol{\theta} \tag{8.4}$$

with $p((\boldsymbol{x}_k, y_k) \mid \boldsymbol{\theta}) = p_\varepsilon(y_k - \boldsymbol{x}_k^\top \boldsymbol{\theta})$, and $p_{\boldsymbol{\theta}_j}(\boldsymbol{\theta}; k-1)$ the prior pdf of $\boldsymbol{\theta}_j$.

Then, the posterior pdf of the parameter vector $\boldsymbol{\theta}_{\hat{q}_k}$, $p_{\boldsymbol{\theta}_{\hat{q}_k}}(\boldsymbol{\theta}; k)$, computed as $p_{\boldsymbol{\theta}_{\hat{q}_k}}(\boldsymbol{\theta}; k \mid (\boldsymbol{x}_k, y_k))$, is updated via Bayes' rule (see (A.5)), as

$$p_{\boldsymbol{\theta}_{\hat{q}_k}}(\boldsymbol{\theta}; k) = \frac{p_\varepsilon(y_k - \boldsymbol{x}_k^\top \boldsymbol{\theta})\, p_{\boldsymbol{\theta}_{\hat{q}_k}}(\boldsymbol{\theta}; k-1)}{\int_{\Theta_{\hat{q}_k}} p_\varepsilon(y_k - \boldsymbol{x}_k^\top \boldsymbol{\theta})\, p_{\boldsymbol{\theta}_{\hat{q}_k}}(\boldsymbol{\theta}; k-1)\, d\boldsymbol{\theta}},$$

while the pdfs of the others remain unchanged: $p_{\boldsymbol{\theta}_j}(\boldsymbol{\theta}; k) = p_{\boldsymbol{\theta}_j}(\boldsymbol{\theta}; k-1)$, $j \neq \hat{q}_k$.

A practical implementation of this approach has to devise a scheme for the computation of the pdfs. To that end, particle filtering is well adapted. The basic idea is to numerically approximate the pdf of $\boldsymbol{\theta}_j$ (defined over a dense set) with a pdf supported over a finite set of M points, $\{\boldsymbol{\theta}_j^{i,k}\}_{i=1}^M$, called particles:

$$p_{\theta_j}(\boldsymbol{\theta}; k) \approx \hat{p}_{\theta_j}(\boldsymbol{\theta}; k) = \sum_{i=1}^{M} w_j^{i,k} \delta\left(\boldsymbol{\theta} - \boldsymbol{\theta}_j^{i,k}\right), \tag{8.5}$$

where $w_j^{i,k} = \mathbb{P}(\theta = \theta_j^{i,k}) > 0$, with $\sum_{i=1}^{M} w_j^{i,k} = 1$, weights the contribution of the particle $\boldsymbol{\theta}_j^{i,k}$ and δ is the Dirac delta function (see (A.1)). Then (8.4) can be approximated by

$$p((\boldsymbol{x}_k, y_k) \mid q_k = j) \approx \sum_{i=1}^{M} w_j^{i,k-1} p_\varepsilon(y_k - \boldsymbol{x}_k^\top \boldsymbol{\theta}_j^{i,k-1}). \tag{8.6}$$

At each iteration, the particles are resampled and the weights reestimated to provide a new approximation of the parameter vector pdf. Using the approximations (8.5) and (8.6), the algorithm becomes applicable with a small computational load.

One of the main advantages of the Bayesian approach is that it allows prior knowledge on the distribution of the parameters to be directly included in the algorithm by simply setting $p_{\theta_j}(\boldsymbol{\theta}; 0) = p_{\theta_j}^0(\boldsymbol{\theta})$. On the other hand, the quality of the estimates highly depends on these initial guesses for the prior distributions, due to the convergence of the algorithm to a local optimum.

8.1.2 Model Tracking

Here, we first present a recursive algebraic procedure which extends the approach of Sect. 6.1.1 to identify SARX models. Then, we describe how to estimate the mode with a general scheme for signal segmentation and the detection of abrupt changes.

8.1.2.1 Algebraic Approach

Consider first a single, nonhybrid, ARX system without noise $y_k = \boldsymbol{x}_k^\top \theta$, with $\boldsymbol{x}_k = \begin{bmatrix} -y_{k-1} & \dots & -y_{k-n_a} & u_{k-1} & \dots & u_{k-n_b} \end{bmatrix}^\top$ (2.11), and $\boldsymbol{\theta} = \begin{bmatrix} a_1 & \dots & a_{n_a} & b_1 & \dots & b_{n_b} \end{bmatrix}^\top$ (2.12), $\boldsymbol{\theta} \in \mathbb{R}^d$, $d = n_a + n_b$, for given n_a and n_b. It is easy to write the *equation error* as

$$\boldsymbol{\beta}^\top z_k = 0, \tag{8.7}$$

with $\boldsymbol{\beta} = \begin{bmatrix} \boldsymbol{\theta}^\top & 1 \end{bmatrix}^\top$ and $z_k = \begin{bmatrix} \boldsymbol{x}_k^\top & -y_k \end{bmatrix}^\top$.

By applying the normalized least mean squares (NLMS) algorithm (2.59) to this equation error, we can recursively estimate $\boldsymbol{\beta}$ by

$$\hat{\boldsymbol{\beta}}_{k+1} = \left(\boldsymbol{I}_{d+1} - \frac{\mu \boldsymbol{\Pi}_{d+1} z_k z_k^\top}{1 + \mu \|\boldsymbol{\Pi}_{d+1} z_k\|_2^2} \right) \hat{\boldsymbol{\beta}}_k,$$

where $\hat{\beta}_k$ is the estimate at the time k, $\mu > 0$ the tuning parameter, and $\boldsymbol{\Pi}_{d+1} = \begin{bmatrix} \boldsymbol{I}_d & \boldsymbol{0}_d \\ \boldsymbol{0}_d^\top & 0 \end{bmatrix}$.

The equation error identifier is exponentially convergent to the true parameter vector $\boldsymbol{\beta}$ if, as commonly, the ARX model is minimal (see Remark 2.2 on p. 30), and if the input sequence is *persistently exciting*, i.e., if there is an $S \in \mathbb{N}$ and $\rho_1, \rho_2 > 0$ such that, for all $i \geq n_b$,

$$\rho_1 \boldsymbol{I}_d \prec \sum_{k=i}^{i+S-n_a+1} \boldsymbol{u}_k \boldsymbol{u}_k^\top \prec \rho_2 \boldsymbol{I}_d,$$

where $\boldsymbol{u}_k = \begin{bmatrix} u_{k-n_b} & \dots & u_{k+n_a-1} \end{bmatrix}^\top$ (see (B.16)).

Returning to SARX systems, the Hybrid Decoupling Constraint (HDC), in the form of the homogeneous polynomial of degree s in $d+1$ variables (6.4),

$$P_s(\boldsymbol{z}_k) = \boldsymbol{h}^\top \boldsymbol{\nu}_s(\boldsymbol{z}_k) = 0,$$

where $\boldsymbol{h} \in \mathbb{R}^{M_s}$, with $M_s = \binom{s+d}{s} = \frac{(s+d)!}{s!d!}$, is considered instead of (8.7). Thus, by applying the recursive scheme above, $\boldsymbol{h}$ can be recursively estimated by

$$\hat{\boldsymbol{h}}_{k+1} = \left(\boldsymbol{I}_{M_s} - \frac{\mu \boldsymbol{\Pi}_{M_s} \boldsymbol{\nu}_s(\boldsymbol{z}_k) \boldsymbol{\nu}_s^\top(\boldsymbol{z}_k)}{1 + \mu \|\boldsymbol{\Pi}_{M_s} \boldsymbol{\nu}_s(\boldsymbol{z}_k)\|_2^2} \right) \hat{\boldsymbol{h}}_k,$$

with $\boldsymbol{\Pi}_{M_s} = \begin{bmatrix} \boldsymbol{I}_{M_s-1} & \boldsymbol{0}_{M_s-1} \\ \boldsymbol{0}_{M_s-1}^\top & 0 \end{bmatrix}$, and the parameter vector $\boldsymbol{\beta}$ by

$$\hat{\boldsymbol{\beta}}_k = \frac{\mathrm{D}\boldsymbol{\nu}_s^\top(\boldsymbol{z}_k) \hat{\boldsymbol{h}}_k}{\boldsymbol{a}_d^\top \mathrm{D}\boldsymbol{\nu}_s^\top(\boldsymbol{z}_k) \hat{\boldsymbol{h}}_k},$$

where $\mathrm{D}\boldsymbol{\nu}_s(\boldsymbol{z}_k)$ is the Jacobian of $\boldsymbol{\nu}_s$ at $\boldsymbol{z}_k$ (see (B.19)), and $\boldsymbol{a}_d = \begin{bmatrix} 0 & \dots & 0 & 1 \end{bmatrix}^\top \in \mathbb{R}^{d+1}$.

The exponential convergence of the estimates to the true values is now guaranteed if there is an $S \in \mathbb{N}$ and $\rho_1, \rho_2 > 0$ such that, for all $i \geq \max\{n_a, n_b\}$,

$$\rho_1 \boldsymbol{I}_{M_s-1} \prec \sum_{k=i}^{i+S} \boldsymbol{\Gamma}_{M_s} \boldsymbol{\nu}_s(\boldsymbol{z}_k) \boldsymbol{\nu}_s^\top(\boldsymbol{z}_k) \boldsymbol{\Gamma}_{M_s}^\top \prec \rho_2 \boldsymbol{I}_{M_s-1},$$

with $\boldsymbol{\Gamma}_{M_s} = \begin{bmatrix} \boldsymbol{I}_{M_s-1} & \boldsymbol{0}_{M_s-1} \end{bmatrix} \in \mathbb{R}^{(M_s-1) \times M_s}$.

This approach can be extended to deal with the cases of unknown submodel orders and unknown number of submodels s, with associated conditions of persistence of excitation. However, the recursive method inherits from the batch version the complexity issues related to the high dimensionality of $\boldsymbol{h}$ induced by the polynomial embedding.

8.1.2.2 Segmentation

For recursive identification of switched ARX models, there are several similar approaches derived from signal processing and general recursive identification, more precisely tracking of time-varying systems and signal segmentation.

The idea is to estimate recursively the parameters of one model able to track the system response to the input and detect the abrupt changes in these parameters which correspond to changes of mode for hybrid systems. This gives the switching sequence and the time segments corresponding to the different modes. Then the parameter values of the submodels can be obtained as the last values of the model before the switches or reestimated from the data of the time segments. This shows that the different tasks, recursive parameter estimation, abrupt changes detection, and submodel parameter estimation, can be performed together when the current data point is processed, or sequentially, yielding successively for a whole data set the single model parameter series, the time segments of the modes separated by the changes, and then the parameters of the submodels.

For recursive parameter estimation, time-varying ARX SISO models, extending standard ARX models (2.9), are considered as

$$y_k = \boldsymbol{x}_k^\top \boldsymbol{\theta}_k + \varepsilon_k, \tag{8.8}$$

with the regression vector (2.11) and the parameter vector similar to (2.12), but time-varying.

More or less standard recursive identification algorithms, not specifically designed for tracking parameter abrupt changes, such as recursive least squares (RLS) (2.60), can be employed. A possible description of model (8.8) with jumping parameters is the state-space model

$$\begin{cases} \boldsymbol{\theta}_k = \boldsymbol{\theta}_{k-1} + \boldsymbol{\omega}_k \\ y_k = \boldsymbol{x}_k^\top \boldsymbol{\theta}_k + \varepsilon_k, \end{cases} \tag{8.9}$$

where $\boldsymbol{\omega}_k$ takes value $\boldsymbol{\nu}_k$ with probability p and zero otherwise, with $\boldsymbol{\nu}_k$ a white Gaussian noise with zero mean and covariance matrix $\boldsymbol{R}_1$. As $\boldsymbol{\omega}_k$ is not Gaussian, the Kalman filter (2.58) does not provide the optimal solution.

To improve the estimation, the Adaptive Forgetting through Multiple Models (AFMM) algorithm makes running in parallel a fixed number M of models. Each model is updated independently by an algorithm of Kalman filter type, and its posterior probability is calculated. The least likely model is killed at each sample, while the most likely one gives birth to a new one by allowing for a possible jump from it, and the covariance matrix of the parameter change is set to $\boldsymbol{R}_1$. The time-varying estimate is formed as a Gaussian mixture of the M models with weights equal to their posterior probability. There are nevertheless a number of design parameters for this method. In addition to the number of models M, there are p, the probability that the model exhibits an abrupt change at any given time, the covariance matrix $\boldsymbol{R}_1$, the variance of the innovations R_2, and the initial values of the parameters, $\boldsymbol{\theta}_0$, and their covariance matrix, $\boldsymbol{P}_0$ (see (2.58)).

For the detection of abrupt changes in the parameters, various statistical tests can also be applied. As usual, these tests imply a trade-off between false detections and change misses and inherently induce a delay in the detection. This can be illustrated with a very simple example, where the test variable is $t = \|\boldsymbol{\theta}_k - \boldsymbol{\theta}_{k-1}\|_2$ and a change is detected if $t > a$. A too small value for the threshold a leads to false alarms, when changes in the parameter vector are due to noise, whereas a too large value leads to nondetection of parameter switches. To improve the trade-off, the two successive values of the parameter vector in the test variable can be calculated from two sliding windows, but this increases the detection delay and makes the estimation of the change instant more complicated.

8.2 State-Space Models

Several arguments are in favor of the choice of the state-space (SS) form for dynamical models. First, they offer a unified treatment for SISO and MIMO systems. Then, they allow the analysis of useful control properties, such as controllability and observability, and offer a strong theory for process control, observer design, fault detection, and isolation.

Recall the state-space form of discrete-time linear hybrid models from (4.9):

$$\begin{cases} \mathbf{x}_{k+1} = \boldsymbol{A}_{q_k}\mathbf{x}_k + \boldsymbol{B}_{q_k}\boldsymbol{u}_k + \boldsymbol{w}_k \\ \mathbf{y}_k = \boldsymbol{C}_{q_k}\mathbf{x}_k + \boldsymbol{D}_{q_k}\boldsymbol{u}_k + \boldsymbol{v}_k, \end{cases} \tag{8.10}$$

where $\mathbf{x}_k \in \mathbb{R}^{n_x}$, $\boldsymbol{u}_k \in \mathbb{R}^{n_u}$, and $\mathbf{y}_k \in \mathbb{R}^{n_y}$ are, respectively, the continuous state, input, and output vectors, $\boldsymbol{w}_k \in \mathbb{R}^{n_x}$ and $\boldsymbol{v}_k \in \mathbb{R}^{n_y}$ are noise/error terms, and $q_k \in \{1, \ldots, s\}$ is the discrete mode. The matrices $\boldsymbol{A}_j \in \mathbb{R}^{n_x \times n_x}$, $\boldsymbol{B}_j \in \mathbb{R}^{n_x \times n_u}$, $\boldsymbol{C}_j \in \mathbb{R}^{n_y \times n_x}$ and $\boldsymbol{D}_j \in \mathbb{R}^{n_y \times n_u}$ describe each submodel j. For affine models, a constant term is added in each equation.

An equivalent formulation for (8.10) is

$$\begin{cases} \mathbf{x}_{k+1} = \sum_{j=1}^{s} \beta_{jk}\left(\boldsymbol{A}_j\mathbf{x}_k + \boldsymbol{B}_j\boldsymbol{u}_k\right) + \boldsymbol{w}_k \\ \mathbf{y}_k = \sum_{j=1}^{s} \beta_{jk}\left(\boldsymbol{C}_j\mathbf{x}_k + \boldsymbol{D}_j\boldsymbol{u}_k\right) + \boldsymbol{v}_k, \end{cases} \tag{8.11}$$

where the β_{jk}'s are binary variables encoding the assignment of the kth point to the jth submodel, such that, for any k, $\beta_{jk} \in \{0, 1\}$, $j = 1, \ldots, s$, and $\sum_{j=1}^{s} \beta_{jk} = 1$.

Besides, the predictors of the s submodels are given as

$$\begin{cases} \hat{\mathbf{x}}_{j(k+1)} = \boldsymbol{A}_j\hat{\mathbf{x}}_{jk} + \boldsymbol{B}_j\boldsymbol{u}_k \\ \hat{\mathbf{y}}_{jk} = \boldsymbol{C}_j\hat{\mathbf{x}}_{jk} + \boldsymbol{D}_j\boldsymbol{u}_k \end{cases}, \quad j = 1, \dots, s. \tag{8.12}$$

For switched state-space models, the discrete state q_k is an independent variable. For piecewise affine (PWA) or piecewise linear state-space models, the discrete state q_k depends on the continuous state and input vectors, $\mathbf{x}_k$ and $\boldsymbol{u}_k$:

$$q_k = j \quad \text{if and only if} \begin{bmatrix} \mathbf{x}_k \\ \boldsymbol{u}_k \end{bmatrix} \in \Omega_j, \quad j = 1, \dots, s,$$

where Ω_j, $j \in \{1, \dots, s\}$, are regions that form a partition of the state-input domain $\Omega \subset \mathbb{R}^{n_\mathbf{x}+n_u}$. Particularly, a (complete) polyhedral partition is described as:

$$\Omega_j = \left\{ \begin{bmatrix} \mathbf{x} \\ \boldsymbol{u} \end{bmatrix} \in \Omega : \boldsymbol{H}_j \begin{bmatrix} \mathbf{x} \\ \boldsymbol{u} \\ 1 \end{bmatrix} \leq \mathbf{0} \right\}, \quad j = 1, \dots, s, \text{ s.t.} \begin{cases} \cup_{j=1}^{s} \Omega_j = \Omega, \\ \Omega_i \cap \Omega_j = \emptyset, \forall i \neq j, \end{cases}$$

where the maximum number of rows for the $\boldsymbol{H}_j$'s is $s - 1$. Strict inequalities can replace inequalities for some regions in order to avoid that the boundaries belong to two adjacent regions. Note that, depending on the context, only the state $\mathbf{x}$ or the input $\boldsymbol{u}$ can define the regions. This polyhedral partition for state-space hybrid models is the counterpart of the one given in (4.12) for PWARX/PWA input–output models.

8.2.1 Problem Statement

The general identification problem of linear hybrid state-space models can be set as follows.

Problem 8.1 Given only an input–output data set $\{(\boldsymbol{u}_k, \mathbf{y}_k)\}_{k=1}^{N}$ generated by a system in the form (8.10), estimate the switching sequence $\{q_k\}_{k=1}^{N}$, the number s of discrete states, i.e., the number of submodels, the model order $n_\mathbf{x}$, all of the system parameter matrices $\boldsymbol{A}_j, \boldsymbol{B}_j, \boldsymbol{C}_j, \boldsymbol{D}_j$, $j = 1, \dots, s$, and states $\{\mathbf{x}_k\}$, and, if the model is piecewise linear, the regions Ω_j, $j = 1, \dots, s$.

All the entries of the matrices $\boldsymbol{A}_j \in \mathbb{R}^{n_\mathbf{x} \times n_\mathbf{x}}$, $\boldsymbol{B}_j \in \mathbb{R}^{n_\mathbf{x} \times n_u}$, $\boldsymbol{C}_j \in \mathbb{R}^{n_\mathbf{y} \times n_\mathbf{x}}$, $\boldsymbol{D}_j \in \mathbb{R}^{n_\mathbf{y} \times n_u}$ can be regrouped in a vector $\boldsymbol{\theta}_j \in \mathbb{R}^d$, for a more compact notation. Then, as in (4.19) for I/O models, Problem 8.1 can be formulated for switched systems, with given number s of submodels and submodel order $n_\mathbf{x}$, as a Mixed-Integer NonLinear Program (MINLP):

$$\begin{aligned}
&\min_{\{\boldsymbol{\theta}_j \in \mathbb{R}^d\}_{j=1}^s, \{\beta_{jk}\}} \frac{1}{N} \sum_{k=1}^{N} \sum_{j=1}^{s} \beta_{jk} \|\mathbf{y}_k - \hat{\mathbf{y}}_{jk}\|_2^2 \\
&\text{s.t. } \beta_{jk} \in \{0, 1\}, \; j = 1, \ldots, s, \; k = 1, \ldots, N, \\
&\sum_{j=1}^{s} \beta_{jk} = 1, \; k = 1, \ldots, N,
\end{aligned} \tag{8.13}$$

where the β_{jk}'s are defined as in (8.11) and the $\hat{\mathbf{y}}_{jk}$'s as in (8.12).

In addition to the general ill-posedness issues for hybrid systems identification detailed in Sect. 4.2.2, there are other difficulties specific to hybrid systems in SS form. In Sect. 2.2, we mentioned, already for regular, single mode SS models, that the state is most often not measured. Second, the state vector is defined up to a similarity transformation. The models can have numerous parameters and may be overparameterized. Also, the model order is often unknown, as can be the models' orders for I/O models. This order can be nonetheless estimated by subspace methods (see Sect. 2.2.2.2).

When considering hybrid systems, with multiple submodels, additional issues arise. In Sects. 4.1 and 8.2, and Problem 8.1, the submodels have a common order $n_{\mathbf{x}}$, meaning that only the parameters of the model can jump, but the structure cannot change. Alternatively, we can consider different orders $n_{\mathbf{x}j}$ for the subsystems. In that case, the problem reduces to estimating the switching sequence and the number of submodels, as the identification problem can be dealt with by the methods of Sect. 2.2.2. Conversely, if a same order is sought for the submodels, the state basis must be kept for these ones. Moreover, the identification of SS models requires data in sufficient number. In subspace methods, this number is the data block size used to build the Hankel matrices (see (2.53)). So, a minimum dwell time is required in each mode, and state-space modeling is not applicable to fast switching systems.

8.2.2 *Off-Line Approaches*

Solving Problem 8.1 is deemed computationally intractable. From there, different approaches try to find suboptimal solutions to a more or less relaxed problem, with various assumptions. Some are common such as a minimum dwell time or the persistence of excitation. Others restrain the complete ambition, such as requiring that the switching times are known or that the full state can be accessed. Yet others depend on the considered approach, such as the stability of the submodels.

8.2.2.1 From I/O to SS Identification

A straightforward approach is to perform the identification of submodels in I/O form as developed in Chap. 6 and then convert them into SS models. Indeed, any switched affine ARX (SARX) (I/O) model with s modes, possibly MIMO, has an equivalent switched affine SS model with s modes (the contrary is true, but not with the same number of modes). This has been shown in Sect. 2.2.1 for SISO systems. The initial I/O model can be identified off-line or recursively, without requiring a minimum dwell time. Then, the SS model can be obtained directly or from the identified switching sequence.

8.2.2.2 Change Detection Techniques

Another approach is to use first a change detection technique and identify the different submodels on the delimitated time segments, with the techniques described in Sect. 2.2.2. Note that in this case, the state, which is generally unknown, cannot be used for partitioning the time periods of the submodels. Note also that the change detection techniques are basically recursive, while the following parameter estimation process can be batch.

Stable Systems

The following change detection technique can be applied to estimate the switching times, with the assumption that the submodels are stable and a minimum dwell time allows the change detection and parameter estimation. A residual is calculated on a sliding window and used in a statistical test to determine the switching instants. Then, the parameter matrices of each submodel are identified via a subspace method. The comparison of the identified parameters allows us to combine the similar local models. Merging of the submodels and estimation of the parameters are repeated alternatively until the number of models converges.

Integrating model (8.10) from $k-i$ to k, for submodel j, as (2.41) in the noiseless case, gives

$$\begin{aligned} \mathbf{y}_k =& \boldsymbol{C}_j \boldsymbol{A}_j^{i-1} \boldsymbol{x}_{k-i} + \boldsymbol{C}_j \boldsymbol{A}_j^{i-2} \boldsymbol{B}_j \boldsymbol{u}_{k-i} + \cdots + \boldsymbol{C}_j \boldsymbol{B}_j \boldsymbol{u}_{k-1} + \boldsymbol{D}_j \boldsymbol{u}_k \\ &+ \boldsymbol{C}_j \boldsymbol{A}_j^{i-2} \boldsymbol{w}_{k-i} + \cdots + \boldsymbol{C}_j \boldsymbol{w}_{k-1} + \boldsymbol{v}_k. \end{aligned}$$

With the assumption that the submodels are stable, the term $\boldsymbol{C}_j \boldsymbol{A}_j^{i-1}$ can be neglected:

$$\begin{aligned} \mathbf{y}_k \approx& \boldsymbol{C}_j \boldsymbol{A}_j^{i-2} \boldsymbol{B}_j \boldsymbol{u}_{k-i} + \cdots + \boldsymbol{C}_j \boldsymbol{B}_j \boldsymbol{u}_{k-1} + \boldsymbol{D}_j \boldsymbol{u}_k \\ &+ \boldsymbol{C}_j \boldsymbol{A}_j^{i-2} \boldsymbol{w}_{k-i} + \cdots + \boldsymbol{C}_j \boldsymbol{w}_{k-1} + \boldsymbol{v}_k. \end{aligned}$$

Then, the signals are stacked on a sliding window of length $i + l \geq \tau$, with τ the dwell time, yielding

$$\boldsymbol{Y}_l \approx \boldsymbol{H}^d \boldsymbol{U}_l + \boldsymbol{H}^s \boldsymbol{W}_l + \boldsymbol{V}_l$$

with

$$\boldsymbol{Y}_l = \begin{bmatrix} \mathbf{y}_k \ \mathbf{y}_{k+1} \ \cdots \ \mathbf{y}_{k+l-1} \end{bmatrix} \in \mathbb{R}^{n_y \times l}, \ \boldsymbol{V}_l = \begin{bmatrix} \boldsymbol{v}_k \ \boldsymbol{v}_{k+1} \ \ldots \ \boldsymbol{v}_{k+l-1} \end{bmatrix} \in \mathbb{R}^{n_y \times l},$$

$$U_l = \begin{bmatrix}\tilde{u}_k \ \tilde{u}_{k+1} \ \dots \ \tilde{u}_{k+l-1}\end{bmatrix} \in \mathbb{R}^{i.n_u \times l}, \ W_l = \begin{bmatrix}\tilde{w}_k \ \tilde{w}_{k+1} \ \dots \ \tilde{w}_{k+l-1}\end{bmatrix} \in \mathbb{R}^{i.n_x \times l},$$

where

$$\tilde{u}_k = \begin{bmatrix} u_{k-i+1} \\ u_{k-i+2} \\ \vdots \\ u_k \end{bmatrix}, \quad \tilde{w}_k = \begin{bmatrix} w_{k-i+1} \\ w_{k-i+2} \\ \vdots \\ w_k \end{bmatrix},$$

$$H^d = \begin{bmatrix} CA_j^{i-2}B_j \ CA_j^{i-3}B_j \ \dots \ C_jB_j \ D_j \end{bmatrix} \in \mathbb{R}^{n_y \times i.n_u},$$

$$H^s = \begin{bmatrix} CA_j^{i-2} \ CA_j^{i-3} \ \dots \ C_j \ 0 \end{bmatrix} \in \mathbb{R}^{n_y \times i.n_x}.$$

As in subspace methods (see (2.56)), U_l is eliminated by projection of the terms in this equation on the orthogonal complement of the row space of U_l:

$$Y_l \Pi_{U_l}^{\perp} \approx H^s W_l \Pi_{U_l}^{\perp} + V_l \Pi_{U_l}^{\perp},$$

with $\Pi_{U_l}^{\perp} = I_l - U_l^{\top}(U_l U_l^{\top})^{-1} U_l$, and thus $U_l \Pi_{U_l}^{\perp} = 0$.

Therefore, choosing any column of this residual matrix, the last for instance, i.e., multiplying by the indicator vector $a = \begin{bmatrix}0 \dots 0 \ 1\end{bmatrix}^{\top}$, gives the residual

$$e_l = Y_l \Pi_{U_l}^{\perp} a = H^s W_l \Pi_{U_l}^{\perp} a + V_l \Pi_{U_l}^{\perp} a.$$

For a given mode of the hybrid system, this residual, which is a linear combination of matrices of zero mean Gaussian white noises (see Appendix A.3.1.1), has zero mean. But when the mode changes in the interval of the sliding window, it can be shown rather easily that this residual is no longer zero mean. This can be detected, for instance, by a statistical test applied to $e_l^{\top} R^{-1} e_l$, with R the variance matrix of e_l in the absence of switching. Nevertheless, due to the detection delay, the switching instant cannot be precisely determined. Thus, with $t_{\max}$ the maximum delay, the $t_{\max}$ points before the detection time are excluded.

Then, on the delimitated time segments, the submodel parameters are estimated by a subspace method and the dubious points are reassigned. Finally, the similar submodels are merged.

Tracking the Rank of the Observability Subspaces

We now consider an alternative method which does not require the submodels to be stable, but instead that they only involve measurement noise. Consider Eq. (2.56), which becomes

$$Y_f \Pi_{U_f}^{\perp} = \mathcal{O}_i X_i \Pi_{U_f}^{\perp} + N_f \Pi_{U_f}^{\perp},$$

for a moving window of n data with the terms defined as in (2.54), particularly $Y_f = \begin{bmatrix}\tilde{y}_i \ \tilde{y}_{i+1} \ \cdots \ \tilde{y}_{i+n-1}\end{bmatrix}$. When a switching occurs during the interval, the dimension of the observability subspaces changes, and this can be detected by inspecting the singular values of $Y_f \Pi_{U_f}^{\perp}$.

8.2.2.3 Recovering a Common State Basis

Suppose that the parameter matrices of the submodels and the associated data segments, i.e., the mode sequence $\{q_k\}$, are given. This could be obtained by a change detection technique mentioned above, followed by the identification of parameter matrices on the delimited segments by a subspace method and merging of the similar submodels based on some distance. As the parameter matrices have been identified separately, they are defined up to a similarity transformation which is different for each local model (see Sect. 2.2.1) and we need to recover a common state basis. The state transformations can be obtained by comparing, at each transition, the states of the model before and after the transition.

More precisely, from the submodel matrices, at a transition time τ_i from submodel m to submodel n, it is possible to estimate the final state given by submodel m, $\hat{\mathbf{x}}^m_{\tau_i}$, and the initial state given by submodel n, $\hat{\mathbf{x}}^n_{\tau_i}$, and relate these states with

$$\hat{\mathbf{x}}^m_{\tau_i} = \boldsymbol{T}_{mn}\hat{\mathbf{x}}^n_{\tau_i}.$$

With $\tau_1, \tau_2, \ldots$ the different transition times, the transformation matrix $\boldsymbol{T}_{mn}$ can then be correctly estimated via (B.27) and (B.31) as

$$\boldsymbol{T}_{mn} = \left[\hat{\mathbf{x}}^m_{\tau_1}\ \hat{\mathbf{x}}^m_{\tau_2}\ \ldots\right]\left[\hat{\mathbf{x}}^n_{\tau_1}\ \hat{\mathbf{x}}^n_{\tau_2}\ \ldots\right]^+,$$

where $\left[\hat{\mathbf{x}}^n_{\tau_1}\ \hat{\mathbf{x}}^n_{\tau_2}\ \ldots\right]^+$ is the pseudo-inverse of $\left[\hat{\mathbf{x}}^n_{\tau_1}\ \hat{\mathbf{x}}^n_{\tau_2}\ \ldots\right]$, if this matrix is of full column rank $n_{\mathbf{x}}$. This means that a sufficiently large number of transitions is required for which the states at the transition are linearly independent.

The submodel n is then updated as follows:

$$\begin{aligned}
\hat{\boldsymbol{A}}_n &\leftarrow \boldsymbol{T}_{mn}\hat{\boldsymbol{A}}_n\boldsymbol{T}_{mn}^{-1}\\
\hat{\boldsymbol{B}}_n &\leftarrow \boldsymbol{T}_{mn}\hat{\boldsymbol{B}}_n\\
\hat{\boldsymbol{C}}_n &\leftarrow \hat{\boldsymbol{C}}_n\boldsymbol{T}_{mn}^{-1},\\
\hat{\boldsymbol{D}}_n &\leftarrow \hat{\boldsymbol{D}}_n,
\end{aligned}$$

in order to bring the state of model n in the same basis as model m. The process is applied for all the different transitions, thus for all the submodels.

8.2.2.4 Canonical Representation

The problems related to the multiplicity of possible state space bases can be circumvented by using a canonical representation (see Sect. 2.2.1.4). For SISO or MISO systems, obtaining an observability canonical form is standard, but this is not the case in the general case of MIMO systems.

Under the assumptions that the system is observable and the matrix $\boldsymbol{A}$ has a characteristic polynomial equal to its minimal polynomial (see Remark 2.2), this can be done as follows. The idea is to replace one output of the original system by a linear combination of all the outputs in order to be able to find a similarity transformation to an observability canonical form. Consider a single model, of the form (2.34), with full matrices $\boldsymbol{A} \in \mathbb{R}^{n_x \times n_x}$, $\boldsymbol{B} \in \mathbb{R}^{n_x \times n_u}$, $\boldsymbol{C} \in \mathbb{R}^{n_y \times n_x}$, and $\boldsymbol{D} \in \mathbb{R}^{n_y \times n_u}$. With $\boldsymbol{h}^\top = \begin{bmatrix} h_1 \dots h_{n_y} \end{bmatrix}$ generated randomly from a uniform distribution, it holds with probability 1 that the observability matrix $\mathcal{O}_{n_x}(\boldsymbol{A}, \boldsymbol{h}^\top \boldsymbol{C})$ built as in (2.40) with $\boldsymbol{h}^\top \boldsymbol{C}$ instead of $\boldsymbol{C}$ is full rank, i.e., $\text{rank}(\mathcal{O}_{n_x}(\boldsymbol{A}, \boldsymbol{h}^\top \boldsymbol{C})) = n_x$. Thus, introducing a square matrix $\boldsymbol{H} \in \mathbb{R}^{n_y \times n_y}$ defined as the identity matrix with one row, for instance here the first, equal to $\boldsymbol{h}^\top$, leads to the transformed model:

$$\bar{\mathbf{y}}_k = \boldsymbol{H}\mathbf{y}_k = \boldsymbol{H}\left(\boldsymbol{C}\mathbf{x}_k + \boldsymbol{D}\boldsymbol{u}_k\right) = \bar{\boldsymbol{C}}_k + \bar{\boldsymbol{D}}\boldsymbol{u}_k.$$

With the similarity transformation $\boldsymbol{T} = \mathcal{O}_{n_x}(\boldsymbol{A}, \boldsymbol{h}^\top \boldsymbol{C})$, this model can be then cast into an observability canonical form such that

$$\tilde{\boldsymbol{A}} = \boldsymbol{T}\boldsymbol{A}\boldsymbol{T}^{-1} = \begin{bmatrix} 0 & 1 & 0 & \dots & 0 \\ 0 & 0 & 1 & \dots & 0 \\ \vdots & \vdots & \vdots & \ddots & \vdots \\ 0 & 0 & 0 & \dots & 1 \\ -a_0 & -a_1 & -a_2 & \dots & -a_{n_x-1} \end{bmatrix}, \qquad \tilde{\boldsymbol{B}} = \boldsymbol{T}\boldsymbol{B},$$

$$\tilde{\boldsymbol{C}} = \boldsymbol{H}\boldsymbol{C}\boldsymbol{T}^{-1} = \begin{bmatrix} [1\ 0 \dots 0] \\ \boldsymbol{c}_2^\top \\ \vdots \\ \boldsymbol{c}_{n_y}^\top \end{bmatrix}, \qquad \tilde{\boldsymbol{D}} = \boldsymbol{H}\boldsymbol{D},$$

where the a_i's are the coefficients of the characteristic polynomial of $\boldsymbol{A}$, $P(\boldsymbol{A}) = a_0 + a_1 \mathrm{q} + \dots + a_{n_x-1}\mathrm{q}^{n_x-1} + \mathrm{q}^{n_x}$, and the $\boldsymbol{c}_i^\top$'s, $i = 2, \dots, n_y$, are rows of $\boldsymbol{C}$.

This canonical form, which is identifiable (see footnote 4 on p. 31), can then be used for the different submodels.

8.2.2.5 Small Sets Identification and Classification

The canonical form above can be used in a particular procedure for switched linear SS system identification. Indeed, contrarily to many approaches which need the submodels to be active longer than the dwell time, this procedure starts from local models identified from small data sets. More precisely, five steps are carried on. The first step consists in data division into small sequential sets of w points, where w is sufficient to perform model identification and at the same time less than or equal to the dwell time of the system. The other steps are successively local model identification for each small set, clustering of the estimated local models using the Markov parameters (see (2.42)), classification of the original data in the identified submodels, and final identification of the submodels. The identification is performed by a MOESP method, and, as previously mentioned, a similarity transformation allows us to obtain models in canonical form, maintaining the control of the state basis. For clustering the Markov parameters of the estimated local models, an algorithm such as k-means (Algorithm 2) could have been used. Note nevertheless that the k-means algorithm needs to fix the number of clusters, here the number s of submodels. Instead, alternative clustering methods can be considered, with a focus on their ability to automatically determine the number of clusters and their sensitivity to outliers, which can result from models identified from sets containing submodel switchings.

8.2.2.6 Block-Coordinate Descent

The MINLP (8.13) is a very difficult optimization problem that involves both binary variables and nonlinear terms in the cost function. It could be tackled using a heuristic in the spirit of the block-coordinate descent methods seen in Sects. 6.1.3 and 6.2.2, which alternate between the assignment of data points to modes (computing the β_{jk}'s) and the estimation of the submodels. Here, a subspace method can be employed for the latter. Recall that, on the one hand, these methods cannot be guaranteed to yield an optimal solution, but, on the other hand, they can be computationally efficient.

8.2.2.7 Bounded Switchings for the Segmentation of Data Ordered in Time

Keeping the block-coordinate descent strategy in mind for dealing with mixed-integer programming problems, we now describe an alternative approach dedicated to the segmentation of data ordered in time, as considered in Sect. 6.2.4.4 but with SS submodels.

First, notice that when the system (8.11) switches between time k and time $k+1$, $|\beta_{j(k+1)} - \beta_{jk}|$ is zero except for exactly two values of j corresponding to the mode at time k and the mode at time $k+1$. Thus, the total number of switchings over the sequence of length N is

$$\frac{1}{2}\sum_{k=1}^{N-1}\sum_{j=1}^{s} |\beta_{j(k+1)} - \beta_{jk}|.$$

Therefore, we can estimate the submodels while minimizing the number of switchings by solving

$$\min_{\{\theta_j \in \mathbb{R}^d\}_{j=1}^s, \beta \in \{0,1\}^{s\times N}} \sum_{k=1}^{N}\sum_{j=1}^{s} \beta_{jk}\|\mathbf{y}_k - \hat{\mathbf{y}}_{jk}\|_2^2 + \frac{\lambda}{2}\sum_{k=1}^{N-1}\sum_{j=1}^{s} |\beta_{j(k+1)} - \beta_{jk}|,$$

$$\text{s.t.} \sum_{j=1}^{s} \beta_{jk} = 1, \quad k = 1, \dots, N,$$

where $\boldsymbol{\theta}_j$ gathers all the parameters of the jth submodel, $\hat{\mathbf{y}}_{jk}$ is its predicted output, and $\lambda > 0$ is a regularization hyperparameter influencing the number of switchings in the solution. Then, by introducing $s(N-1)$ slack variables ζ_{jk} and using a similar trick as in (2.32), this can be reformulated as[1]

$$\min_{\substack{\{\boldsymbol{\theta}_j \in \mathbb{R}^d\}_{j=1}^s,\\ \boldsymbol{\beta} \in \{0,1\}^{s\times N},\\ \boldsymbol{\zeta} \in \{0,1\}^{s\times(N-1)}}} \sum_{k=1}^{N}\sum_{j=1}^{s} \beta_{jk}\|\mathbf{y}_k - \hat{\mathbf{y}}_{jk}\|_2^2 + \frac{\lambda}{2}\sum_{k=1}^{N-1}\sum_{j=1}^{s}\zeta_{jk}, \tag{8.14}$$

$$\text{s.t.} \sum_{j=1}^{s}\beta_{jk} = 1, \quad k = 1,\dots,N,$$

$$-\zeta_{jk} \leq \beta_{j(k+1)} - \beta_{jk} \leq \zeta_{jk}, \quad j = 1,\dots,s,\ k = 1,\dots,N-1.$$

Hence, in a block-coordinate descent approach in which the $\hat{\mathbf{y}}_{jk}$'s are kept constant during the classification step, this amounts to a binary linear program with respect to the β_{jk}'s and ζ_{jk}'s. Such optimization problems are known to be $\mathcal{NP}$-hard with a typical complexity that grows exponentially with the number of variables. Therefore, such an approach is intrinsically limited to small data set sizes N. Yet, generic solvers might still handle (8.14) more efficiently than when dealing directly with (8.13).

In order to ease the tuning of the hyperparameters, we can also consider the alternative of fixing a bound S on the total number of switchings and a constrained form for the classification problem:

$$\min_{\substack{\{\boldsymbol{\theta}_j \in \mathbb{R}^d\}_{j=1}^s,\\ \boldsymbol{\beta} \in \{0,1\}^{s\times N},\\ \boldsymbol{\zeta} \in \{0,1\}^{s\times(N-1)}}} \sum_{k=1}^{N}\sum_{j=1}^{s} \beta_{jk}\|\mathbf{y}_k - \hat{\mathbf{y}}_{jk}\|_2^2,$$

$$\text{s.t.} \sum_{j=1}^{s}\beta_{jk} = 1, \quad k = 1,\dots,N,$$

$$-\zeta_{jk} \leq \beta_{j(k+1)} - \beta_{jk} \leq \zeta_{jk}, \quad j = 1,\dots,s,\ k = 1,\dots,N-1,$$

$$\frac{1}{2}\sum_{k=1}^{N-1}\sum_{j=1}^{s}\zeta_{jk} \leq S.$$

The advantage here is that S can take only integer values, which can be tested more thoroughly than the real values of λ in (8.14). In addition, by fixing S, we can also set s to $S+1$ without loss of generality.

[1]Note that since the β_{jk}'s are binary, the feasible values of the ζ_{jk}'s are in $\{0, 1\}$.

Notes

Input–Output Models

In the framework of parallel identifiers, the recursive approach alternating mode detection and parameter update is due to [1]. As an extension, the submodel recognition including the cluster covariance matrices is inspired from [2, 3], where, for MIMO PWA systems, parameter update is performed using an inverse QR factorization approach. The Bayesian approach for hybrid system identification is due to [4], where the pdfs are approximated by particle filters (see [5] for a more detailed overview of these approaches).

The recursive algebraic procedure to identify SARX models has been proposed in [6–8] extending the condition of persistence of excitation for recursive ARX model identification of [9].

Reference [10] proposed a sparse estimation approach for the recursive identification of PWARX models.

Regarding the identification of SOE models, [11] adapted the *batch* clustering-based technique of [12] by replacing the least squares estimation of the ARX submodels by the estimation of OE submodels, using instrumental variables. Several works deal with recursive approaches. An extension of [1] cited above, which detects the mode mismatches and resets the variance matrices of the corresponding parameter vectors, is applied to OE submodels in [13]. In [14], a recursive identification algorithm for SOE systems with bounded noise is presented with convergence properties.

Already in the review of [15], different aspects of model adaption and signal tracking are discussed and the importance of prior assumptions about the parameter variations (random walk, jump changes, Markov chain) is highlighted to yield efficient algorithms based on a trade-off between tracking ability and noise rejection. For detection of abrupt changes, one can refer to [16]. One can see also [17], for a more complete treatment, and many applications. The Adaptive Forgetting through Multiple Models (AFMM) algorithm was proposed by [18].

State-Space Models

Various works deal with off-line identification of hybrid SS systems. The equivalence between switched affine ARX (SARX) (I/O) models and switched affine SS models is explored in [19].

In the segmentation of the signals according to the submodel changes, the change detection technique for stable systems is due to [20]; the one based on checking the dimension of the observability subspaces as well as the common basis recovering can be found in [21–23].

The procedure based on small sets division is detailed in [24], with the similarity transformation borrowed from [25].

For the coordinate descent algorithm, see [26]. An approach bounding the total number of switchings for segmentation is presented in [27] for SS models and [28] for I/O models.

Few recursive approaches have been developed. One can refer to [29, 30].

References

1. Bako, L., Boukharouba, K., Duviella, E., Lecoeuche, S.: A recursive identification algorithm for switched linear/affine models. Nonlinear Anal.: Hybrid Syst. **5**(2), 242–253 (2011)
2. Breschi, V., Bemporad, A., Piga, D.: Identification of hybrid and linear parameter varying models via recursive piecewise affine regression and discrimination. In: Proceedings of of the 2016 European Control Conference (ECC), Aalborg, Denmark, pp. 2632–2637 (2016)
3. Breschi, V., Piga, D., Bemporad, A.: Identification of hybrid and linear parameter varying models via recursive piecewise affine regression and discrimination. Automatica **73**, 155–162 (2016)
4. Juloski, A.Lj., Weiland, S., Heemels, W.P.M.H. : A Bayesian approach to identification of hybrid systems. IEEE Trans. Autom. Control **50**(10), 1520–1533 (2005)
5. Arulampalam, M.S., Maskell, S., Gordon, N., Clapp, T.: A tutorial on particle filters for online nonlinear/non-Gaussian Bayesian tracking. IEEE Trans. Signal Process. **50**(2), 174–188 (2002)
6. Vidal, R., Anderson, B.D.O.: Recursive identification of switched ARX hybrid models: exponential convergence and persistence of excitation. In: Proceedings of the 43rd IEEE Conference on Decision and Control (CDC), Paradise Island, The Bahamas, pp. 32–37 (2005)
7. Hashambhoy, Y., Vidal, R.: Recursive identification of switched ARX models with unknown number of models and unknown orders. In: Proceedings of the 44th IEEE Conference on Decision and Control and European Control Conference (CDC-ECC), Seville, Spain, pp. 6115–6121 (2005)
8. Vidal, R.: Recursive identification of switched ARX systems. Automatica **44**(9), 2274–2287 (2008)
9. Anderson, B.D.O., Johnson Jr., C.R.: Exponential convergence of adaptive identification and control algorithms. Automatica **18**(1), 1–13 (1982)
10. Mattsson, P., Zachariah, D., Stoica, P.: Recursive identification method for piecewise arx models: a sparse estimation approach. IEEE Trans. Signal Process. **64**(19), 5082–5093 (2011)
11. Canty, N., O'Mahony, T., Cychowski, M.T.: An output error algorithm for piecewise affine system identification. Control Eng. Pract. **20**(4), 444–452 (2012)
12. Ferrari-Trecate, G., Muselli, M., Liberati, D., Morari, M.: A clustering technique for the identification of piecewise affine systems. Automatica **39**(2), 205–217 (2003)
13. Wang, J., Chen, T.: Online identification of switched linear output error models. In: Proceedings of the 2011 IEEE International Symposium on Computer-Aided Control System Design (CACSD), Denver, CO, USA, pp. 1379–1384 (2011)
14. Goudjil, A., Pouliquen, M., Pigeon, E., Gehan, O., Targui, B.: Recursive output error identification algorithm for switched linear systems with bounded noise. In: Proceedings of the 20th IFAC World Congress, Toulouse, France, IFAC-PapersOnLine, vol. 50(1), pp. 14,112–14,117 (2017)
15. Ljung, L., Gunnarsson, S.: Adaptation and tracking in system identification - a survey. Automatica **26**(1), 7–21 (1990)
16. Basseville, M., Nikiforov, I.V.: Detection of Abrupt Changes - Theory and Application. Prentice-Hall, Upper Saddle River (1993)
17. Gustafsson, F.: Adaptive Filtering and Change Detection. Wiley, New York (2000)
18. Andersson, P.: Adaptive forgetting in recursive identification through multiple models. Int. J. Control **42**(5), 1175–1193 (1985)
19. Weiland, S., Juloski, A.Lj., Vet, B.: On the equivalence of switched affine models and switched ARX models. In: Proceedings of the 45th IEEE Conference on Decision and Control (CDC), San Diego, CA, USA, pp. 2614–2618 (2006)
20. Pekpe, K.M., Mourot, G., Gasso, K., Ragot, J.: Identification of switching systems using change detection technique in the subspace framework. In: Proceedings of the 43rd IEEE Conference on Decision and Control (CDC), Paradise Island, The Bahamas, pp. 3720–3725 (2004)
21. Verdult, V., Verhaegen, M.: Subspace identification of piecewise linear systems. In: Proceedings of the 43rd IEEE Conference on Decision and Control (CDC), Paradise Island, Bahamas, pp. 3838–3843 (2004)

22. Borges, G.A., Verdult, V., Verhaegen, M., Ayala Botto, M.: A switching detection method based on projected subspace classification. In: Proceedings of the 44th IEEE Conference on Decision and Control, and the European Control Conference (CDC-ECC), Seville, Spain, pp. 344–349 (2005)
23. Borges, G.A., Verdult, V., Verhaegen, M.: Iterative subspace identification of piecewise linear systems. In: Proceedings of the 14th IFAC Symposium on System Identification (SYSID), Newcastle, Australia, IFAC Proceedings Volumes, vol. 39(1), pp. 368–373 (2006)
24. Lopes, R.V., Borges, G.A., Ishihara, Ja.Y.: New algorithm for identification of discrete-time switched linear systems. In: Proceedings of the 2013 American Control Conference (ACC), Washington, DC, USA, pp. 6219–6224 (2013)
25. Mercère, G., Bako, L.: Parameterization and identification of multivariable state-space systems: a canonical approach. Automatica **47**(8), 1547–1555 (2011)
26. Bertsekas, D.P.: Nonlinear Programming. Athena Scientific (1999)
27. Sefidmazgi, M.G., Kordmahalleh, M.M., Homaifar, A., Karimoddini, A., Tunstel, E.: A bounded switching approach for identification of switched MIMO systems. In: Proceedings of the 2016 IEEE International Conference on Systems, Man, and Cybernetics (SMC), Budapest, Hungary, pp. 4743–4748 (2016)
28. Sefidmazgi, M.G., Kordmahalleh, M.M., Homaifar, A., Karimoddini, A.: Switched linear system identification based on bounded-switching clustering. In: Proceedings of the 2015 American Control Conference (ACC), Chicago, IL, USA, pp. 1806–1811 (2015)
29. Bako, L., Mercère, G., Lecoeuche, S.: On-line structured subspace identification with application to switched linear systems. Int. J. Control **82**(8), 1496–1515 (2009)
30. Chen, D., Bako, L., Lecoeuche, S.: Recursive sparse learning method: application to jump markov linear systems. In: Proceedings of the 18th IFAC World Congress, Milano, Italy, IFAC Proceedings Volumes, vol. 44(1), pp. 3198–3203 (2011)

Chapter 9
Nonlinear Hybrid System Identification (⋆)

This chapter deals with the estimation of nonlinear hybrid models, where the submodels are given by nonlinear functions of the regressors. More precisely, we consider arbitrary and unknown nonlinearities and focus on nonparametric submodels. Note that in the parametric case, linearly parametrized nonlinearities can be handled by the methods of the previous chapters with a straightforward extension of the regression vector (e.g., by using one regressor for each nonlinear term).

Therefore, the chapter will make extensive use of the material introduced in Sect. 2.4.2, whose content should be kept in mind.

As will be seen, most approaches consider submodels given by kernel expansions as in (2.72) or (2.73). Since these models are linear with respect to their parameters, α_k, they can be introduced in the methods for switched linear or PWA models, with however a few technical issues as discussed below. This is a typical feature of the so-called kernel methods in machine learning: introducing kernel functions in a linear model to produce a nonlinear model without breaking the algorithmic advantages of the linear method. However, when doing so one must remember the crucial role of regularization in preventing the overfitting phenomenon, as discussed in Sect. 2.4.2.1.

Computational issues are also of primary importance in this context. In particular, the number of parameters to optimize in kernel models is typically proportional to the number of data, which can become prohibitively large for methods that rely on global optimization strategies, such as the continuous optimization approach. A workaround is to fix the structure of the model beforehand with a small set of preselected support vectors. For this purpose, several structure selection schemes will be exposed in this chapter. Another issue is that of the amount of memory required to store the so-called kernel matrix (2.65), which usually constitutes the starting point of kernel model estimation. A lot of work has been done in the machine learning community to bypass the memory limitation and produce efficient algorithms for large data sets. Therefore, when possible, it is desirable to express hybrid model estimation methods as standard machine learning problems, for which such algorithms exist. We will see in this chapter that this is the case for the error sparsification approach.

F. Lauer and G. Bloch, *Hybrid System Identification*,
Lecture Notes in Control and Information Sciences 478,
https://doi.org/10.1007/978-3-030-00193-3_9

Chapter Outline

Sections 9.1 and 9.2 present the extension to nonlinear switched system identification of the continuous optimization approach of Sect. 6.1.2 and the error sparsification approach of Sect. 6.2.3, respectively. Then, Sect. 9.3 discusses modifications of the sum-of-norms approach of Sect. 7.3.2 for learning piecewise smooth models including nonlinear pieces.

9.1 Continuous Optimization Approach for Switched Nonlinear Models

Submodels in the kernel form (2.73) can be introduced in the product-of-errors (PE) framework of Sect. 6.1.2 with an additional regularization term. This leads to the nonlinear PE estimator for nonlinear hybrid systems expressed as the solution to

$$\min_{\{\boldsymbol{\alpha}_j \in \mathbb{R}^N\}_{j=1}^s, \boldsymbol{b} \in \mathbb{R}^s} \sum_{k=1}^N \prod_{j=1}^s \ell\left(y_k - \sum_{i=1}^N \alpha_{ji} K_j(\boldsymbol{x}_i, \boldsymbol{x}_k) - b_j \right) + \lambda \sum_{j=1}^s R(\boldsymbol{\alpha}_j), \quad (9.1)$$

where ℓ is a smooth loss function and $R(\boldsymbol{\alpha}_j)$ is the regularizer weighted by $\lambda > 0$ and acting on the parameters $\boldsymbol{\alpha}_j = \begin{bmatrix} \alpha_{j1} & \dots & \alpha_{jN} \end{bmatrix}^\top$ of the submodel f_j. For instance, the model complexity can be measured by the ℓ_1-norm of the parameter vector, i.e., $R(\boldsymbol{\alpha}_j) = \|\boldsymbol{\alpha}_j\|_1$. This regularizer penalizes nonsmooth functions and ensures sparsity as a certain number of parameters α_{jk} will tend toward zero. Regularization over the ℓ_2-norm of the parameter vectors, i.e., $R(\boldsymbol{\alpha}_j) = \|\boldsymbol{\alpha}_j\|_2^2$ is also possible, but may result in less sparse models.

For submodels in kernel form (2.73), the nonconvex optimization program (9.1) involves a large number of variables associated with the number of *potential* support vectors (SVs). Since the kernel submodels consider all the data points $\boldsymbol{x}_k$, $k = 1, \dots, N$, as potential SVs, the number of variables α_{jk} and b_j is $s(N + 1)$. Thus, solving this problem for large N with a global optimization approach such as the one of the MCS solvers (see Sect. 6.1.2.1) may be prohibitively time-consuming.

9.1.1 *Overall Procedure*

A number of methods can be devised to reduce the number of parameters in (2.73) *before* starting the optimization. Let

$$\mathcal{S}_j = \{\boldsymbol{x}_{jk}\}_{k=1}^{M_j} \quad (9.2)$$

denote the set of M_j SVs retained for the jth reduced-size submodel

$$\tilde{f}_j(\boldsymbol{x}) = \sum_{k=1}^{M_j} \tilde{\alpha}_{jk} K_j(\boldsymbol{x}_{jk}, \boldsymbol{x}) + b_j. \tag{9.3}$$

The $(M_j + 1)$ parameters of submodel $\tilde{f}_j$ are now given by $\tilde{\boldsymbol{\alpha}}_j = \left[\tilde{\alpha}_{j1} \ \dots \ \tilde{\alpha}_{jM_j}\right]^\top$ and b_j.

With these notations, the complete identification procedure is as follows.

1. Find the structure of each submodel $\tilde{f}_j(\boldsymbol{x})$ as in (9.3) by applying one of the structure selection methods presented below.
2. Train the hybrid model by solving

$$\min_{\{\tilde{\boldsymbol{\alpha}}_j \in \mathbb{R}^{M_j}\}_{j=1}^s, \boldsymbol{b} \in \mathbb{R}^s} \sum_{k=1}^N \prod_{j=1}^s \ell\left(y_k - \tilde{f}_j(\boldsymbol{x}_k)\right) + \lambda \sum_{j=1}^s \frac{\tilde{\boldsymbol{\alpha}}_j^\top \tilde{\boldsymbol{\alpha}}_j}{M_j}. \tag{9.4}$$

3. Estimate the mode $\hat{q}_k$ for each data point by

$$\hat{q}_k = \operatorname*{argmin}_{j \in [s]} \ell(y_k - \tilde{f}_j(\boldsymbol{x}_k)), \ k = 1, \dots, N, \tag{9.5}$$

   and classify the data into s subsets accordingly.
4. Reestimate the submodels with a nonlinear estimator applied independently to each data subset.

Note that the reduced-size submodels (9.3) are based on an intrinsically sparse representation of the data, hence the choice of the smooth squared ℓ_2-norm regularization over the low-dimensional parameter vectors $\tilde{\boldsymbol{\alpha}}_j$ in (9.4) instead of an ℓ_1-norm regularization (see the discussion in Sect. 2.4.2.4). Note also the normalization by the dimension M_j of $\tilde{\boldsymbol{\alpha}}_j$.

The final optimization program (9.4) involves only $\sum_{j=1}^s (M_j + 1)$ variables instead of $s(N + 1)$ as in (9.1). This allows the complexity of the procedure to scale only linearly with respect to the number of training data N (through the summation in the error term), as for the original continuous optimization method of Sect. 6.1.2.

In Step 1 of this procedure, the nature of the problem and of the optimization program (9.1) calls for structure selection methods that can apply *without* knowledge of the target values y_k. Indeed, these values cannot be assigned to a submodel ahead of Step 3 and *before* optimizing the parameters, which are too numerous to allow for this optimization without applying structure selection first. In particular, the few structure selection methods described below satisfy this requirement and avoid solving multiple instances of (9.4).

After the classification of the data in Step 3, the submodels can be reestimated in Step 4 by considering s independent problems with s distinct data sets, to which any classical nonlinear estimation method can be applied. The sparsity and accuracy of the final model thus depend on the properties of this particular method. For instance, one can use the classical kernel ridge regression introduced in Sect. 2.4.2.

9.1.2 A Few Methods for Structure Selection

The following describes three methods that can be used to reduce the size of the submodels before the optimization. All these approaches can be used to build reduced-size kernel hybrid models, since they are only based on the input data and do not use the target output, which is undetermined in this context due to the unknown switches of the hybrid system.

In the following, $\boldsymbol{K}_j$ will denote the kernel matrix of mode j with components $(\boldsymbol{K}_j)_{ik} = K_j(\boldsymbol{x}_i, \boldsymbol{x}_k)$ and $\boldsymbol{K}_{\mathcal{S}_j}$ its submatrix built from the rows and columns corresponding to the SVs in $\mathcal{S}_j = \{\boldsymbol{x}_{jk}\}_{k=1}^{M_j}$, i.e.,

$$\boldsymbol{K}_{\mathcal{S}_j} = \begin{bmatrix} K_j(\boldsymbol{x}_{j1}, \boldsymbol{x}_{j1}) & \dots & K_j(\boldsymbol{x}_{j1}, \boldsymbol{x}_{jM_j}) \\ \vdots & \ddots & \vdots \\ K_j(\boldsymbol{x}_{jM_j}, \boldsymbol{x}_{j1}) & \dots & K_j(\boldsymbol{x}_{jM_j}, \boldsymbol{x}_{jM_j}) \end{bmatrix}. \tag{9.6}$$

Also note that in the procedures proposed below, a data point $\boldsymbol{x}_k$ originally generated by a particular mode can be considered as a SV for another mode. The main idea here is to capture only the general distribution of the data in the feature space $\mathcal{X}_\phi$ in order to ensure sufficient support for the model. However, when applying this approach as a preliminary step for piecewise smooth model estimation, where a particular submodel is only active in a given region of input space, the procedures also select SVs outside of this region. In this case, how to obtain sparser representations remains an open issue.

9.1.2.1 Entropy Maximization

One of the aims of an SV selection algorithm is to obtain a sufficiently sparse model structure while ensuring a sufficient coverage of the feature space. Inspired by the fixed-size least squares SVM (LS-SVM), such an SV selection algorithm can be defined through the maximization of the quadratic Rényi entropy H, which quantifies the diversity, uncertainty, or randomness of a system. For a particular mode j, we approximate H by

$$H_j \approx -\log \frac{1}{M_j^2} \boldsymbol{1}^\top \boldsymbol{K}_{\mathcal{S}_j} \boldsymbol{1}, \tag{9.7}$$

where $\boldsymbol{K}_{\mathcal{S}_j}$ is given by (9.6). This criterion can be maximized (at least locally) with a simple randomized procedure, that iteratively exchanges a data point from $\mathcal{S}_j$ with one of the remaining points while testing if this increases the criterion (9.7).

In this procedure, the numbers of SVs, $\{M_j\}_{j=1}^{s}$, are hyperparameters that must be fixed a priori. For Gaussian RBF kernels with bandwidth parameter σ_j, the numbers M_j can be set according to the heuristic

$$M_j = \left\lfloor \frac{1}{\sigma_j} \max_{i \in [d]} \left(\max_{k \in [N]} x_{ki} - \min_{k \in [N]} x_{ki} \right) \right\rfloor, \tag{9.8}$$

where $\lfloor \cdot \rfloor$ denotes the integer part of its argument and x_{ki} is the ith component of $\boldsymbol{x}_k$.

This heuristic is not optimal in the sense of minimizing the generalization error, but it ensures sufficient support for the model over the whole input space. The numbers M_j in (9.8) strongly depend on the bandwidths σ_j, since more SVs are needed to cover the whole input space with a smaller bandwidth. In practice, the values of σ_j can influence the quality of the model as they control the smoothness of the submodels. Proper tuning of these values may require multiple trials or prior knowledge on the relative smoothness of the subsystems in the model. However, suboptimal numbers M_j are sufficient to obtain rough mode estimates $\hat{q}_k$ and a data classification to reestimate the submodels in Step 4 of the general procedure. If these refined submodels are learned by SVR techniques, for instance, then the final number of SVs is automatically determined.

9.1.2.2 Feature Vector Selection

In the feature vector selection (FVS) method, the selection of support vectors aims at finding a suitable set of basis vectors in the feature space $\mathcal{X}_\phi$ that spans the data subspace.

If we let $\boldsymbol{\theta}_j = \sum_{k=1}^N \alpha_{jk} \boldsymbol{\phi}_j(\boldsymbol{x}_k)$, then the kernel expansion (2.72) can be rewritten in terms of inner products in feature space to yield a linear form with respect to $\boldsymbol{\theta}_j$ as

$$f_j(\boldsymbol{x}) = \sum_{k=1}^N \alpha_{jk} \langle \boldsymbol{\phi}_j(\boldsymbol{x}_k), \boldsymbol{\phi}_j(\boldsymbol{x}) \rangle + b_j = \langle \boldsymbol{\theta}_j, \boldsymbol{\phi}_j(\boldsymbol{x}) \rangle + b_j. \tag{9.9}$$

The vector $\boldsymbol{\theta}_j$ is represented by means of a set of N vectors $\{\boldsymbol{\phi}_j(\boldsymbol{x}_k)\}_{k=1}^N$, and there are N parameters α_{jk} to be determined. In practice, the dimension of the subspace which contains the whole nonlinearly mapped data set in feature space is significantly lower than N and equal to the numerical rank of the kernel matrix $\boldsymbol{K}_j$.

Thus, in order to reduce the number of parameters, one can express $\boldsymbol{\theta}_j$ from a reduced set of basis vectors $\{\boldsymbol{\phi}_j(\boldsymbol{x}_{jk})\}_{k=1}^{M_j}$, where $\boldsymbol{x}_{jk} \in \mathcal{S}_j$ and with, typically, $M_j \ll N$.

In comparison to the previous method, M_j is not fixed *a priori*, but simply corresponds to the dimension of the smallest subspace containing the data in feature space.

The set $\mathcal{S}_j$ (9.2) induces the set of basis vectors for the data subspace in feature space. For any $\boldsymbol{x}$ in input space, this basis can be used to produce an approximation of $\boldsymbol{\phi}_j(\boldsymbol{x})$ as the closest point to the subspace. In the feature vector selection, the set of SVs, $\mathcal{S}_j$, is thus estimated as the one that minimizes the reconstruction error between this approximation and $\boldsymbol{\phi}_j(\boldsymbol{x})$ over the entire training set. This is equivalent to finding the set $\mathcal{S}_j$ which maximizes the following criterion:

$$J(\mathcal{S}_j) = \frac{1}{N}\sum_{k=1}^{N} \frac{\boldsymbol{k}_{\mathcal{S}_j k}^\top \boldsymbol{K}_{\mathcal{S}_j}^{-1} \boldsymbol{k}_{\mathcal{S}_j k}}{K_j(\boldsymbol{x}_k, \boldsymbol{x}_k)}, \tag{9.10}$$

where $\boldsymbol{k}_{\mathcal{S}_j k} = \left[K_j(\boldsymbol{x}_{j1}, \boldsymbol{x}_k) \dots K_j(\boldsymbol{x}_{jM_j}, \boldsymbol{x}_k)\right]^\top$.

Various algorithms have been proposed for the maximization. For instance, one can apply a randomized procedure in which data points are iteratively added to the set $\mathcal{S}_j$ and retained if this increases $J(\mathcal{S}_j)$. The procedure ends when $\boldsymbol{K}_{\mathcal{S}_j}$ is no longer invertible, at which point the number M_j is automatically estimated as the dimension of the basis, i.e., $M_j = |\mathcal{S}_j|$.

9.1.2.3 Kernel Principal Component Regression (KPCR)

We start by describing kernel principal components methods in a general and standard setting (with a single kernel function and mode), before extending these concepts to hybrid models.

Kernel Principal Component Methods

As in the principal component analysis (PCA), one of the different goals of the *kernel principal component analysis* (KPCA) is the dimensionality reduction of a data set. KPCA considers the nonlinear *centered* mapping $\boldsymbol{x} \mapsto \phi(\boldsymbol{x}) \in \mathcal{X}_\phi \subset \mathbb{R}^L$, where $\mathbb{R}^L$ is a high-dimensional feature space. For N observations $\boldsymbol{x}_k$, the mapped data are regrouped in the matrix $\boldsymbol{\Phi} \in \mathbb{R}^{N\times L}$ as

$$\boldsymbol{\Phi} = \left[\phi(\boldsymbol{x}_1) \dots \phi(\boldsymbol{x}_k) \dots \phi(\boldsymbol{x}_N)\right]^\top.$$

PCA seeks directions, known as the principal component directions, with direction vectors $\boldsymbol{u}_i \in \mathbb{R}^L$, on which the data points $\boldsymbol{x}_k$ are orthogonally projected with a maximal dispersion (sample variance) under the constraints that the $\boldsymbol{u}_i$'s are mutually orthogonal ($\forall i \neq i'$, $\boldsymbol{u}_i^\top \boldsymbol{u}_{i'} = 0$) and unitary ($\forall i$, $\boldsymbol{u}_i^\top \boldsymbol{u}_i = 1$) vectors.

KPCA applies a similar scheme, but on the mapped data points $\phi(\boldsymbol{x}_k)$ stacked in $\boldsymbol{\Phi}$ in order to extract *nonlinear* principal component directions. With the sample covariance $\boldsymbol{C} \in \mathbb{R}^{L\times L}$ of the features,

$$\boldsymbol{C} = \frac{1}{N}\sum_{k=1}^{N} \phi(\boldsymbol{x}_k)\phi(\boldsymbol{x}_k)^\top,$$

this corresponds to the eigendecomposition

$$N\boldsymbol{C} = \boldsymbol{\Phi}^\top\boldsymbol{\Phi} = \boldsymbol{U}\boldsymbol{\Lambda}\boldsymbol{U}^\top,$$

with $\boldsymbol{U} = \left[\boldsymbol{u}_1 \dots \boldsymbol{u}_L\right] \in \mathbb{R}^{L\times L}$, and $\boldsymbol{\Lambda} = \operatorname{diag}(\left[\lambda_1 \dots \lambda_L\right]) \in \mathbb{R}^{L\times L}$.

In order to avoid computations in the high-dimensional space $\mathcal{X}_\phi$, KPCA performs the dual eigendecomposition of the kernel matrix $\boldsymbol{K} \in \mathbb{R}^{N\times N}$ (2.65), with $K_{kk'} = \langle \phi(\boldsymbol{x}_k), \phi(\boldsymbol{x}_{k'})\rangle = K(\boldsymbol{x}_k, \boldsymbol{x}_{k'})$ (2.76):

$$\boldsymbol{K} = \boldsymbol{\Phi}\boldsymbol{\Phi}^\top = \boldsymbol{V}\boldsymbol{\Lambda}\boldsymbol{V}^\top,$$

with $\boldsymbol{V} = \begin{bmatrix}\boldsymbol{v}_1 \dots \boldsymbol{v}_N\end{bmatrix} \in \mathbb{R}^{N\times N}$ and $\boldsymbol{\Lambda} = \text{diag}(\begin{bmatrix}\lambda_1 \dots \lambda_N\end{bmatrix}) \in \mathbb{R}^{N\times N}$. The mapped data are assumed to be centered in feature space, i.e., $\sum_{k=1}^N \phi(\boldsymbol{x}_k) = \boldsymbol{0}$. If not, the kernel matrix $\boldsymbol{K}$ must be replaced by

$$\boldsymbol{K}' = \left(\boldsymbol{I}_N - \frac{1}{N}\boldsymbol{1}_{N\times N}\right)\boldsymbol{K}\left(\boldsymbol{I}_N - \frac{1}{N}\boldsymbol{1}_{N\times N}\right). \tag{9.11}$$

It can be noticed that $N\boldsymbol{C}\boldsymbol{\Phi}^\top\boldsymbol{v}_i = \boldsymbol{\Phi}^\top\boldsymbol{\Phi}\boldsymbol{\Phi}^\top\boldsymbol{v}_i = \boldsymbol{\Phi}^\top\boldsymbol{K}\boldsymbol{v}_i = \lambda_i\boldsymbol{\Phi}^\top\boldsymbol{v}_i$, implying that $(\boldsymbol{\Phi}^\top\boldsymbol{v}_i, \lambda_i)$ is an eigenvector–eigenvalue pair for $N\boldsymbol{C}$. Furthermore, we have $\|\boldsymbol{\Phi}^\top\boldsymbol{v}_i\|_2^2 = \boldsymbol{v}_i^\top\boldsymbol{\Phi}\boldsymbol{\Phi}^\top\boldsymbol{v}_i = \lambda_i$. Thus, a *normalized* eigenvector of $N\boldsymbol{C}$ is $\boldsymbol{u}_i = \lambda_i^{-\frac{1}{2}}\boldsymbol{\Phi}^\top\boldsymbol{v}_i$, or in matrix form

$$\boldsymbol{U} = \boldsymbol{\Phi}^\top\boldsymbol{V}\boldsymbol{\Lambda}^{-\frac{1}{2}}.$$

There are $r = \text{rank}(\boldsymbol{\Phi}\boldsymbol{\Phi}^\top) = \text{rank}(\boldsymbol{\Phi}^\top\boldsymbol{\Phi}) \le \min\{N, L\}$ nonlinear principal components corresponding to $\lambda_i \ne 0$, i.e., the number of singular values in the SVD (see Appendix B.2.8.2).

Moreover, as in PCA, only $m < r$ directions can be sufficient to describe the structure of the data $\boldsymbol{\Phi}$, allowing for a dimensionality reduction. The eigenvectors $\boldsymbol{u}_i$ with the corresponding eigenvalues λ_i are arranged in decreasing order, $\lambda_1 \ge \lambda_2 \dots \ge \lambda_N$. Then, for a given $\rho \in [0, 1]$, the cumulative energy content ratio can be used to estimate m as the smallest number such that

$$\frac{\sum_{i=1}^m \lambda_i}{\sum_{i=1}^N \lambda_i} \ge \rho, \tag{9.12}$$

with $\sum_{i=1}^N \lambda_i = \text{tr}(\boldsymbol{K})$ by (B.35). Note that with a Gaussian RBF kernel K as in (2.69), $\text{tr}(\boldsymbol{K}) = N$ due to the fact that $K(\boldsymbol{x}, \boldsymbol{x}) = 1$ for all $\boldsymbol{x}$. Thus, the m first nonlinear principal components are given by

$$\tilde{\boldsymbol{\Phi}} = \begin{bmatrix}\boldsymbol{\Phi}\boldsymbol{u}_1 \dots \boldsymbol{\Phi}\boldsymbol{u}_m\end{bmatrix} = \boldsymbol{\Phi}\tilde{\boldsymbol{U}} = \boldsymbol{\Phi}\boldsymbol{\Phi}^\top\tilde{\boldsymbol{V}}\tilde{\boldsymbol{\Lambda}}^{-\frac{1}{2}} = \boldsymbol{K}\tilde{\boldsymbol{V}}\tilde{\boldsymbol{\Lambda}}^{-\frac{1}{2}},$$

with the reduced matrices $\tilde{\boldsymbol{U}} = \begin{bmatrix}\boldsymbol{u}_1 \dots \boldsymbol{u}_m\end{bmatrix}$, $\tilde{\boldsymbol{V}} = \begin{bmatrix}\boldsymbol{v}_1 \dots \boldsymbol{v}_m\end{bmatrix}$, and $\tilde{\boldsymbol{\Lambda}} = \text{diag}(\begin{bmatrix}\lambda_1 \dots \lambda_m\end{bmatrix})$. These projections of the original data points $\{\boldsymbol{x}_k\}_{k=1}^N$ along the m first principal component directions are computed from the kernel matrix of these points.

For a point $\boldsymbol{x}$, the projection $\tilde{\phi}(\boldsymbol{x}) \in \mathbb{R}^m$ is expressed as

$$\tilde{\phi}(\boldsymbol{x}) = \tilde{\boldsymbol{U}}^\top \phi(\boldsymbol{x}) = \tilde{\boldsymbol{\Lambda}}^{-\frac{1}{2}} \tilde{\boldsymbol{V}}^\top \boldsymbol{\Phi} \phi(\boldsymbol{x}) = \boldsymbol{A}\boldsymbol{k}(\cdot, \boldsymbol{x}), \tag{9.13}$$

with $\boldsymbol{A} = \left[\frac{\boldsymbol{v}_1}{\sqrt{\lambda_1}} \cdots \frac{\boldsymbol{v}_m}{\sqrt{\lambda_m}}\right]^\top \in \mathbb{R}^{m \times N}$ and $\boldsymbol{k}(\cdot, \boldsymbol{x}) = \left[K(\boldsymbol{x}_1, \boldsymbol{x}) \dots K(\boldsymbol{x}_N, \boldsymbol{x})\right]^\top$.

Finally, the reduced *kernel principal component regression* (KPCR) model uses this projection as

$$\tilde{f}(\boldsymbol{x}) = \boldsymbol{\beta}^\top \tilde{\phi}(\boldsymbol{x}) + b = \boldsymbol{\beta}^\top \boldsymbol{A}\boldsymbol{k}(\cdot, \boldsymbol{x}) + b, \tag{9.14}$$

with the coefficient vector $\boldsymbol{\beta} \in \mathbb{R}^m$.

Reducing the Size of Hybrid Models

Following the KPCR approach described above, the number of optimization variables in (9.1) can be reduced by using only several principal components of the kernel matrix which are sufficient to account for most of the structure in the data.

Formally, for a particular mode j, we are interested in finding the kernel principal components that can represent all data points associated with this mode. However, as the discrete state q_k (determining to which mode belongs a data point) is unknown for the training data, we have to compute the kernel principal components from the whole data set $\mathcal{S} = \{\boldsymbol{x}_k\}_{k=1}^N$ for each mode. Note nevertheless that these principal components can be different from one mode to another if the kernel functions K_j are different.

Let $\boldsymbol{\Phi}_j$ be the $(N \times L)$ matrix whose kth row is the vector $\phi_j^\top(\boldsymbol{x}_k)$ of the observation $\boldsymbol{x}_k$ mapped into the L-dimensional feature space $\mathcal{X}_{\phi_j}$, and $\boldsymbol{K}_j = \boldsymbol{\Phi}_j \boldsymbol{\Phi}_j^\top$ the associated kernel matrix. We assume that the mapped data are centered in feature space, i.e., $\sum_{k=1}^N \phi_j(\boldsymbol{x}_k) = \boldsymbol{0}$. If not, the kernel matrix $\boldsymbol{K}_j$ must be replaced by $\boldsymbol{K}'_j$ as $\boldsymbol{K}$ by $\boldsymbol{K}'$ in (9.11).

The eigendecomposition of $\boldsymbol{K}_j$, $\boldsymbol{K}_j = \boldsymbol{V}_j \boldsymbol{\Lambda}_j \boldsymbol{V}_j^\top$, gives the eigenvector and eigenvalue matrices $\boldsymbol{V}_j = \left[\boldsymbol{v}_{j,1} \dots \boldsymbol{v}_{j,N}\right] \in \mathbb{R}^{N \times N}$, and $\boldsymbol{\Lambda}_j = \text{diag}(\left[\lambda_{j,1} \dots \lambda_{j,N}\right]) \in \mathbb{R}^{N \times N}$. Then, the eigenvalues $\lambda_{j,i}$ in $\boldsymbol{\Lambda}_j$ are sorted in decreasing order with the associated eigenvectors $\boldsymbol{v}_{j,i}$ in $\boldsymbol{V}_j$ and the first M_j nonlinear principal components are retained for a desired percentage of data variance, as in (9.12).

As in (9.13), the projection $\tilde{\phi}_j(\boldsymbol{x}) \in \mathbb{R}^{M_j}$ along the m first principal component directions is expressed as

$$\tilde{\phi}_j(\boldsymbol{x}) = \boldsymbol{A}_j \boldsymbol{k}_j(\cdot, \boldsymbol{x}),$$

with $\boldsymbol{A}_j = \left[\frac{\boldsymbol{v}_{j,1}}{\sqrt{\lambda_{j,1}}} \cdots \frac{\boldsymbol{v}_{j,m}}{\sqrt{\lambda_{j,m}}}\right]^\top \in \mathbb{R}^{M_j \times N}$ and $\boldsymbol{k}_j(\cdot, \boldsymbol{x}) = \left[K_j(\boldsymbol{x}_1, \boldsymbol{x}) \dots K_j(\boldsymbol{x}_N, \boldsymbol{x})\right]^\top$, and, as in (9.14), the reduced-size KPCR submodel as

$$\tilde{f}_j(\boldsymbol{x}) = \boldsymbol{\beta}_j^\top \tilde{\phi}_j(\boldsymbol{x}) + b_j = \boldsymbol{\beta}_j^\top \boldsymbol{A}_j \boldsymbol{k}(\cdot, \boldsymbol{x}) + b_j.$$

Thus, one obtains a submodel with only $M_j + 1$ parameters that need to be estimated. However, the resulting model needs to retain the N original data points instead of M_j as in (9.3). Indeed, computing its output for a new input $\boldsymbol{x}$ involves the vector $\boldsymbol{k}_j(\cdot, \boldsymbol{x}) \in \mathbb{R}^N$. In addition, the eigenvalue decomposition of a too large kernel matrix $\boldsymbol{K}_j$ can be prohibitive. To avoid these issues, the kernel matrix can be approximated by a low-rank matrix $\tilde{\boldsymbol{K}}_j$ via the Nyström method.

9.2 Error Sparsification Approach for Switched Nonlinear Models

The error sparsification approach described in Sect. 6.2.3 can be extended to estimate nonlinear submodels. First recall that this approach works iteratively by estimating the submodels one by one instead of all at once. At each iteration, a submodel for the mode that dominates the data set is identified as one that yields zero (or close to zero) errors over many points. Then, the corresponding data points are removed from the data set and the process is repeated until all data points are satisfactorily approximated by one of the submodels.

The basic idea to extend this approach to nonlinear submodels is to replace the linear submodels by expansions over a set of basis functions. Here, we consider kernel expansions as in (2.72) with their structure either fixed in advance or resulting from the application of the Representer Theorem 2.1 with a nonparametric function class. We first describe the method for a general regularizer $R(f)$ and nonlinear model f before detailing these two cases.

For a given function class $\mathcal{H}$, the nonlinear submodel f of the first mode is estimated by solving

$$\min_{f \in \mathcal{H}} \sum_{k=1}^{N} w_k |y_k - f(\boldsymbol{x}_k)| + \lambda R(f), \tag{9.15}$$

where $w_k > 0$ is a positive weight with a similar interpretation as in the reweighting scheme of Sect. 6.2.3.2. By defining the error vector $\boldsymbol{e} \in \mathbb{R}^N$ with components $e_k = y_k - f(\boldsymbol{x}_k)$, we see that the first term is merely $\|\boldsymbol{W}\boldsymbol{e}\|_1$, where $\boldsymbol{W} = \operatorname{diag}\left(\begin{bmatrix} w_1 & \dots & w_N \end{bmatrix}^\top\right)$. As in Sect. 6.2.3, by minimizing this term, we aim at recovering a model that yields the sparsest error vector and thus that should approximate the behavior of the dominant mode. Then, the procedure is repeated with a reduced data set (and thus with a decreased value of N in the above).

In comparison with the approach described in the previous Sect. 9.1, the optimization programs will all be convex. Beside the classical benefits of convex optimization, this also leverages the limitation on the number of variables. This is particularly advantageous here, as it allows for the simultaneous estimation of the model structure and parameters (as in classical nonlinear regression by, e.g., kernel ridge regression, see Sect. 2.4.2.3). Therefore, the error sparsification approach does not require the preliminary step of structure selection discussed in Sect. 9.1.2 for the continuous optimization approach.

9.2.1 *Choice of the Regularizer*

We now discuss two particular choices for the regularizer $R(f)$ when the function class $\mathcal{H}$ is an RKHS as in (2.67):

$$\mathcal{H} = \left\{ f \in \mathbb{R}^{\mathcal{X}} : f = \sum_{i=1}^{m} \alpha_i K(\boldsymbol{x}_i, \cdot),\ m \in \mathbb{N}, \alpha_i \in \mathbb{R}, \boldsymbol{x}_i \in \mathcal{X}, \|f\|_{\mathcal{H}} < +\infty \right\}.$$

9.2.1.1 Capacity Control Regularization

Following the standard approach for learning in an RKHS (see Sect. 2.4.2), the regularizer can be chosen as the squared RKHS norm, $R(f) = \frac{1}{2}\|f\|_{\mathcal{H}}^2$. In this case, the nonlinear submodels are estimated by solving the convex optimization problem

$$\min_{f \in \mathcal{H}} \sum_{k=1}^{N} w_k |y_k - f(\boldsymbol{x}_k)| + \frac{\lambda}{2}\|f\|_{\mathcal{H}}^2. \tag{9.16}$$

The solution is in the form of (2.72), i.e.,

$$f = \sum_{k=1}^{N} \alpha_k K(\boldsymbol{x}_k, \cdot), \tag{9.17}$$

by application of a straightforward extension of the Representer Theorem 2.1 to weighted loss functions. Indeed, in the proof of this theorem, the only required property for the data term is that it depends on f only via the function values $f(\boldsymbol{x}_k)$, $k = 1, \dots, N$. Here, the purpose of the additional weights w_k is to increase the sparsity of the error vector, as will be illustrated in Example 9.2.

9.2.1.2 Sparsity-Inducing Regularization

Consider now a regularization scheme explicitly targeting sparse models, i.e., models that can be written as linear combinations of a few basis functions. Given a predefined large set of m basis functions f_i, this could be written as the search for a model

$$f = \sum_{i=1}^{m} \alpha_i f_i \tag{9.18}$$

with as few nonzero parameters $\alpha_i \in \mathbb{R}$ as possible. Therefore, it first seems tempting to choose a regularizer that explicitly counts the number of nonzeros, i.e., $R(f) = \|\boldsymbol{\alpha}\|_0$ with $\boldsymbol{\alpha} = \begin{bmatrix}\alpha_1 \dots \alpha_m\end{bmatrix}^\top$. However, such regularizers are not suitable in general for two reasons. First, they are highly nonconvex and nonsmooth, which leads to algorithmic difficulties for their minimization. Second, they do not really penalize nonsmooth models and thus might be less efficient against overfitting. Indeed, imposing a small ℓ_0-norm for $\boldsymbol{\alpha}$ does not constrain the magnitude of its nonzero entries, which could result in very large derivatives for f. Therefore, we instead consider the ℓ_1-norm regularizer,

$$R(f) = \|\boldsymbol{\alpha}\|_1, \tag{9.19}$$

which, in addition to favoring sparse solutions, also constrains the magnitude of the parameters.

In order to mimic the kernel-based models used in the previous Sect. 9.2.1.1, we here choose the basis functions f_i as the kernel functions computed at the data points. In this case, the model (9.18) indeed coincides with (9.17) and the estimation problem amounts to solving linear programs. Specifically, let $\boldsymbol{K}$ be the kernel matrix as in (2.65), $\boldsymbol{y} = \begin{bmatrix}y_1 \dots y_N\end{bmatrix}^\top$ and $\boldsymbol{W} = \operatorname{diag}\left(\begin{bmatrix}w_1 \dots w_N\end{bmatrix}^\top\right)$. Then, with the regularizer in (9.19), the problem (9.15) can be written as

$$\min_{\boldsymbol{\alpha}\in\mathbb{R}^N} \|\boldsymbol{W}(\boldsymbol{y} - \boldsymbol{K}\boldsymbol{\alpha})\|_1 + \lambda\|\boldsymbol{\alpha}\|_1, \tag{9.20}$$

which can easily be cast into a linear program as discussed in Sect. 2.1.3.1 (see (2.31)–(2.32)).

9.2.2 *Sparsity Versus Compressibility*

Sparsity is defined with respect to the number of zeros in a vector, and in particular in the error vector for the considered approach. As a result, with noisy data, the error vector cannot be sparse, or, more precisely, should not be sparse. Indeed, contrary to the linear case, with highly flexible nonlinear models, the error vector can be sparse. But this is not desired as this corresponds to a nonlinear model that has learned the noise in the data rather than the underlying meaningful dependency between the input and the output.

Compressibility extends the notion of sparsity to noisy cases. A vector $\boldsymbol{e} \in \mathbb{R}^N$ is said to be compressible when the number of nonsignificant entries is large compared to N. Specifically, let this number be measured by an ℓ_0-pseudo-norm (see (B.2)) operating on thresholded entries: for some $\epsilon > 0$,

$$\forall \boldsymbol{e} \in \mathbb{R}^N, \quad \|\boldsymbol{e}\|_{0,\epsilon} = |\{k \in [N] : |e_k| > \epsilon\}| = \sum_{k=1}^{N} \mathbb{1}_{|e_k|>\epsilon}$$
$$= \|\max\{\boldsymbol{0}, |\boldsymbol{e}| - \epsilon\}\|_0 ,$$

where the max and absolute value apply entrywise.

Maximizing the compressibility can thus be set as the minimization of this pseudo-norm, for which the ℓ_1-relaxation is

$$\|\boldsymbol{e}\|_{1,\epsilon} = \sum_{k=1}^{N} \max\{0, |e_k| - \epsilon\}.$$

Reshaping the error sparsification approach as an error compression approach then amounts, for instance, to replacing the ℓ_1-loss in (9.16) with the ϵ-insensitive loss (2.78). This leads to

$$\min_{f\in\mathcal{H}} \sum_{k=1}^{N} w_k \max\{0, |y_k - f(\boldsymbol{x}_k)| - \epsilon\} + \frac{\lambda}{2}\|f\|_{\mathcal{H}}^2. \tag{9.21}$$

Interestingly, except for the weights w_k, this formulation is exactly that of the original support vector regression (SVR) (see Sect. 2.4.2.5) and coincides with weighted SVR formulations proposed in a different context where the weights depend on the confidence on the data points. This observation has a number of consequences.

On the practical side, note that though (9.21) is convex, solving it requires to compute and store in memory the kernel matrix $\boldsymbol{K}$ of size $N \times N$, which quickly becomes impractical when $N > 10000$ (this comment also applies to (9.20)). However, efficient solvers able to deal with such large data sets have been developed for SVR and can readily be used for hybrid system identification. These are usually based on decomposition techniques in which the optimization is carried only with respect to a few variables at each iteration.

On the theoretical side, we can draw a relationship between the sparsity (or compressibility) of the error vector and the sparsity of the parameter vector $\boldsymbol{\alpha}$ in the final form (2.72) of the solution:

$$f(\boldsymbol{x}) = \sum_{k=1}^{N} \alpha_k K(\boldsymbol{x}_k, \boldsymbol{x}).$$

In particular, we know from (2.79) that the only nonzero parameters α_k in this SVR solution are those corresponding to data indexes k such that

$$|y_k - f(\boldsymbol{x}_k)| \geq \epsilon.$$

These points are the ones that are outside the tube of insensitivity or exactly on its border. Thus,

$$\|\boldsymbol{\alpha}\|_0 = \|\boldsymbol{y} - \boldsymbol{K}\boldsymbol{\alpha}\|_{0,\epsilon} + \sum_{k=1}^{N} \mathbb{1}_{|y_k - \boldsymbol{k}_k^\top \boldsymbol{\alpha}| = \epsilon}.$$

where $\boldsymbol{k}_k^\top$ is the kth row of $\boldsymbol{K}$.

We now give two examples on a large data set to illustrate the application of the method and the effect of the reweighting scheme.

Example 9.1 (*Switching nonlinear regression with SVR*) This example considers the approximation of two overlapping nonlinear functions (a sinusoid and a quadratic) from a set of 3000 data points with Gaussian noise of standard deviation $\sigma_\varepsilon = 0.5$. The submodels are estimated by solving (9.21) for a Gaussian RBF kernel K of parameter $\sigma = 0.5$ and ϵ set to $\epsilon = 1.0043\sigma_\varepsilon$, in accordance with standard recommendations.

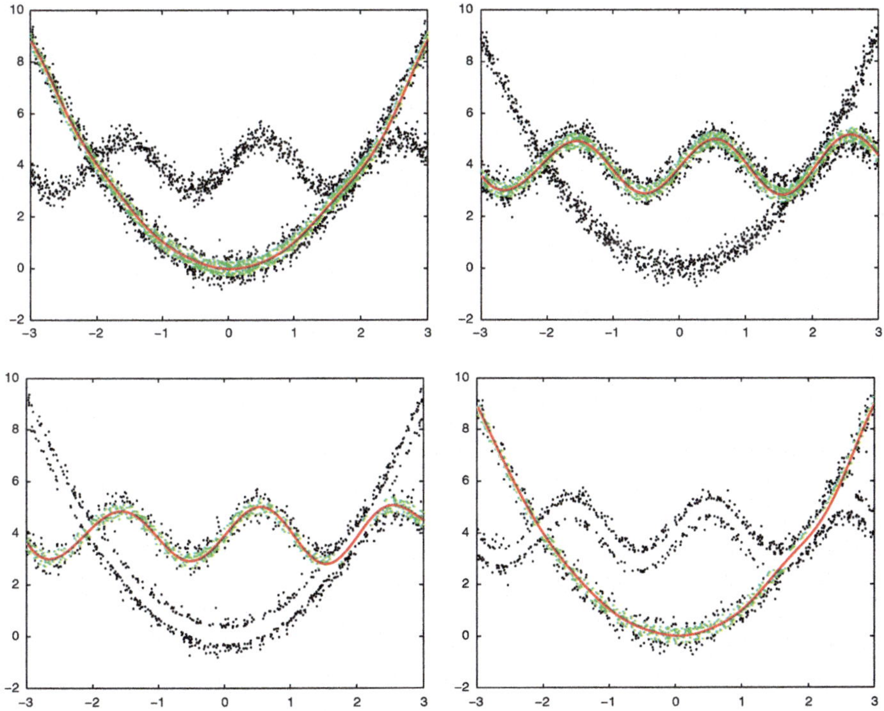

Fig. 9.1 Example 9.1: illustration of the nonlinear error sparsification method depending on which one of the quadratic (left column) or the sinusoidal (right column) modes dominates the data set. The first row of plots shows for each case the result of the first iteration estimating the first submodel, while the submodels estimated in the second iteration after removing points within ϵ of the first submodel are plotted in the second row. [1] © 2013 ACM, Inc. Reprinted by permission. https://doi.org/10.1145/2461328.2461336

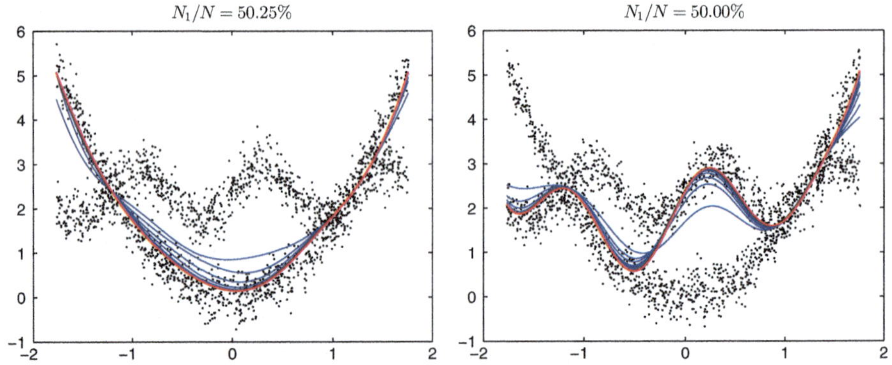

Fig. 9.2 Example 9.2: iterations of the reweighting process for $N_1/N = 50.25\%$ (left) and $N_1/N = 50\%$ (right), where N_1 is the number of points generated by the quadratic. The thin blue lines are the results of intermediate iterations leading to the final solution plotted as a thick red line. [1] © 2013 ACM, Inc. Reprinted by permission. https://doi.org/10.1145/2461328.2461336

The first row of Fig. 9.1 shows the first submodel obtained when either one of the two functions dominates the other in terms of the number of data points. In both cases, the method correctly estimates the submodel corresponding to the dominating mode. Then, after removing the points close to this submodel, a second submodel is estimated. In particular, a sufficient fraction of data is eliminated to allow for the recovery of the second submodel. However, a significant fraction (a bit less than 1/3) of the data remains unassigned at the end of the procedure. This is due to the Gaussian noise which is unbounded and generates many points outside of the tube of insensitivity. In such a case, the algorithm can be stopped either after a given number of submodels have been estimated or when a sufficiently small fraction of the data are left unassigned.

Example 9.2 (*Effect of the reweighting scheme*) In Example 9.1, a single iteration with the initial weights $w_k = 1$ already yielded a satisfactory discrimination between the two modes due to the large fraction of points associated with the dominating one (about 66%). Figure 9.2 shows the influence of reweighting in more challenging cases, i.e., when this fraction is closer to 50%. For 50.25%, the first iteration is not very accurate, but ten iterations of the reweighting scheme provide a good approximation of the target submodel. However, for exactly 50% of data of each mode, the estimated model switches between the two target submodels and fails to discriminate between the modes.

9.3 Sum-of-Norms Approach for Piecewise Smooth Models

Recall the parameter sparsification approach of Sect. 6.2.4 and its adaptation to the PWA case in Sect. 7.3.2. In general terms, this approach first considers N local models which are learned together so that each one approximates a single data point and that the variations between the models are minimized. For PWA models, these variations are measured only between models assigned to data points that are close in the regression space $\mathcal{X}$. We will now extend this approach to PWS models with nonlinear submodels.

9.3.1 Controlling the Variations of Nonlinear Local Models

We aim at learning a collection of N local models $\{f_k\}_{k=1}^N$ with as few different functions as possible. The difference between two functions f_k and f_j can be measured by a functional norm. In particular, we will use functions from an RKHS $\mathcal{H}$ and measure their differences with the distance induced by the RKHS norm (2.68): $\|f_j - f_k\|_{\mathcal{H}}$. The last ingredient one should not forget when dealing with such rich function spaces is regularization, which here can be naturally implemented with the squared RKHS norms of the local models. Putting all this together, the extension of (7.10) to the nonlinear case becomes

$$\min_{\{f_k \in \mathcal{H}\}_{k=1}^N} \sum_{k=1}^N \ell(y_k - f_k(\boldsymbol{x}_k)) + \gamma \sum_{k=1}^N \|f_k\|_{\mathcal{H}}^2 + \lambda \sum_{j=1}^N \sum_{k=1}^N w_{jk} \|f_j - f_k\|_{\mathcal{H}}, \quad (9.22)$$

where $\gamma > 0$ is the hyperparameter that controls the complexity of the local models f_k while λ controls the complexity of the global PWS model in terms of the variations of the local ones. The weights w_{jk} are used in an iteratively reweighting scheme to increase the sparsity of the variations (see Sect. 6.2.3.2).

A fundamental difference between (9.22) and the other versions such as (7.10) using parametrized linear local models is that the variables $\{f_k\}_{k=1}^N$ are functions of $\mathcal{H}$ and not vectors of $\mathbb{R}^d$. However, a finite-dimensional formulation of (9.22) is obtained thanks to the following theorem, which extends the representer theorem of Sect. 2.4.2.3. Indeed, the formulation in Theorem 2.1 does not include multiple functions to be learned, nor allows for variational terms involving the norm of the difference between these functions.

Theorem 9.1 (Extended representer theorem) *Any solution* $\{f_k^*\}_{k=1}^N$ *to* (9.22) *is a collection of functions that all lie in the span of the kernel functions taken at the data points:*

$$\forall k \in \{1, \ldots, N\}, \quad f_k^* \in \operatorname{span}\left(\{K(\boldsymbol{x}, \cdot)\}_{\boldsymbol{x} \in \mathcal{X}_N}\right),$$

where $\mathcal{X}_N = \{\boldsymbol{x}_k\}_{k=1}^N$ *contains all regression vectors.*

Proof Let $\mathcal{S} = \operatorname{span}\left(\{K(\boldsymbol{x}, \cdot)\}_{\boldsymbol{x} \in \mathcal{X}_N}\right)$ denote the subspace of interest in $\mathcal{H}$ and $\mathcal{S}^\perp$ its orthogonal complement. Then, every function $f_k \in \mathcal{H}$ can be decomposed into a sum of two orthogonal components as

$$f_k = f_k^{\mathcal{S}} + f_k^\perp, \quad f_k^{\mathcal{S}} \in \mathcal{S}, \; f_k^\perp \in \mathcal{S}^\perp, \quad \mathcal{S} \perp \mathcal{S}^\perp. \quad (9.23)$$

Note that in this case, the function values, $f_k(\boldsymbol{x}_k) = f_k^S(\boldsymbol{x}_k) + f_k^\perp(\boldsymbol{x}_k)$, only depend on the components f_k^S, since

$$f_k^\perp \in \mathcal{S}^\perp \Rightarrow f_k^\perp \perp \mathcal{S} \Rightarrow \langle f_k^\perp, K(\boldsymbol{x}_k, \cdot)\rangle_\mathcal{H} = 0$$

and, by the reproducing property of K (see Definition 2.2),

$$f_k^\perp(\boldsymbol{x}_k) = \langle f_k^\perp, K(\boldsymbol{x}_k, \cdot)\rangle_\mathcal{H} = 0.$$

This implies $f_k(\boldsymbol{x}_k) = f_k^S(\boldsymbol{x}_k)$, and thus that $\ell(y_k - f_k(\boldsymbol{x}_k)) = \ell(y_k - f_k^S(\boldsymbol{x}_k)), k = 1, \dots, N$, in the data term of (9.22).

Regarding the complexity control term, note that for all $f_k \in \mathcal{H}$,

$$\begin{aligned}
\|f_k\|_\mathcal{H}^2 &= \langle f_k, f_k\rangle_\mathcal{H} \\
&= \langle f_k^S, f_k^S\rangle_\mathcal{H} + \langle f_k^\perp, f_k^\perp\rangle_\mathcal{H} + 2\langle f_k^S, f_k^\perp\rangle_\mathcal{H} \\
&= \|f_k^S\|_\mathcal{H}^2 + \|f_k^\perp\|_\mathcal{H}^2,
\end{aligned}$$

due to the orthogonality between f_k^S and $f_k^\perp$.

For the variational term, we have, $\forall (j, k) \in [N]^2$,

$$\begin{aligned}
\|f_j - f_k\|_\mathcal{H} &= \|f_j^S - f_k^S + f_j^\perp - f_k^\perp\|_\mathcal{H} \\
&= \sqrt{\|f_j^S - f_k^S\|_\mathcal{H}^2 + \|f_j^\perp - f_k^\perp\|_\mathcal{H}^2 + 2\langle f_j^S - f_k^S, f_j^\perp - f_k^\perp\rangle_\mathcal{H}}.
\end{aligned}$$

Besides, $\langle f_j^S - f_k^S, f_j^\perp - f_k^\perp\rangle_\mathcal{H} = \langle f_j^S, f_j^\perp\rangle_\mathcal{H} - \langle f_j^S, f_k^\perp\rangle_\mathcal{H} - \langle f_k^S, f_j^\perp\rangle_\mathcal{H} + \langle f_k^S, f_k^\perp\rangle_\mathcal{H} = -\langle f_j^S, f_k^\perp\rangle_\mathcal{H} - \langle f_k^S, f_j^\perp\rangle_\mathcal{H}$. But since all f_k^S belong to $\mathcal{S}$ and all $f_k^\perp$ are orthogonal to that subspace, we have $\forall (j, k) \in [N]^2$, $\langle f_j^S, f_k^\perp\rangle_\mathcal{H} = 0$, leading to

$$\|f_j - f_k\|_\mathcal{H} = \sqrt{\|f_j^S - f_k^S\|_\mathcal{H}^2 + \|f_j^\perp - f_k^\perp\|_\mathcal{H}^2}$$

and

$$\|f_j - f_k\|_\mathcal{H} \geq \|f_j^S - f_k^S\|_\mathcal{H}.$$

Let $J\left(\{f_k\}_{k=1}^N\right)$ denote the cost functional of (9.22). Then, for any set of functions, $\{f_k\}_{k=1}^N \in \mathcal{H}^N$, decomposed as in (9.23), the partial results on the data, regularization, and variational terms lead to

$$\begin{aligned}
J\left(\{f_k\}_{k=1}^N\right) - J\left(\{f_k^S\}_{k=1}^N\right) &= \lambda \sum_{k=1}^N w_{jk}\left(\|f_j - f_k\|_\mathcal{H} - \|f_j^S - f_k^S\|_\mathcal{H}\right) + \gamma \sum_{k=1}^N \|f_k^\perp\|_\mathcal{H}^2 \\
&\geq 0.
\end{aligned}$$

In addition, if $f_k^\perp \neq 0$ for some $k \in [N]$, then $\|f_k^\perp\|_\mathcal{H} > 0$ and

$$J\left(\{f_k\}_{k=1}^N\right) > J\left(\{f_k^S\}_{k=1}^N\right).$$

Hence, any minimizer, $\{f_k^*\}_{k=1}^N$, of (9.22) admits a decomposition (9.23) with $f_k^{\perp *} = 0$, $k = 1, \dots, N$, which concludes the proof.

By applying Theorem 9.1, we can replace the f_k's in (9.22) by the linear combinations of kernel functions

$$f_k = \sum_{i=1}^N \alpha_{ki} K(\boldsymbol{x}_i, \cdot),$$

with weights $\alpha_{ki} \in \mathbb{R}$ to be estimated. Recalling the definition of the RKHS norm (2.68), this yields the finite-dimensional and convex optimization problem

$$\begin{aligned} \min_{\{\boldsymbol{\alpha}_k \in \mathbb{R}^N\}_{k=1}^N} \quad & \sum_{k=1}^N \ell\left(y_k - \boldsymbol{\alpha}_k^\top \boldsymbol{k}_k\right) + \gamma \sum_{k=1}^N \boldsymbol{\alpha}_k^\top \boldsymbol{K} \boldsymbol{\alpha}_k \\ & + \lambda \sum_{j=1}^N \sum_{k=1}^N w_{jk} \sqrt{(\boldsymbol{\alpha}_j - \boldsymbol{\alpha}_k)^\top \boldsymbol{K} (\boldsymbol{\alpha}_j - \boldsymbol{\alpha}_k)}, \end{aligned} \tag{9.24}$$

where $\boldsymbol{\alpha}_k = \left[\alpha_{k1} \dots \alpha_{kN}\right]^\top$, $\boldsymbol{K}$ is the kernel matrix (2.65), and $\boldsymbol{k}_k$ is its kth column.

For the squared or absolute loss functions, Problem (9.24) can be rewritten as a second-order cone program, suitable for general purpose solvers, by computing the Cholesky factorization $\boldsymbol{K} = \boldsymbol{R}^\top \boldsymbol{R}$ as in (B.33). For instance, with $\ell(e) = |e|$, this yields

$$\begin{aligned} \min_{\{\boldsymbol{\alpha}_k \in \mathbb{R}^N, \boldsymbol{t}_k \in \mathbb{R}^N, \mu_k \in \mathbb{R}\}_{k=1}^N} \quad & \sum_{k=1}^N e_k + \gamma \sum_{k=1}^N \mu_k + \lambda \sum_{j=1}^N \sum_{k=1}^N t_{jk} \\ \text{s.t.} \quad & -e_k \leq y - \boldsymbol{k}_k^\top \boldsymbol{\alpha}_k \leq e_k, \quad k = 1, \dots, N, \\ & \left\| \begin{matrix} (1-\mu_k)/2 \\ \boldsymbol{R}\boldsymbol{\alpha}_k \end{matrix} \right\|_2 \leq (1+\mu_k)/2, \quad k = 1, \dots, N, \\ & w_{jk} \|\boldsymbol{R}(\boldsymbol{\alpha}_j - \boldsymbol{\alpha}_k)\|_2 \leq t_{jk}, \quad j = 1, \dots, N, \ k = 1, \dots, N. \end{aligned} \tag{9.25}$$

9.3.2 Clustering Local Models in the RKHS

After solving the convex optimization problem above, the classification of the data into modes should be obtained by clustering the local models. However, compared to the original approach of Sect. 7.3.3, the local models now live in the functional space $\mathcal{H}$ and the classical k-means algorithm of Sect. 3.2, which clusters finite-dimensional feature vectors φ_k in $\mathbb{R}^p$ for some $p < \infty$, has to be extended to this setting.

Specifically, k-means is designed to minimize the sum of squared Euclidean distances,

$$\sum_{j=1}^{s} \sum_{\varphi_k \in G_j} d(\varphi_k, \overline{\varphi}_j)^2 = \sum_{j=1}^{s} \sum_{\varphi_k \in G_j} \|\varphi_k - \overline{\varphi}_j\|_2^2,$$

with respect to the means $\overline{\varphi}_j$ of s groups $\{G_j\}_{j=1}^{s}$ of feature vectors in $\mathbb{R}^p$. In order to cluster functions of the RKHS $\mathcal{H}$, the distances must be computed in $\mathcal{H}$ with

$$d_{\mathcal{H}}(f_k, \overline{f}_j)^2 = \|f_k - \overline{f}_j\|_{\mathcal{H}}^2 = \langle f_k - \overline{f}_j, f_k - \overline{f}_j \rangle_{\mathcal{H}}.$$

Since all f_k belong to the span of $\{K(\boldsymbol{x}_k, \cdot)\}_{k=1}^{N}$, the mean functions $\overline{f}_j$ also belong to this subspace of $\mathcal{H}$. For instance, for the jth group G_j, the mean function is

$$\overline{f}_j = \frac{1}{|G_j|} \sum_{f_i \in G_j} f_i = \frac{1}{|G_j|} \sum_{i: f_i \in G_j} \sum_{k=1}^{N} \alpha_{ik} K(\boldsymbol{x}_k, \cdot) = \sum_{k=1}^{N} \overline{\alpha}_{jk} K(\boldsymbol{x}_k, \cdot)$$

with the mean parameters

$$\overline{\alpha}_{jk} = \frac{1}{|G_j|} \sum_{i: f_i \in G_j} \alpha_{ik}.$$

Using the factorization $\boldsymbol{K} = \boldsymbol{R}^\top \boldsymbol{R}$, this simplifies the computations as

$$d_{\mathcal{H}}(f_k, \overline{f}_j)^2 = (\boldsymbol{\alpha}_k - \overline{\boldsymbol{\alpha}}_j)^\top \boldsymbol{K} (\boldsymbol{\alpha}_k - \overline{\boldsymbol{\alpha}}_j) = \|\boldsymbol{R}(\boldsymbol{\alpha}_k - \overline{\boldsymbol{\alpha}}_j)\|_2^2.$$

Thus, the k-means Algorithm 2 of Sect. 3.2 can be applied in a straightforward manner with the Euclidean distance and feature vectors $\varphi_k = \boldsymbol{R}\boldsymbol{\alpha}_k \in \mathbb{R}^N$ in order to cluster the f_k's and produce the final classification of the data points. Additional information on the location of the local models can also be concatenated to these feature vectors, as discussed in Sect. 7.3.3.

Example 9.3 (Piecewise smooth regression) Figure 9.3 presents an example of PWS regression with two modes. Here, the sum-of-norms approach yields the correct classification of the $N = 100$ data points into two groups thanks to the variational terms that draw the local models f_k close to each other (while allowing for large but rare differences). Indeed, Fig. 9.4 shows that this leads to a rather easy clustering problem with two well-separated groups of local models. The mean models for each group, $\overline{f}_1$ and $\overline{f}_2$, are also plotted in Fig. 9.3. In practice, the correct classification of the data points implies that we could easily obtain even more accurate models in a final step by learning them separately on each subset of the data.

Figure 9.3 also shows the outcome of the switching regression method of Sect. 9.2, which also estimates the submodels in an RKHS via convex optimization, but without taking into account the piecewise nature of the target function. The result is rather unsatisfactory: A single nonlinear submodel is built to approximate most of the data

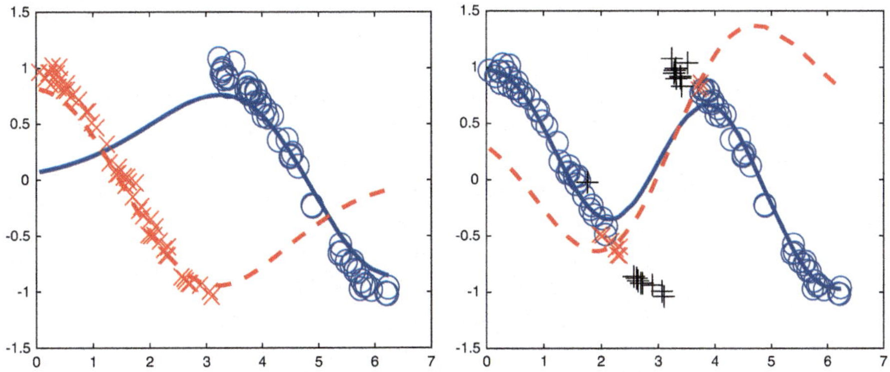

Fig. 9.3 Example 9.3 (PWS regression). Left: data classified (as $\circ$ or $\times$) by the sum-of-norms approach and the mean functions $\overline{f}_1$ (–) and $\overline{f}_2$ (– –). Right: classification and models obtained by the switching nonlinear regression method of Sect. 9.2 with points left unassigned (+)

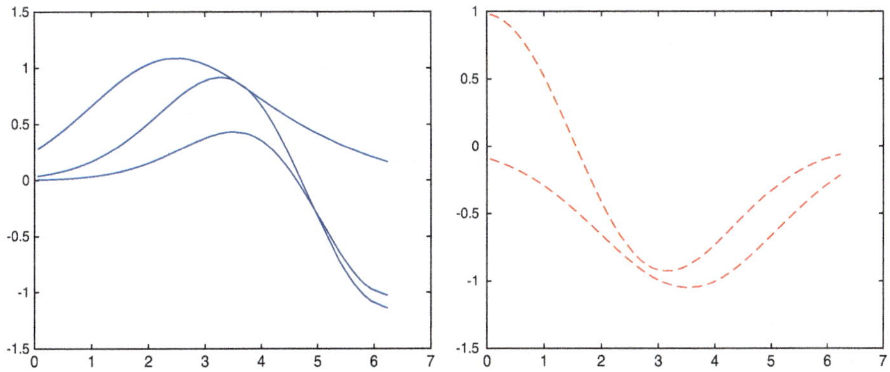

Fig. 9.4 $N = 100$ functions f_k obtained by solving (9.22), classified in two groups (left and right plots) by the k-means algorithm operating in the RKHS. The one hundred f_k's solution to (9.22) forms a set of only five different functions, which can easily be separated into two groups to recover the correct classification of the data points

points except those close to the jump of the target function. In addition, the second submodel cannot accurately estimate this jump, which leaves a number of data points unassigned and calls for further iterations with additional submodels. This contrasts with the piecewise *affine* case where switching regression techniques can be used to estimate the mode as discussed in Sect. 7.1. The reason why such a strategy is not so easily applicable in the PWS case is that a single nonlinear model can easily approximate many points across multiple modes, whereas this is usually not possible when constraining the submodels to be linear.

Notes

A first attempt at nonlinear hybrid system identification can be found in [2], with a preliminary version of the product-of-errors method. Later on, the continuous optimization approach was extended to nonlinear kernel-based submodels in [3] and its various methods for the selection of the support vectors were discussed in [4]. This paper extended a number of methods to the hybrid case. In particular, the LS-SVM method and its fixed-size variant are described in [5], the FVS method is due to [6, 7], kernel principal component regression to [8], and Nyström approximations of the kernel matrix to [9]. Eigendecompositions in feature space used in some of these methods are based on the work of [10] and detailed in [11].

The error sparsification approach of [12] was extended to the nonlinear case in [13] with a sparsity-inducing regularization. The choice of different regularization terms was discussed in [1], which also proposed a version with the ϵ-insensitive loss for measuring the sparsity of the error vector. The extension of the representer theorem (Theorem 2.1) to weighted loss functions discussed in Sect. 9.2.1 was already included in the results of [14]. Support vector regression and the ϵ-insensitive loss are due to [15], while linear programming variants with sparsity-inducing regularization were proposed by [16, 17] and weighted versions appeared, for instance, in [18]. The most famous decomposition technique for building efficient SVR solvers is the sequential minimal optimization (SMO) of [19, 20]. "Optimal" values for the parameter ϵ, as the one considered in Example 9.1, were studied in [21, 22].

The sum-of-norms approach was extended to PWS systems in [23] from which we took the proof of the Representer Theorem 9.1. In the case where the index k provides the ordering of the data in time, replacing the variational term in (9.22) with $\sum_{k=2}^{N} \|f_k - f_{k-1}\|_{\mathcal{H}}$ yields a method similar in spirit to [24] for segmenting nonlinear ARX systems over time.

Software

The MATLAB code for the continuous optimization framework available at https://members.loria.fr/FLauer/files/COFSR/COFSR.zip also features nonlinear submodels with the selection method based on the maximization of the entropy. For the error sparsification approach, LibSVM [25] (http://www.csie.ntu.edu.tw/~cjlin/libsvm/) provides an efficient implementation of support vector regression with the ϵ-insensitive loss, including the possibility to deal with weighted instances. The convex programs appearing in the sum-of-norms approach can be solved in MATLAB using the CVX toolbox [26, 27] available at http://cvxr.com/cvx.

References

1. Le, V.L., Lauer, F., Bako, L., Bloch, G.: Learning nonlinear hybrid systems: from sparse optimization to support vector regression. In: Proceedings of the 16th ACM/IEEE International Conference on Hybrid Systems: Computation and Control (HSCC), Philadelphia, PA, USA, pp. 33–42 (2013)

2. Lauer, F., Bloch, G.: Switched and piecewise nonlinear hybrid system identification. In: Proceedings of the 11th International Workshop on Hybrid Systems: Computation and Control (HSCC), St. Louis, MO, USA. Lecture Notes in Computer Science, vol. 4981, pp. 330–343 (2008)
3. Lauer, F., Bloch, G., Vidal, R.: Nonlinear hybrid system identification with kernel models. In: Proceedings of the 49th IEEE International Conference on Decision and Control (CDC), Atlanta, GA, USA, pp. 696–701 (2010)
4. Le, V.L., Bloch, G., Lauer, F.: Reduced-size kernel models for nonlinear hybrid system identification. IEEE Trans. Neural Netw. **22**(12), 2398–2405 (2011)
5. Suykens, J.A.K., Van Gestel, T., De Brabanter, J., De Moor, B., Vandewalle, J.: Least Squares Support Vector Machines. World Scientific, Singapore (2002)
6. Baudat, G., Anouar, F.: Feature vector selection and projection using kernels. Neurocomputing **55**(1–2), 21–38 (2003)
7. Cawley, G.C., Talbot, N.L.: Reduced rank kernel ridge regression. Neural Process. Lett. **16**(3), 293–302 (2002)
8. Rosipal, R., Girolami, M., Trejo, L.J., Cichocki, A.: Kernel PCA for feature extraction and de-noising in nonlinear regression. Neural Comput. Appl. **10**(3), 231–243 (2001)
9. Williams, C.K.I., Seeger, M.: Using the Nyström method to speed up kernel machines. In: Leen, T.K., Diettrich, T.G., Tresp, V. (eds.) Advances in Neural Information Processing Systems 13, pp. 682–688 (2001)
10. Schölkopf, B., Smola, A., Müller, K.R.: Nonlinear component analysis as a kernel eigenvalue problem. Neural Comput. **10**(5), 1299–1319 (1998)
11. Shawe-Taylor, J., Cristianini, N.: Kernel Methods for Pattern Analysis. Cambridge University Press, Cambridge (2014)
12. Bako, L.: Identification of switched linear systems via sparse optimization. Automatica **47**(4), 668–677 (2011)
13. Bako, L., Boukharouba, K., Lecoeuche, S.: An ℓ_0-ℓ_1 norm based optimization procedure for the identification of switched nonlinear systems. In: Proceedings of the 49th IEEE Conference on Decision and Control (CDC), Atlanta, GA, USA, pp. 4467–4472 (2010)
14. Schölkopf, B., Herbrich, R., Smola, A.J.: A generalized representer theorem. In: Proceedings of the 14th Annual Conference on Computational Learning Theory (COLT), Amsterdam, The Netherlands. Lecture Notes in Computer Science, vol. 2111, pp. 416–426 (2001)
15. Vapnik, V.N.: The Nature of Statistical Learning Theory. Springer, Berlin (1995)
16. Smola, A., Schölkopf, B., Rätsch, G.: Linear programs for automatic accuracy control in regression. In: Proceedings of the 9th International Conference on Artificial Neural Networks (ICANN), Edinburgh, UK, pp. 575–580 (1999)
17. Mangasarian, O.L., Musicant, D.R.: Large scale kernel regression via linear programming. Mach. Learn. **46**(1–3), 255–269 (2002)
18. Tay, F.E.H., Cao, L.: Modified support vector machines in financial time series forecasting. Neurocomputing **48**(1–4), 847–861 (2002)
19. Platt, J.C.: Fast training of support vector machines using sequential minimal optimization. In: Schölkopf, B., Burges, C., Burges, A. (eds.) Advances in Kernel Methods: Support Vector Learning, pp. 185–208. MIT Press, Cambridge (1999)
20. Shevade, S.K., Keerthi, S.S., Bhattacharyya, C., Murthy, K.R.K.: Improvements to the SMO algorithm for SVM regression. IEEE Trans. Neural Netw. **11**(5), 1188–1193 (2000)
21. Smola, A., Murata, N., Schölkopf, B., Müller, K.R.: Asymptotically optimal choice of ε-loss for support vector machines. In: Proceedings of the 8th International Conference on Artificial Neural Networks (ICANN), Skövde, Sweden, Springer, pp. 105–110 (1998)
22. Kwok, J.T., Tsang, I.W.: Linear dependency between ϵ the input noise in ϵ-support vector regression. IEEE Trans. Neural Netw. **14**(3), 544–553 (2003)
23. Lauer, F., Bloch, G.: Piecewise smooth system identification in reproducing kernel Hilbert space. In: Proceedings of the 53rd IEEE Conference on Decision and Control (CDC), Los Angeles, CA, USA, pp. 6498–6503 (2014)

24. Falck, T., Ohlsson, H., Ljung, L., Suykens, J.A.K., De Moor, B.: Segmentation of times series from nonlinear dynamical systems. In: Proceedings of the 18th IFAC World Congress, Milano, Italy, IFAC Proceedings Volumes, vol. 44(1), pp. 13,209–13,214 (2011)
25. Chang, C.C., Chih-Jen, L.: LIBSVM: a library for support vector machines. ACM Trans. Intell. Syst. Technol. **2**(3), 27:1–27:27 (2001). http://www.csie.ntu.edu.tw/~cjlin/libsvm/
26. Grant, M., Boyd, S.: CVX: Matlab software for disciplined convex programming, version 2.1 (2017). http://cvxr.com/cvx
27. Grant, M.C., Boyd, S.P.: Graph implementations for nonsmooth convex programs. In: Blondel, V., Boyd, S.P., Kimura, H. (eds.) Recent Advances in Learning and Control. Lecture Notes in Control and Information Sciences, vol. 2, pp. 95–110. Springer, Berlin (2008). http://stanford.edu/~boyd/papers/pdf/graph_dcp.pdf

Chapter 10
Outlook

In this monograph, after an overview of classical system identification and classification, the different forms of hybrid models were given and their related identification problems were exposed. The particularities of hybrid system identification were emphasized, among which the close intertwining of regression and classification tasks and the combinatorial character of the problem are of primary importance. This character limits to small-scale problems the use of exact methods described in Chap. 5. Thus, for switched and piecewise affine systems, several suboptimal methods were presented to handle more complex cases (with more modes or a larger data dimensionality) in Chaps. 6, 7, and 9 for nonlinear submodels.

These methods form a representative sample among the numerous approaches proposed over the last 15 years. Given such a large choice of methods, the practitioner could ask for the determination of the one method to apply in a specific context. However, this demand cannot be easily answered. In any particular context, the most suitable method might depend on numerous factors, including the amount of noise, the number of data, the dimension of the data, the number of modes, whether this number is fixed/known or not, etc. Furthermore, since most of the practical methods only offer heuristic solutions without optimality guarantees, no final statement can be made regarding the superiority of one over the others in all circumstances. This is the reason why we chose to present several approaches in each chapter without highlighting one in particular.

That said, a baseline method that can be quickly applied for switched system identification in practice is the k-LinReg algorithm presented in Sect. 6.1.3. It is computationally efficient and has been observed to yield reasonable accuracy in many practical cases. For PWA regression, the same algorithm could also be used to define a baseline by following the strategy depicted in Sect. 7.1, but with the possible pitfalls mentioned there. Alternatively, the clustering-based method of [1] described in Sect. 7.3.1 provides a solid basis for a first attempt at PWA system identification. Implementations of these methods are freely available from https://members.loria.fr/FLauer/files/klinreg/ for k-LinReg and from http://sisdin.unipv.it/lab/personale/pers_hp/ferrari/HIT_toolbox.php for the clustering-based technique.

F. Lauer and G. Bloch, *Hybrid System Identification*,
Lecture Notes in Control and Information Sciences 478,
https://doi.org/10.1007/978-3-030-00193-3_10

In this book, we also refrained from reporting on numerical experiments for reasons similar to those mentioned above. While such experiments could help to emphasize a few particularities of some methods, we believe that a thorough empirical comparison of all the methods is doomed to fail due to the numerous factors coming into play and that could favor one method or the other. Despite these comments, there remains the need for a solid empirical study that can help to characterize the specific conditions under which each method is likely to yield satisfactory results.

The following lists a few other open issues for hybrid system identification. In addition to these, classical open issues in the fields of system identification and machine learning are also relevant.

10.1 Other Forms of Hybrid Models

We mention below open issues related to input–output (I/O) models, nonlinear submodels, and continuous-time models.

10.1.1 Input–Output Models

For the identification of I/O hybrid models, the main part of the book (except Chap. 8) focuses on ARX submodels in a (pure) regression framework. However, when dealing with dynamical systems, even regular (nonhybrid) ones, the assumptions behind this setting do not always hold in practice (see Remark 2.1), which can lead to biased estimates of the model parameters.

This calls for the development of hybrid system identification methods that can deal with submodels in ARMAX, OE, or BJ form, as described in Sect. 2.1. Currently, only a few examples can be found in the literature for OE submodels, such as a batch approach proposed in [2] and a recursive one working in a bounded-error framework in [3, 4]. In [5], the Bayesian approach described for ARX submodels in Sect. 8.1.1.2 is applied to OE submodels, while [6] considers switched systems in an error in variables framework.

10.1.2 Nonlinear Submodels

A chapter is devoted in this book to nonlinear hybrid system identification, more precisely hybrid systems with nonlinear submodels. Although involving such submodels in hybrid systems is contradictory to the original idea of using linear submodels to extend linear control design to nonlinear systems, real switched or piecewise-defined hybrid systems can behave nonlinearly within a mode. This field has not received much attention yet, and there is still much work to be done, related to machine learning for black-box identification of the submodels or the inclusion of predefined nonlinearly parametrized structures.

10.1.3 Continuous-Time Models

Identification of continuous-time models was initiated in the middle of the twentieth century. However, the estimation of continuous-time models from discrete-time data was only reconsidered rather recently [7, 8]. Software packages are now available [9].

Few recent works deal with the identification of continuous-time hybrid models, either arbitrarily switched [10] or piecewise affine [11], and this has to be developed by capitalizing on the various direct or indirect nonhybrid approaches for linear or nonlinear systems [12].

10.2 Computational Complexity and Exact Methods

Chap. 5 essentially shows that most optimization problems encountered in hybrid system identification were $\mathcal{NP}$-hard. But, are these problems *strongly* $\mathcal{NP}$-hard?

The distinction lies in the choice of the model of computation: A problem is strongly $\mathcal{NP}$-hard if it is $\mathcal{NP}$-hard under the unary (instead of binary) encoding (see Sect. 5.2.1.1). Since our proofs of $\mathcal{NP}$-hardness are based on a reduction of the Partition problem, which is known to be only weakly $\mathcal{NP}$-hard, the question remains open. Beyond the genuine interest in answering theoretical questions, showing that the problems are only weakly $\mathcal{NP}$-hard might lead to exact algorithms that run in pseudo-polynomial time. Such algorithms have a polynomial-time complexity with respect to the number of inputs and their magnitudes; thus, they remain "reasonable" save only for instances involving very large inputs (see [13] for more details).

From a more practical viewpoint, future work could develop new heuristic methods inspired by the exact algorithms described in Chap. 5. Preliminary results in that direction were obtained in [14] by substituting a random sampling for the complete enumeration of subsets of points defining the classification.

Finally, it seems that the global optimization of PWA models in an efficient manner with respect to the number of data remains elusive.

10.3 Statistical Guarantees

Statistical analysis of the hybrid models and estimation methods discussed in this book is scarce in the literature. This is in part due to the inherent difficulty of hybrid system identification and the fact that research first focused on the nontrivial task of deriving computationally efficient and accurate estimation methods for the minimization of the error on the data. However, bounds on the expected error (risk) of piecewise smooth and switching models were recently derived by [15, 16] in a statistical learning framework. These bounds are uniform over the model class, which means that they hold for any model and not only for the global minimizers of the empirical error. Thus, they are particularly well suited for piecewise/switching regression, where most estimation methods cannot be guaranteed to yield a global solution. However, these results consider a static regression context, where the observations are assumed to be independent. Making them hold in the nonindependent case would be of primary importance for the identification of hybrid dynamical systems.

10.4 Model Selection

Model selection is an important issue for hybrid system identification, all the more since it occurs at two different levels. On the one hand, the selection of the most appropriate submodel structure directly influences the accuracy of the model, especially in the nonlinear case. On the other hand, in many practical cases, the number of modes must also be estimated from the data, which can be seen as the selection of a suitable structure for the overall hybrid model.

The first level of model selection remains a central problem in machine learning despite a large body of works in the literature. Inspired by this field, the second level could be tackled on the basis of statistical guarantees as discussed above via the *Structural Risk Minimization* principle. The idea is to minimize an upper bound on the expected error instead of the error on the data. Such error bounds are usually made of two terms: a data fit term and a confidence term. While the first decreases with the complexity of the model (e.g., the number of modes), the confidence term typically grows with the complexity, so that minimizing the sum of the two can help us find a satisfactory trade-off between them in an automated manner.

Regarding the estimation of the number of modes, an interesting connection can also be made between switching regression and the standard clustering problem (see [16]), for which approaches based on a notion of algorithmic stability show promising results for tuning the number of groups, see [17].

10.5 Applications

In the introduction (Chap. 1), several applications for hybrid system modeling have been enumerated, showing that many fields can benefit from the research described in this book. Yet, new applications can be expected, particularly with piecewise-defined models for system control (see [18] for one recent example). But other domains dealing with static data showing multiple tendencies or aggregated from multiple sources might also give rise to interesting use cases for switching or piecewise smooth regression techniques.

References

1. Ferrari-Trecate, G., Muselli, M., Liberati, D., Morari, M.: A clustering technique for the identification of piecewise affine systems. Automatica **39**(2), 205–217 (2003)
2. Canty, N., O'Mahony, T., Cychowski, M.T.: An output error algorithm for piecewise affine system identification. Control Eng. Pract. **20**(4), 444–452 (2012)
3. Wang, J., Chen, T.: Online identification of switched linear output error models. In: Proceedings of the 2011 IEEE International Symposium on Computer-Aided Control System Design (CACSD), Denver, CO, USA, pp. 1379–1384 (2011)
4. Goudjil, A., Pouliquen, M., Pigeon, E., Gehan, O., Targui, B.: Recursive output error identification algorithm for switched linear systems with bounded noise. In: Proceedings of the 20th IFAC World Congress, Toulouse, France, IFAC-PapersOnLine, vol. 50(1), pp. 14,112–14,117 (2017)
5. Juloski, A.Lj., Weiland, S.: A bayesian approach to the identification of piecewise linear output error models. In: Proceedings of the 14th IFAC Symposium on System Identification (SYSID), Newcastle, Australia, IFAC Proceedings Volumes, vol. 39(1), pp. 374–379 (2006)
6. Feng, C., Lagoa, C.M., Ozay, N., Sznaier, M.: Hybrid system identification: an SDP approach. In: Proceedings of the 49th IEEE International Conference on Decision and Control (CDC), Atlanta, GA, USA, pp. 1546–1552 (2010)
7. Rao, G.P., Unbehauen, H.: Identification of continuous-time systems. IEE Proc. Control Theory Appl. **153**(2), 185–200 (2006)
8. Garnier, H., Wang, L. (eds.): Identification of Continuous-time Models from Sampled Data. Springer, Berlin (2008)
9. Ljung, L.: Experiments with identification of continuous time models. In: Proceedings of the 15th IFAC Symposium on System Identification (SYSID), Saint-Malo, France, IFAC Proceedings Volumes, vol. 42(10), pp. 1175–1180 (2009)
10. Keshvari-Khor, H., Karimpour, A., Pariz, N.: Identification of continuous-time switched linear systems from low-rate sampled data. IET Control Theory A. **12**(14), 1964–1973 (2018)
11. Kersting, S., Buss, M.: Recursive estimation in piecewise affine systems using parameter identifiers and concurrent learning. In: International Journal of Control (2017) in Press
12. Guo, Y., Guo, L.Z., Billings, S.A., Wei, H.L.: Identification of continuous-time models for nonlinear dynamic systems from discrete data. Int. J. Syst. Sci. **47**(12), 3044–3054 (2016)
13. Garey, M.R., Johnson, D.S.: Computers and Intractability: A Guide to the Theory of NP-Completeness. W.H. Freeman and Co (1979)
14. Lauer, F. On the exact minimization of saturated loss functions for robust regression and subspace estimation. Pattern Recogn. Lett. **112**, 317–323 (2018)
15. Lauer, F.: Error bounds for piecewise smooth and switching regression (2017). arXiv:1707.07938
16. Lauer, F.: Error bounds with almost radical dependence on the number of components for multi-category classification, vector quantization and switching regression. In: French Conference on Machine Learning (FCML/CAp), Rouen, France (2018)
17. von Luxburg, U.: Clustering stability: an overview. Found. Trends Mach. Learn. **2**(3), 235–274 (2010)
18. Schirrer, A., Mayrhofer, J., Ritzberger, D.: Efficient piecewise-affine coupled-system feedforward control of a nonlinear elastomer coupling. In: Proceedings of the 9th Vienna International Conference on Mathematical Modelling (MATHMOD), Vienna, Austria, IFAC-PapersOnLine, vol. 51(2), pp. 861–866 (2018)

Appendix A
Basics of Probability

This appendix recalls basic definitions and facts from probability, such as random variables, probability distributions, mean and variance.

Remark A.1 Symbols used in this appendix usually refer to generic objects and do not bear the same meaning as in the rest of the book.

A.1 Probability Space

Given a random experiment, the set of all possible outcomes Ω is called the *sample space*. A σ-algebra on Ω is a set Σ of subsets of Ω such that

(i) $\Omega \in \Sigma$;
(ii) $\forall \mathcal{S} \in \Sigma$, $(\Omega \setminus \mathcal{S}) \in \Sigma$;
(iii) if $\forall i \in \mathbb{N}$, $\mathcal{S}_i \in \Sigma$, then $\cup_{i \in \mathbb{N}} \mathcal{S}_i \in \Sigma$.

Elements of Σ are refered to as *events*. A *probability measure* $\mathbb{P}$ on (Ω, Σ) is a function from Σ into $[0, 1]$ yielding the probability of events such that $\mathbb{P}(\emptyset) = 0$, $\mathbb{P}(\Omega) = 1$ and, for countably many disjoint sets $\mathcal{S}_i$,

$$\mathbb{P}\left(\bigcup_i \mathcal{S}_i\right) = \sum_i \mathbb{P}(\mathcal{S}_i).$$

A triplet $(\Omega, \Sigma, \mathbb{P})$ such that all of the above holds is known as a *probability space*.

A.2 Random Variables

A (real) *random variable* X is a function associating real numbers $X(\omega)$ to the outcomes $\omega \in \Omega$ of a random experiment. If X can take only a finite number of values,

F. Lauer and G. Bloch, *Hybrid System Identification*,
Lecture Notes in Control and Information Sciences 478,
https://doi.org/10.1007/978-3-030-00193-3

then it is a *discrete* random variable, and otherwise a *continuous* one. Random variables are characterized via their probability distribution which yields the probability that the variable takes a value in a given subset of $\mathbb{R}$.

More generally, given a set $\mathcal{X}$ and a σ-algebra $\Sigma_{\mathcal{X}}$ on $\mathcal{X}$, the *probability distribution* P_X of a random variable X taking values in $\mathcal{X}$ is a function

$$P_X : \Sigma_{\mathcal{X}} \to [0, 1]$$

such that, for all events $\mathcal{A} \in \Sigma_{\mathcal{X}}$,

$$P_X(\mathcal{A}) = \mathbb{P}(\{\omega \in \Omega \ : \ X(\omega) \in \mathcal{A}\}) = \mathbb{P}(X \in \mathcal{A}).$$

A random variable $X \in \mathcal{X}$ has a *probability density function* (pdf) $p_X : \mathcal{X} \to [0, +\infty)$ if

$$\forall \mathcal{A} \in \Sigma_{\mathcal{X}}, \quad P_X(\mathcal{A}) = \int_{\mathcal{A}} p_X(x) dx.$$

The pdf of a discrete random variable X taking values in $\{x_i\}_{i=1}^n$ can be expressed as

$$p_X(x) = \sum_{i=1}^n \mathbb{P}(X = x_i)\delta(x - x_i), \tag{A.1}$$

where the Dirac delta function $\delta(x)$, more rigorously Dirac distribution or Dirac measure, equals zero everywhere except at zero and is such that $\int_{-\infty}^{+\infty} \delta(x)dx = 1$.

A.2.1 Mean and Variance

The *mean* (or *expectation*, or *expected value*) of a random variable X of pdf p_X is defined as

$$\mathbb{E}X = \int_{\mathcal{X}} x p_X(x) dx \tag{A.2}$$

whenever the integral exists. For a function $f : \mathcal{X} \to \mathbb{R}$ of this random variable, we have

$$\mathbb{E}f(X) = \int_{\mathcal{X}} f(x) p_X(x) dx. \tag{A.3}$$

The *variance* of a random variable X is the quantity

$$\sigma_X^2 = \mathbb{E}(X - \mathbb{E}X)^2, \tag{A.4}$$

σ_X being its *standard deviation*.

A.2.2 Common Distributions

The uniform distribution P_X on a finite set $\mathcal{X}$ of cardinality n is such that

$$\forall x \in \mathcal{X}, \quad P_X(x) = \frac{1}{n}.$$

Conversely, the uniform distribution on an interval $[a, b] \subset \mathbb{R}$ is such that

$$p_X(x) = \begin{cases} \frac{1}{b-a}, & \text{if } x \in [a, b] \\ 0, & \text{otherwise.} \end{cases}$$

A random variable X is said to have Gaussian distribution of mean μ and variance σ^2 if its pdf is given by

$$p_X(x) = \frac{1}{\sigma\sqrt{2\pi}} \exp\left(\frac{-(x-\mu)^2}{2\sigma^2}\right).$$

A.3 Random Pairs and Random Vectors

A random pair (X, Y) taking values in $\mathcal{X} \times \mathcal{Y}$ is characterized by its *joint probability distribution*

$$P_{X,Y} : \Sigma_{\mathcal{X}} \times \Sigma_{\mathcal{Y}} \to [0, 1]$$

such that

$$P_{X,Y}(\mathcal{A}, \mathcal{B}) = \mathbb{P}(X \in \mathcal{A}, Y \in \mathcal{B}).$$

The distribution P_X of X and P_Y of Y are refered to as the *marginal* distributions and a similar terminology is used for pdfs. If it exists, the joint pdf $p_{X,Y} : \mathcal{X} \times \mathcal{Y} \to [0, +\infty)$ is such that

$$\forall (\mathcal{A}, \mathcal{B}) \in \Sigma_{\mathcal{X}} \times \Sigma_{\mathcal{Y}}, \quad P_{X,Y}(\mathcal{A}, \mathcal{B}) = \int_{\mathcal{A}} \int_{\mathcal{B}} p_{X,Y}(x, y) dx dy.$$

This definition and the ones below can be generalized to any finite number n of random variables X_i, $i = 1, \dots, n$, to yield a random vector X taking values $\boldsymbol{x} \in \mathcal{X} \subseteq \mathbb{R}^n$. In particular, for a random vector X, its mean,

$$\mathbb{E}X = \begin{bmatrix} \mathbb{E}X_1 \\ \vdots \\ \mathbb{E}X_n \end{bmatrix},$$

is a vector of $\mathbb{R}^n$ and the square matrix

$$\begin{aligned} \boldsymbol{C}_X &= \mathbb{E}\left[(X - \mathbb{E}X)(X - \mathbb{E}X)^\top\right] \\ &= \begin{bmatrix} (X_1 - \mathbb{E}X_1)^2 & \dots & (X_1 - \mathbb{E}X_1)(X_n - \mathbb{E}X_n) \\ \vdots & \ddots & \vdots \\ (X_1 - \mathbb{E}X_1)(X_n - \mathbb{E}X_n) & \dots & (X_n - \mathbb{E}X_n)^2 \end{bmatrix} \end{aligned}$$

is known as its variance-covariance matrix (or just covariance matrix for short).

Note that random vectors are usually written in uppercase non-bold letters, contrary to real vectors (their values) which are written in lowercase bold letters. However, in cases where more importance is given to the fact that they are vectors, the same symbol (in lowercase and bold) is used for both random vectors and their values.

A.3.1 Independence

Two continuous random variables X and Y are said *independent* if and only if their joint pdf is given by the product of their marginal pdfs:

$$\forall (x, y) \in \mathcal{X} \times \mathcal{Y}, \quad p_{X,Y}(x, y) = p_X(x) p_Y(y).$$

In this case, we have

$$\mathbb{E}[XY] = \mathbb{E}[X]\mathbb{E}[Y].$$

A.3.1.1 White Noise

A random (or stochastic) process $\{\varepsilon_k\}$ in discrete time k is a *white noise* if $\mathbb{E}[\varepsilon_k] = 0$ (zero mean), $\mathbb{E}[\varepsilon_k^2] = \sigma_\varepsilon^2$ (constant variance) and the variables ε_k, ε_l, $\forall k \neq l$, are independent.

If the variables ε_k are identically distributed, the independence condition reduces to

$$\forall k \neq l, \quad \mathbb{E}[\varepsilon_k \varepsilon_l] = 0.$$

This can be generalized to n dimensional vector-valued random processes $\{\boldsymbol{\varepsilon}_k\}$, for which the white noise condition becomes $\mathbb{E}[\boldsymbol{\varepsilon}_k] = \boldsymbol{0}$ (zero mean), $\mathbb{E}[\boldsymbol{\varepsilon}_k \boldsymbol{\varepsilon}_k^\top] = \text{diag}(\left[\sigma_1 \dots \sigma_n\right]^\top)$ (a constant diagonal covariance matrix), and, if the vectors $\boldsymbol{\varepsilon}_k$ are identically distributed, $\forall k \neq l$, $\mathbb{E}[\boldsymbol{\varepsilon}_k \boldsymbol{\varepsilon}_l^\top] = \boldsymbol{0}$.

A.3.2 Conditional Probability

The conditional probability of an event $\mathcal{A}$ given $\mathcal{B}$ is, for $\mathbb{P}(\mathcal{B}) \neq 0$,

$$\mathbb{P}(\mathcal{A}|\mathcal{B}) = \frac{\mathbb{P}(\mathcal{A}, \mathcal{B})}{\mathbb{P}(\mathcal{B})}.$$

For a pair (X, Y) of continuous random variables, the conditional pdf of X given Y is defined for all y such that $p_Y(y) \neq 0$ as

$$p_{X|Y}(x|y) = \frac{p_{X,Y}(x, y)}{p_Y(y)}.$$

The *Bayes rule* states that

$$p_{Y|X}(y|x) = \frac{p_{X|Y}(x|y)p_Y(y)}{p_X(x)} \tag{A.5}$$

and is easily derived from the definitions of the conditional pdfs involved.

Appendix B
Basics of Linear Algebra

This appendix collects basic definitions and facts from linear algebra, such as vectors, matrices, vector spaces, eigenvalue/singular value decomposition.

Remark B.1 Symbols used in this appendix usually refer to generic objects and do not bear the same meaning as in the rest of the book.

B.1 Vectors

Throughout the book, vectors of $\mathbb{R}^n$ are written in lowercase and bold letters. We denote $\boldsymbol{x}$ a column vector such as

$$\boldsymbol{x} \in \mathbb{R}^n \Leftrightarrow \boldsymbol{x} = \begin{bmatrix} x_1 \\ \vdots \\ x_n \end{bmatrix}, \; x_i \in \mathbb{R}, \; i = 1, \dots, n.$$

The vector $\in \mathbb{R}^n$ with all entries equal to zero is denoted by $\mathbf{0}$ or $\mathbf{0}_n$, and the one $\in \mathbb{R}^n$ with all entries equal to one by $\mathbf{1}$ or $\mathbf{1}_n$.

The ℓ_p-norms on $\mathbb{R}^n$ are defined for $p \in [1, \infty)$ as

$$\|\boldsymbol{x}\|_p = \left(\sum_{i=1}^{n} |x_i|^p \right)^{1/p}, \tag{B.1}$$

and for $p = \infty$ as

$$\|\boldsymbol{x}\|_\infty = \max_{i \in [n]} |x_i|.$$

Any norm on $\mathbb{R}^n$ is a convex function. However, the ℓ_0-pseudo-norm, which counts the number of nonzero entries and is defined as

F. Lauer and G. Bloch, *Hybrid System Identification*,
Lecture Notes in Control and Information Sciences 478,
https://doi.org/10.1007/978-3-030-00193-3

$$\|\boldsymbol{x}\|_0 = |\{i \in [n] : x_i \neq 0\}|, \tag{B.2}$$

is not a norm (since it does not satisfy the homogeneity condition that $\forall \lambda \in \mathbb{R}$, $\forall \boldsymbol{x} \in \mathbb{R}^n$, $\|\lambda \boldsymbol{x}\| = |\lambda| \|\boldsymbol{x}\|$) and is nonconvex.

B.2 Matrices

Throughout the book, matrices are written in uppercase and bold letters. A matrix $\boldsymbol{A} \in \mathbb{R}^{m \times n}$ is an array of m rows and n columns with real entries $(\boldsymbol{A})_{ij} = A_{ij}$.

B.2.1 Basic Facts on Matrices

The *matrix transpose* $\boldsymbol{A}^\top \in \mathbb{R}^{n \times m}$ of $\boldsymbol{A} \in \mathbb{R}^{m \times n}$ has entries given by

$$\left(\boldsymbol{A}^\top\right)_{ij} = A_{ji}, \quad i = 1, \dots, n, \; j = 1, \dots, m. \tag{B.3}$$

A vector $\boldsymbol{x} \in \mathbb{R}^n$ can be treated as a single-column matrix. Then, its transpose is the row vector

$$\boldsymbol{x}^\top = \left[x_1 \dots x_n\right].$$

A matrix $\boldsymbol{A} \in \mathbb{R}^{m \times n}$ can be written such as

$$\boldsymbol{A} = \left[\boldsymbol{c}_1 \dots \boldsymbol{c}_n\right] = \left[\boldsymbol{r}_1 \dots \boldsymbol{r}_m\right]^\top,$$

with the columns (column vectors) $\boldsymbol{c}_j$, $j = 1, \dots, n$, and the rows (row vectors) $\boldsymbol{r}_i^\top$, $i = 1, \dots, m$.

A *square* matrix $\boldsymbol{A}$ has as many rows as columns, i.e., $m = n$.

The *trace* of a square matrix $\boldsymbol{A} \in \mathbb{R}^{n \times n}$ is the sum of the entries on its main diagonal:

$$\operatorname{tr}(\boldsymbol{A}) = \sum_{i=1} A_{ii}. \tag{B.4}$$

The *Frobenius norm* of a matrix $\boldsymbol{A} \in \mathbb{R}^{m \times n}$ is defined by

$$\|\boldsymbol{A}\|_F = \sqrt{\sum_{i=1}^{m} \sum_{j=1}^{n} A_{ij}^2}. \tag{B.5}$$

A *diagonal* matrix $\boldsymbol{A}$ is a square matrix with zero entries except on its main diagonal. Such matrices can be obtained by applying the diag$(\cdot)$ operator to vectors:

given a vector $\boldsymbol{a} \in \mathbb{R}^n$,

$$\boldsymbol{A} = \operatorname{diag}(\boldsymbol{a}) = \begin{bmatrix} a_1 & 0 & \dots & 0 \\ 0 & a_2 & \ddots & \\ \vdots & \ddots & \ddots & 0 \\ 0 & & 0 & a_n \end{bmatrix}. \tag{B.6}$$

The diagonal matrix with only ones in its main diagonal is the *identity* matrix $\boldsymbol{I} = \operatorname{diag}(\boldsymbol{1})$. $\boldsymbol{I} \in \mathbb{R}^{n \times n}$ can be denoted as $\boldsymbol{I}_n$.

In a lower (respectively, upper) *triangular* matrix, all entries above (respectively, below) its main diagonal are zeros. A matrix is said to be *strictly triangular* if it is triangular and has only zeros on its main diagonal.

A *symmetric* matrix $\boldsymbol{A}$ is a square matrix such that $\boldsymbol{A}^\top = \boldsymbol{A}$.

A *Jordan block* $\boldsymbol{J}$ is a bidiagonal matrix with the diagonal filled with a single value and the superdiagonal (the diagonal just above the main one) filled with ones:

$$\boldsymbol{J} = \begin{bmatrix} \lambda & 1 & & 0 \\ 0 & \ddots & \ddots & \\ \vdots & & \ddots & 1 \\ 0 & \dots & & \lambda \end{bmatrix}. \tag{B.7}$$

A *Toeplitz matrix* is a matrix whose entries are constant along each diagonal (from left to right). Formally, $\boldsymbol{A} \in \mathbb{R}^{n \times n}$ is a Toeplitz matrix if there exist scalars $a_{-(n-1)}, \dots, a_0, \dots, a_{n-1}$ such that $A_{ij} = a_{j-i}$, thus

$$\boldsymbol{A} = \begin{bmatrix} a_0 & a_1 & a_2 & \dots & \dots & a_{n-1} \\ a_{-1} & a_0 & a_1 & \ddots & & \vdots \\ a_{-2} & a_{-1} & \ddots & \ddots & \ddots & \vdots \\ \vdots & \ddots & \ddots & \ddots & a_1 & a_2 \\ \vdots & & \ddots & a_{-1} & a_0 & a_1 \\ a_{-(n-1)} & \dots & \dots & a_{-2} & a_{-1} & a_0 \end{bmatrix}. \tag{B.8}$$

A *Hankel matrix* is a matrix whose entries are constant along each antidiagonal (from right to left). Formally, $\boldsymbol{A} \in \mathbb{R}^{n \times n}$ is a Hankel matrix if there exist scalars $a_0, \dots, a_{2(n-1)}$ such that $A_{ij} = A_{ji} = a_{i+j-2}$, thus

$$A = \begin{bmatrix} a_0 & a_1 & a_2 & \dots & a_{n-1} \\ a_1 & a_2 & a_3 & \dots & a_n \\ a_2 & a_3 & a_4 & \dots & a_{n+1} \\ \vdots & \vdots & \vdots & & \vdots \\ a_{n-1} & a_n & a_{n+1} & \dots & a_{2(n-1)} \end{bmatrix}. \tag{B.9}$$

For square matrices, the *matrix exponential* is defined as

$$e^{A} = \sum_{i=1}^{\infty} \frac{1}{i!} A^{i}. \tag{B.10}$$

For the matrices $A \in \mathbb{R}^{m\times n}$ and $B \in \mathbb{R}^{p\times q}$, the *Kronecker product* $A \otimes B \in \mathbb{R}^{mp\times nq}$ is a block matrix, where each block is $A_{ij} B$, such as

$$A \otimes B = \begin{bmatrix} A_{11}B & \dots & A_{1n}B \\ \vdots & \ddots & \vdots \\ A_{m1}B & \dots & A_{mn}B \end{bmatrix}. \tag{B.11}$$

The *vectorization operator* applies to matrices $A = \begin{bmatrix} a_1 \; a_2 \; \dots \; a_n \end{bmatrix}$ to yield a vector containing all the matrix entries as

$$\text{vec}(A) = \begin{bmatrix} a_1 \\ \vdots \\ a_n \end{bmatrix}. \tag{B.12}$$

In particular, for general matrices A, B, and C of dimensions $m \times n$, $n \times p$, and $p \times q$, we have

$$\text{vec}(ABC) = (C^{\top} \otimes A)\, \text{vec}(B) \tag{B.13}$$

and

$$\text{vec}(AB) = (B^{\top} \otimes I_m)\, \text{vec}(A) = (I_p \otimes A)\, \text{vec}(B), \tag{B.14}$$

where $\otimes$ is the Kronecker product.

A symmetric matrix $A \in \mathbb{R}^{n\times n}$ is said *positive semi-definite* if

$$\forall x \in \mathbb{R}^n, \quad x^{\top} A x \geq 0. \tag{B.15}$$

A symmetric matrix $A \in \mathbb{R}^{n\times n}$ is said *positive definite*, which is denoted by $A \succ 0$, if

$$\forall x \in \mathbb{R}^n, \quad x^{\top} A x > 0. \tag{B.16}$$

The notation $A \prec B$ means that $(B - A) \succ 0$.

B.2.2 Differential Operators

The *gradient* of the scalar-valued function

$$\begin{aligned} f : \mathbb{R}^n &\to \mathbb{R} \\ \boldsymbol{x} &\mapsto f(\boldsymbol{x}) \end{aligned}$$

is the first derivative of f with respect to $\boldsymbol{x}$:

$$\nabla f = \frac{\partial f}{\partial \boldsymbol{x}} = \begin{bmatrix} \frac{\partial f}{\partial x_1} \\ \vdots \\ \frac{\partial f}{\partial x_n} \end{bmatrix}. \tag{B.17}$$

The second derivative is the *Hessian*

$$\nabla^2 f = \frac{\partial^2 f}{\partial \boldsymbol{x} \partial \boldsymbol{x}^\top} = \frac{\partial^2 f}{\partial \boldsymbol{x}^\top \partial \boldsymbol{x}} = \begin{bmatrix} \frac{\partial^2 f}{\partial x_1 \partial x_1} & \cdots & \frac{\partial^2 f}{\partial x_1 \partial x_n} \\ \vdots & \ddots & \vdots \\ \frac{\partial^2 f}{\partial x_n \partial x_1} & \cdots & \frac{\partial^2 f}{\partial x_n \partial x_n} \end{bmatrix}. \tag{B.18}$$

The *Jacobian* of the vector-valued function

$$\begin{aligned} \boldsymbol{f} : \mathbb{R}^n &\to \mathbb{R}^m \\ \boldsymbol{x} &\mapsto \boldsymbol{f}(\boldsymbol{x}) \end{aligned}$$

is the $m \times n$ dimensional matrix of all the first-order partial derivatives $\frac{\partial f_i}{\partial x_j}$, $i = 1, \ldots, m$, $j = 1, \ldots, n$:

$$\mathrm{D}\boldsymbol{f} = \begin{bmatrix} \frac{\partial \boldsymbol{f}}{\partial x_1} & \cdots & \frac{\partial \boldsymbol{f}}{\partial x_n} \end{bmatrix} = \begin{bmatrix} \frac{\partial f_1}{\partial \boldsymbol{x}^\top} \\ \vdots \\ \frac{\partial f_m}{\partial \boldsymbol{x}^\top} \end{bmatrix} = \begin{bmatrix} \frac{\partial f_1}{\partial x_1} & \cdots & \frac{\partial f_1}{\partial x_n} \\ \vdots & \ddots & \vdots \\ \frac{\partial f_m}{\partial x_1} & \cdots & \frac{\partial f_m}{\partial x_n} \end{bmatrix}. \tag{B.19}$$

B.2.3 Vector Spaces

A *real vector space* $\mathcal{V}$ is a set of vectors with two operations: an addition, $+ : \mathcal{V} \times \mathcal{V} \to \mathcal{V}$, and a scalar multiplication, $\cdot : \mathbb{R} \times \mathcal{V} \to \mathcal{V}$, satisfying a number of properties such as the associativity, commutativity, existence of identity and inverse elements of addition, and the distributivity.

A *subspace* $\mathcal{S}$ of the real vector space $\mathcal{V}$ is a subset of $\mathcal{V}$ with the same operations.

The *span* of a set of vectors $\mathcal{B} = \{\boldsymbol{b}_i\}_{i=1}^n \subset \mathcal{V}$ is the set of all linear combinations of elements of $\mathcal{B}$:

$$\text{span}(\mathcal{B}) = \left\{ \sum_{i=1}^{n} \lambda_i \boldsymbol{b}_i \, : \, \lambda_i \in \mathbb{R} \right\}.$$

A set of vectors $\mathcal{B} = \{\boldsymbol{b}_i\}_{i=1}^n$ is *linearly independent* if and only if

$$\sum_{i=1}^{n} \lambda_i \boldsymbol{b}_i = 0 \implies \lambda_i = 0, \quad i = 1, \dots, n.$$

A *basis* of $\mathcal{V}$ is a set $\mathcal{B} = \{\boldsymbol{b}_i\}_{i=1}^n$ that spans $\mathcal{V}$ and is linearly independent.

B.2.3.1 Inner Product and Hilbert Spaces

An inner product space $\mathcal{V}$ over $\mathbb{R}$ is a real vector space equipped with an inner product $\langle \cdot, \cdot \rangle_{\mathcal{V}} : \mathcal{V} \times \mathcal{V} \to \mathbb{R}$ that is symmetric, linear in its first argument and positive definite (i.e., $\forall x \in \mathcal{V}$, $\langle x, x \rangle_{\mathcal{V}} \geq 0$ and $\langle x, x \rangle_{\mathcal{V}} = 0 \Leftrightarrow x = 0$).

A *Hilbert space* $\mathcal{H}$ is an inner product space that is complete with respect to the distance $d(x, x') = \|x - x'\|_{\mathcal{H}}$ defined in terms of the norm induced by its inner product as

$$\|x\|_{\mathcal{H}} = \sqrt{\langle x, x \rangle_{\mathcal{H}}}.$$

Here, the completeness of $\mathcal{H}$ means that converging sequences of elements in $\mathcal{H}$ converge in $\mathcal{H}$: $\lim_{n \to +\infty} x_n \in \mathcal{H}$ for any sequence (x_n) of elements in $\mathcal{H}$ satisfying $\lim_{n \to +\infty} \sup_{i,j>n} \|x_i - x_j\|_{\mathcal{H}} = 0$.

B.2.3.2 Euclidean Spaces, Range, Null Space and Rank

The Euclidean space $\mathbb{R}^n$ is a Hilbert space with the inner product

$$\langle \boldsymbol{x}, \boldsymbol{x}' \rangle_{\mathbb{R}^n} = \sum_{i=1}^{n} x_i x_i'$$

that can be written as the matrix multiplication $\boldsymbol{x}^\top \boldsymbol{x}'$.

The matrix $\boldsymbol{A} \in \mathbb{R}^{m \times n}$ can be viewed as a linear transformation, $\boldsymbol{x} \mapsto \boldsymbol{y} = \boldsymbol{A}\boldsymbol{x} \in \mathbb{R}^m$, of vectors $\boldsymbol{x} \in \mathbb{R}^n$. In this case, the domain of $\boldsymbol{A}$ is $\mathbb{R}^n$, its *range* (or image) is a subspace of $\mathbb{R}^m$ defined by

$$\text{ran}(\boldsymbol{A}) = \{\boldsymbol{y} \in \mathbb{R}^m \, : \, \boldsymbol{y} = \boldsymbol{A}\boldsymbol{x}\}, \tag{B.20}$$

and its *null space* is the subspace of $\mathbb{R}^n$ defined by

$$\mathrm{null}(\boldsymbol{A}) = \{\boldsymbol{x} \in \mathbb{R}^n : \boldsymbol{A}\boldsymbol{x} = \boldsymbol{0}\}. \tag{B.21}$$

The dimension of $\mathrm{ran}(\boldsymbol{A})$ is called the *rank* of $\boldsymbol{A}$:

$$\mathrm{rank}(\boldsymbol{A}) = \dim(\mathrm{ran}(\boldsymbol{A})), \tag{B.22}$$

and is the maximum number of linearly independent rows or columns of $\boldsymbol{A}$. Conversely, the *nullity* of $\boldsymbol{A}$ is $\mathrm{nullity}(a) = \dim(\mathrm{null}(\boldsymbol{A}))$ and the following relationship holds: $\mathrm{rank}(\boldsymbol{A}) + \mathrm{nullity}(\boldsymbol{A}) = n$. A matrix is said to be of *full rank* if $\mathrm{rank}(\boldsymbol{A}) = \min\{m, n\}$.

The column space of the matrix $\boldsymbol{A} = \begin{bmatrix}\boldsymbol{c}_1 \dots \boldsymbol{c}_n\end{bmatrix} \in \mathbb{R}^{m\times n}$ is the set of all possible linear combinations of the column vectors $\boldsymbol{c}_1, \dots, \boldsymbol{c}_n$, i.e., the span of these vectors, or the set of all possible products $\boldsymbol{A}\boldsymbol{x}$, then the range of $\boldsymbol{A}$.

The row space of the matrix $\boldsymbol{A} = \begin{bmatrix}\boldsymbol{r}_1 \dots \boldsymbol{r}_m\end{bmatrix}^\top \in \mathbb{R}^{m\times n}$ is the set of all possible linear combinations of the row vectors $\boldsymbol{r}_1^\top, \dots, \boldsymbol{r}_m^\top$, i.e., the span of these vectors, which also corresponds to the range of $\boldsymbol{A}^\top$.

B.2.3.3 Function Spaces

A set of real-valued functions on some set $\mathcal{X}$ form a *function space* $\mathcal{F}$ with a vector space structure under the pointwise addition

$$\forall (f, f') \in \mathcal{F}^2, \quad \forall x \in \mathcal{X}, \quad (f + f')(x) = f(x) + f'(x)$$

and scalar multiplication

$$\forall \alpha \in \mathbb{R}, \ \forall f \in \mathcal{F}, \quad \forall x \in \mathcal{X}, \quad (\alpha f)(x) = \alpha f(x).$$

B.2.4 Matrix Inverses and Systems of Linear Equations

The *determinant* of a square matrix $\boldsymbol{A} \in \mathbb{R}^{n\times n}$ can be expressed recursively as

$$\det(\boldsymbol{A}) = \sum_{j=1}^{n} (-1)^{j+1} A_{1j} \det(\boldsymbol{A}_{1j}), \tag{B.23}$$

where $\boldsymbol{A}_{1j}$ is a matrix of dimension $(n-1) \times (n-1)$ obtained by deleting the first row and jth column of $\boldsymbol{A}$, with $\det(\boldsymbol{A}) = A$, for $\boldsymbol{A} = A$, a scalar.

A square matrix $\boldsymbol{A}$ is said to be *invertible* or *nonsingular* if there exists a matrix *inverse* $\boldsymbol{A}^{-1}$ such that

$$\boldsymbol{A}\boldsymbol{A}^{-1} = \boldsymbol{A}^{-1}\boldsymbol{A} = \boldsymbol{I}.$$

In particular, $\left(A^{-1}\right)^{-1} = A$ and, for nonsingular A and B,

$$(AB)^{-1} = B^{-1}A^{-1}$$

and

$$(ABA^{-1})^{-1} = AB^{-1}A^{-1}. \tag{B.24}$$

If $\det(A) = 0$, A is singular.

An *orthogonal* square matrix A is a square matrix such that

$$AA^{\top} = A^{\top}A = I,$$

i.e., A is orthogonal if and only if $A^{-1} = A^{\top}$.

The solution of the system of linear equations

$$Ax = b \tag{B.25}$$

is given, if A is square and nonsingular, by

$$x = A^{-1}b.$$

Otherwise, if $A \in \mathbb{R}^{m\times n}$ is a non-square matrix or if it is singular, the solution (if any) is given by

$$x = A^{+}b, \tag{B.26}$$

where the Moore–Penrose pseudo-inverse $A^{+} \in \mathbb{R}^{n\times m}$ is such that

$$AA^{+}A = A, \quad A^{+}AA^{+} = A^{+}, \quad (AA^{+})^{\top} = AA^{+}, \quad (A^{+}A)^{\top} = A^{+}A.$$

Indeed, if the system (B.25) has a solution x, then $b = Ax = AA^{+}Ax = AA^{+}b$ and the system amounts to $Ax = AA^{+}b$ for which (B.26) gives a solution.

The pseudo-inverse always exists and is unique. In particular, if A is of full column rank, i.e., has n ($n < m$) linearly independent columns, A^{+} is the left inverse:

$$A^{+} = (A^{\top}A)^{-1}A^{\top}. \tag{B.27}$$

Conversely, if A is of full row rank, i.e., has m ($m < n$) linearly independent rows, A^{+} is the right inverse:

$$A^{+} = A^{\top}(AA^{\top})^{-1}. \tag{B.28}$$

In other cases, A^{+} can be computed via the singular value decomposition of A (see Sect. B.2.8.1 below).

The pseudo-inverse also provides the solution to the least squares problem

$$\min_{x \in \mathbb{R}^n} \|\boldsymbol{b} - \boldsymbol{A}\boldsymbol{x}\|_2^2 \tag{B.29}$$

as $\boldsymbol{x}^* = \boldsymbol{A}^+\boldsymbol{b}$.

Indeed, following Sect. 2.1.3.2, the solution is the point at which the gradient of the cost function is zero, which here coincides with the solution of the linear system $\boldsymbol{A}^\top\boldsymbol{A}\boldsymbol{x} = \boldsymbol{A}^\top\boldsymbol{b}$. This solution is in turn given by (B.26) as $\boldsymbol{x}^* = (\boldsymbol{A}^\top\boldsymbol{A})^+\boldsymbol{A}^\top\boldsymbol{b}$, in which the identity $\boldsymbol{A}^+ = (\boldsymbol{A}^\top\boldsymbol{A})^+\boldsymbol{A}^\top$ provides the result.

Finally, matrix extensions of (B.25) and (B.29), i.e., $\boldsymbol{A}\boldsymbol{X} = \boldsymbol{B}$ and $\min_{\boldsymbol{X} \in \mathbb{R}^{n\times p}} \|B - \boldsymbol{A}\boldsymbol{X}\|_F^2$, with $\boldsymbol{B} = \begin{bmatrix}\boldsymbol{b}_1 \dots \boldsymbol{b}_p\end{bmatrix} \in \mathbb{R}^{m\times p}$ and a matrix variable $\boldsymbol{X} = \begin{bmatrix}\boldsymbol{x}_1 \dots \boldsymbol{x}_p\end{bmatrix} \in \mathbb{R}^{n\times p}$, can be solved similarly through the computation of $\boldsymbol{A}^+$ as $\boldsymbol{X} = \boldsymbol{A}^+\boldsymbol{B}$. Indeed, these problems can be decomposed into p independent vector problems (with variable $\boldsymbol{x}_i$ and right-hand side $\boldsymbol{b}_i$) which all share the same matrix $\boldsymbol{A}$. In particular, if $\boldsymbol{A}$ is full column rank, then

$$\boldsymbol{X} = (\boldsymbol{A}^\top\boldsymbol{A})^{-1}\boldsymbol{A}^\top\boldsymbol{B}. \tag{B.30}$$

Transposed versions of these problems, e.g., $\boldsymbol{X}\boldsymbol{A} = \boldsymbol{B} \Leftrightarrow \boldsymbol{A}^\top\boldsymbol{X}^\top = \boldsymbol{B}^\top$ are solved similarly with

$$\boldsymbol{X}^\top = (\boldsymbol{A}^\top)^+\boldsymbol{B}^\top = (\boldsymbol{A}^+)^\top\boldsymbol{B}^\top \quad \Rightarrow \quad \boldsymbol{X} = \boldsymbol{B}\boldsymbol{A}^+. \tag{B.31}$$

B.2.5 Orthogonal Projections

The orthogonal complement $\mathcal{A}^\perp$ of a subspace $\mathcal{A}$ of an inner product space $\mathcal{V}$ is the set of all vectors in $\mathcal{V}$ that are orthogonal to every vector in $\mathcal{A}$:

$$\mathcal{A}^\perp = \{\boldsymbol{x} \in \mathcal{V} : \forall \boldsymbol{y} \in \mathcal{A}, \langle \boldsymbol{x}, \boldsymbol{y}\rangle_{\mathcal{V}} = 0\}.$$

In Euclidean spaces, given general matrices $\boldsymbol{A} \in \mathbb{R}^{m\times n}$ and $\boldsymbol{B} \in \mathbb{R}^{p\times n}$, the *orthogonal projection* of the row space of $\boldsymbol{A}$ on the row space of $\boldsymbol{B}$ is defined as

$$\boldsymbol{A}/\boldsymbol{B} = \boldsymbol{A}\boldsymbol{\Pi}_B = \boldsymbol{A}\boldsymbol{B}^\top(\boldsymbol{B}\boldsymbol{B}^\top)^{-1}\boldsymbol{B}.$$

If $\boldsymbol{B}\boldsymbol{B}^\top$ is singular, the Moore–Penrose pseudo-inverse is used. Similarly the projection of the row space of $\boldsymbol{A}$ on the orthogonal complement of the row space of $\boldsymbol{B}$ is defined as

$$\boldsymbol{A}/\boldsymbol{B}^\perp = \boldsymbol{A}\boldsymbol{\Pi}_B^\perp = \boldsymbol{A}(\boldsymbol{I} - \boldsymbol{\Pi}_B) = \boldsymbol{A}\left(\boldsymbol{I} - \boldsymbol{B}^\top(\boldsymbol{B}\boldsymbol{B}^\top)^{-1}\boldsymbol{B}\right). \tag{B.32}$$

The projections $\boldsymbol{A}/\boldsymbol{B}$ and $\boldsymbol{A}/\boldsymbol{B}^\perp$ decompose the matrix $\boldsymbol{A}$ into two matrices, whose row spaces are mutually orthogonal: $\boldsymbol{A} = \boldsymbol{A}/\boldsymbol{B} + \boldsymbol{A}/\boldsymbol{B}^\perp$. Furthermore, $\boldsymbol{A}\boldsymbol{\Pi}_A = \boldsymbol{A}$ and $\boldsymbol{A}\boldsymbol{\Pi}_A^\perp = \boldsymbol{0}$.

B.2.6 Cholesky Factorization

Any symmetric positive definite matrix $\boldsymbol{A} \in \mathbb{R}^{n \times n}$ admits a *Cholesky factorization*

$$\boldsymbol{A} = \boldsymbol{L}\boldsymbol{L}^\top = \boldsymbol{R}^\top \boldsymbol{R}, \tag{B.33}$$

where $\boldsymbol{L}$ is a lower triangular matrix with only zeros above its diagonal and with positive entries on its diagonal, and, conversely, $\boldsymbol{R} = \boldsymbol{L}^\top$ is an upper triangular matrix. This factorization can be efficiently computed in $\mathcal{O}(n^3)$ operations.

B.2.7 Eigendecomposition

Given a square matrix $\boldsymbol{A} \in \mathbb{R}^{n \times n}$, an *eigvenvalue-eigenvector* pair is a pair $(\lambda, \boldsymbol{u}) \in \mathbb{R} \times \mathbb{R}^n$ with $\boldsymbol{u} \neq \boldsymbol{0}$ such that

$$\boldsymbol{A}\boldsymbol{u} = \lambda \boldsymbol{u}.$$

If $\boldsymbol{A} \in \mathbb{R}^{n \times n}$ is symmetric, then it admits an eigendecomposition

$$\boldsymbol{A} = \boldsymbol{U}\boldsymbol{\Lambda}\boldsymbol{U}^\top \tag{B.34}$$

with a real orthogonal matrix $\boldsymbol{U} \in \mathbb{R}^{n \times n}$ and a diagonal $\boldsymbol{\Lambda} = \operatorname{diag}\left(\left[\lambda_1 \ \ldots \ \lambda_n\right]^\top\right)$ containing the eigenvalues of $\boldsymbol{A}$ of corresponding eigenvectors in the columns of $\boldsymbol{U}$.

The sum of eigenvalues of a matrix is equal to its trace:

$$\sum_{i=1}^{n} \lambda_i = \operatorname{tr}(\boldsymbol{A}). \tag{B.35}$$

B.2.8 Singular Value Decomposition

Any real matrix $\boldsymbol{A} \in \mathbb{R}^{m \times n}$, $m \geq n$, admits a *singular value decomposition* (SVD) as

$$\boldsymbol{A} = \boldsymbol{U}\boldsymbol{S}\boldsymbol{V}^\top \tag{B.36}$$

with orthogonal matrices $\boldsymbol{U} \in \mathbb{R}^{m \times m}$ and $\boldsymbol{V} \in \mathbb{R}^{n \times n}$ and

$$
\boldsymbol{S} = \begin{bmatrix} \sigma_1 & 0 & \dots & 0 \\ 0 & \ddots & & \vdots \\ \vdots & & 0 & \sigma_n \\ 0 & \dots & \dots & 0 \\ \vdots & \ddots & & \\ 0 & & & 0 \end{bmatrix} \in \mathbb{R}^{m \times n},
$$

where $\sigma_1 \geq \sigma_2 \geq \cdots \geq 0$. The uniquely defined σ_i's are called the *singular values* of $\boldsymbol{A}$, while the columns of $\boldsymbol{U}$ (respectively $\boldsymbol{V}$) are called left (respectively right) singular vectors.

The SVD of an $m \times n$ matrix with $m \geq n$ can typically be computed in $\mathcal{O}(m^2 n + mn^2 + n^3)$ operations. This can be reduced to $\mathcal{O}(mn^2 + n^3)$ when computing only the thin SVD, $\boldsymbol{A} = \boldsymbol{U}_1 \boldsymbol{S}_1 \boldsymbol{V}^T$ with $\boldsymbol{U}_1 \in \mathbb{R}^{n \times n}$ not containing the $(m - n)$ last columns of $\boldsymbol{U}$, which is sufficient for most applications.

B.2.8.1 Solving Linear Systems with the SVD

The Moore–Penrose pseudo-inverse of a matrix $\boldsymbol{A} \in \mathbb{R}^{m \times n}$ with $r = \text{rank}(\boldsymbol{A}) < \min\{m, n\}$ can be computed from the SVD as

$$
\boldsymbol{A}^+ = \boldsymbol{V}\boldsymbol{S}^+\boldsymbol{U}^\top,
$$

where

$$
\boldsymbol{S}^+ = \begin{bmatrix} \frac{1}{\sigma_1} & 0 & \dots & 0 & \dots & 0 \\ q\vdots & \ddots & & \vdots & & \vdots \\ \vdots & & \frac{1}{\sigma_r} & \vdots & & \vdots \\ 0 & \dots & \dots & 0 & \dots & 0 \end{bmatrix}
$$

is computed by inverting only the nonzero singular values in $\boldsymbol{S}$ and transposing the resulting matrix.

B.2.8.2 Relationship with the Eigendecomposition

For any matrix $\boldsymbol{A} \in \mathbb{R}^{m \times n}$, $\boldsymbol{A}^\top \boldsymbol{A}$ is a square matrix with the eigendecomposition

$$
\boldsymbol{A}^\top \boldsymbol{A} = \boldsymbol{U}\boldsymbol{\Lambda}\boldsymbol{U}^\top
$$

that can be computed from the SVD of $\boldsymbol{A}$, $\boldsymbol{A} = \tilde{\boldsymbol{U}}\boldsymbol{S}\boldsymbol{V}^\top$. Indeed, since $\tilde{U}$ is orthogonal,

$$
\boldsymbol{A}^\top \boldsymbol{A} = \boldsymbol{V}\boldsymbol{S}^\top \boldsymbol{S}\boldsymbol{V}^\top
$$

and it can be seen that the eigenvalues $\lambda_i \neq 0$ of $\boldsymbol{A}^\top \boldsymbol{A}$ are such that $\lambda_i = \sigma_i^2$, with σ_i the singular values of $\boldsymbol{A}$, and that the corresponding eigenvectors of $\boldsymbol{A}^\top \boldsymbol{A}$ are given by the right singular vectors of $\boldsymbol{A}$.

Index

A

Adaptive Forgetting through Multiple Models (AFMM) algorithm, 191
Affine hybrid model, 81
Akaike information criterion (AIC), 54
Algebraic approach, 142, 189
AutoRegressive Moving Average with eXogenous input (ARMAX) model, 20
AutoRegressive with eXogenous input (ARX) model, 19

B

Bayesian approach, 188
Bayesian information criterion (BIC), 54
Block-structured models, 41
Bounded-error approach, 91, 106, 124, 132, 151
Box-Jenkins (BJ) model, 21
Branch-and-bound optimization, 126

C

Canonical forms, 32
Change detection techniques, 195
Cholesky factorization, 221, 248
Classification, 60, 119, 122, 171
  linear classifiers, 61, 84, 108, 119, 122
  multi-class, 68
  nonlinear, 70
Classification error rate (CE), 94
Clustering, 59, 71, 179, 221
Computational complexity, 109
Controllability, 29
Controllability canonical form, 32
Controllability matrix, 29
Convex polyhedra, 82, 193
Coordinate descent approach, 148, 154
Cross validation, 55

D

Dirac delta function, 189, 234

E

Eigendecomposition, 210, 248
Expectation, 234

F

Feature space, 48, 70, 208, 210
Final prediction error (FPE) criterion, 54
Finite impulse response (FIR) model, 18
FIT criterion, 53
Forgetting factor, 41, 185
Frobenius norm, 240

H

Hammerstein model, 41
Hankel matrix, 241
Hinging hyperplane, 135
Hybrid Decoupling Constraint (HDC), 142, 190
Hybrid Input–Output (I/O) model, 3, 79
Hybrid state-space (SS)
  estimation, 193
  model, 3, 78, 192
Hybrid systems, applications, 6

F. Lauer and G. Bloch, *Hybrid System Identification*,
Lecture Notes in Control and Information Sciences 478,
https://doi.org/10.1007/978-3-030-00193-3

Hybrid systems, general, 3
Hybrid systems, switching mechanism, 3
Hyperparameter, 54, 88, 219
Hyperplane, 62, 69, 84, 97, 119, 144
  optimal separating hyperplane, 63

I
Impulse response, 17
Input–Output (I/O) model, 2, 16

J
Jordan block, 241

K
Kernel function, 45, 70
  Gaussian kernel function, 46, 174, 208, 211, 217
Kernel matrix, 45, 47, 74, 173, 208, 209, 211, 213, 215, 216, 221
Kernel model, 47, 54, 70, 206, 216, 221
Kernel Principal Component Analysis (KPCA), 210
Kernel Principal Component Regression (KPCR), 212
Kernel ridge regression, 47, 207
K-LinReg, 150, 227
K-means, 72, 149, 199, 222
Kronecker product, 242

L
Least squares, 27, 120, 150, 154, 177, 246
Least squares SVM (LS-SVM), 208
Linear programming, 24, 26, 68, 108, 164, 215
Loss function, 22, 51, 153

M
Markov parameters, 30, 33, 198
Matrix exponential, 242
Mean squared error (MSE), 53
Minimal realization, 31
Minimum-of-Errors (ME) approach, 89, 145
MIN PFS problem, 152
Mixed-integer programming, 134, 193
Modal form, 33
Model selection, 53
Multilevel coordinate search (MCS) algorithm, 147, 206

N
Nonlinear ARX (NARX) model, 85
Nonparametric model, 42
Normalized least mean squares (NLMS), 40, 189
Normalized parametric error (NPE), 52
$\mathcal{NP}$-hard, 111, 164

O
Observability, 29
Observability canonical form, 33, 198
Observability matrix, 30
Optimization, 24
  convex optimization, 25, 65, 156, 160, 173, 179, 214
  global optimization, 126
  nonconvex optimization, 71, 146, 206
Orthogonal projection, 247
Output error (OE) model, 21
Overfitting, 43

P
Particle filtering, 188
Piecewise affine (PWA)
  estimation, 108, 119, 134, 186
  model, 79, 82, 169
Piecewise ARX (PWARX)
  see PieceWise Affine (PWA), 82
Piecewise Output Error model, 80
Piecewise smooth (PWS)
  estimation, 93, 219
  model, 82
Piecewise-defined models, 3
Polynomial-time algorithms, 118
Prediction error, 17
Prediction error method (PEM), 22, 34
Principal Component Analysis (PCA), 210
Probability density function, 187, 188, 234
Probability distribution, 234
Product-of-errors (PE) approach, 146, 206
Pseudo-inverse, 246

Q
Quadratic programming, 26, 48, 65, 132, 174

R
Random variable, 233
Recursive identification, 39
Recursive least squares (RLS), 40, 185, 191
Regularization, 44, 50, 61, 206, 219
  local, 172

Renyi entropy, 208
Reproducing kernel Hilbert space, 46
Reweighting techniques, 158, 160, 163, 179, 213, 218, 219
Risk, 42, 60

S

Second-order cone programming, 26, 164, 174, 179, 221
Segmentation, 161, 191
Shift operator, 17
Similarity transformation, 28, 33, 194, 198
Singular value decomposition (SVD), 38, 196, 248
Sparsity, 155, 158, 159, 207, 213, 218, 219
State-space (SS) model, 2, 28
Subspace clustering, 96
Subspace methods, 36
Sum-of-norms approach, 162, 178, 218
Support vector machine (SVM), 63, 69
Support vector regression (SVR), 51, 216
Switched ARX (SARX)
  estimation, 105, 122, 127, 185
  model, 79, 82, 141, 184
Switched nonlinear ARX (SNARX) model, 82
Switched output error (SOE)
  estimation, 183
  model, 80, 82

T

Toeplitz matrix, 241
Transfer function, 17

V

Variance, 234
Vectorization operator, 242

W

White noise, 17, 34, 40, 191, 196, 236
Wiener model, 41

Printed by Printforce, the Netherlands